MODULAR MULTILEVEL CONVERTERS

IEEE Press
445 Hoes Lane
Piscataway, NJ 08854

MODULAR MULTILEVEL CONVERTERS

Analysis, Control, and Applications

Sixing Du
Apparao Dekka
Bin Wu
Navid Zargari

IEEE PRESS SERIES ON POWER ENGINEERING

IEEE PRESS

WILEY

Published by John Wiley & Sons, Inc., Hoboken, New Jersey.
Published simultaneously in Canada.

Library of Congress Cataloging-in-Publication Data is available.

ISBN: 978-1-119-36630-0

Printed in the United States of America.

10 9 8 7 6 5 4 3 2 1

To my parents Lakshmi and Pydiraju, and brother Venkatamuni
[Apparao Dekka]

and

To my parents Lingen and Peifang, and wife Janice
[Bin Wu]

CONTENTS

ABOUT THE AUTHORS

Sixing Du received his B.S. degree in electrical engineering from Taiyuan University of Science and Technology, Taiyuan, China, in 2009, and the M.S. and Ph.D. degrees in electrical engineering from Xi'an Jiaotong University, Xi'an, China, in 2011 and 2014, respectively. He is currently working as a postdoctoral research fellow at the Laboratory for Electric Drive Applications and Research (LEDAR), Ryerson University, Canada.

Dr. Du has published more than 15 journal papers in peer-reviewed Institute of Electrical and Electronics Engineers (IEEE) as the first author, and he holds 7 issued Chinese patents in the fields of power electronics, power quality, and high-voltage direct current (HVDC) transmission. His main research interests include modular multilevel converters (MMC) in power quality, HVDC transmission, and medium-voltage motor drive applications.

Dr. Du has worked closely with Rockwell Automation, Canada, and participated in NSERC/Rockwell Industrial Research Chair program and Mitacs/Rockwell Accelerate program. He has produced more than 10 technical reports on novel MMC topologies and controls. His inventions overcome the major issues of conventional MMC-based drives in zero/low-speed operation, common-mode voltage minimization, and transformer-less utilization. Some of the inventions also facilitate single-stage DC–DC conversion in MVDC/HVDC applications.

Dr. Du received national scholarship for his graduate studies (Ph.D.) from Chinese Ministry of Education in 2013, Student Research Awards from Siemens and Shindengen in 2010 and 2011, and Excellent Ph.D. Graduate Student Award from Xi'an Jiaotong University in 2014. He serves as a reviewer for *IEEE Transactions on Power Electronics*, *IEEE Transactions on Industrial Electronics*, *IEEE Transactions on Industrial Informatics*, *IEEE Transactions on Power Delivery*, *IEEE Transactions on Energy Conversion*, *IET Power Electronics*, and other international journals and conferences.

Apparao Dekka was born in Thanam, Visakhapatnam, India. He received his B.Tech. degree in electrical and electronics engineering from Jawaharlal Nehru Technological University, Hyderabad, India, in 2007, an M.Tech. degree in electrical engineering with specialization in power electronics and power systems from Indian Institute of Technology Bombay, Mumbai, India, in 2010, and Ph.D. degree in electrical and computer engineering from Ryerson University, Toronto, Canada, in 2017. During 2010–2012, he worked as a research engineer at Power Electronics Laboratory, The Petroleum Institute, Abu Dhabi, UAE. He is currently working as a postdoctoral research fellow at the Laboratory for Electric Drive Applications and Research (LEDAR), Ryerson University, Canada. His research interests include high-power converters, pulse width modulation schemes, variable-speed drives, renewable energy systems, HVDC transmission systems, electric vehicles, power quality, energy storage, and model predictive control.

Dr. Dekka worked closely with Rockwell Automation, Centre for Urban Energy, Toronto Hydro, and Connect Canada, and completed several industrial projects in the field of high-power converters, power quality problems, and energy storage. He has published more than 25 peer-reviewed technical papers including 15 journal papers. He has produced more than 13 technical reports for the industry.

Dr. Dekka received national scholarship for his graduate studies (M.Tech.) from the Ministry of Human Resource Development, India, in 2008, Outstanding Reviewer Award for the year 2015 from *IEEE Transactions on Power Electronics* in 2016, Student Paper Travel Award for IEEE-ICIT'16 from *IEEE Industrial Electronics Society* in 2016. During his Ph.D. studies at Ryerson University, he received Best Poster Award from the Department of Electrical and Computer Engineering in 2014, Student Research Awards from IESO, Toronto Hydro, and Connect Canada in 2013 and 2015, and the Research Excellence Award from the Department of Electrical and Computer Engineering, Ryerson University, in 2014, 2015, and 2016. He is a member of IEEE and serves as a reviewer for IEEE/IET journal and conferences.

Bin Wu graduated from Donghua University, Shanghai, China, in 1978, and received his M.A.Sc. and Ph.D. degrees in electrical and computer engineering from the University of Toronto, Canada, in 1989 and 1993, respectively. After being with Rockwell Automation, Canada, from 1992 to 1993, he joined Ryerson University, where he is currently a professor in the Department of Electrical and Computer Engineering and a Senior NSERC/Rockwell Industrial Research Chair (IRC) in Power Electronics and Electric Drives.

Dr. Wu has published more than 400 peer-reviewed technical papers and 4 Wiley-IEEE Press books, and he holds more than 30 issued and pending patents in power electronics, adjustable-speed drives, and renewable energy systems. Dr. Wu is the founder of the Laboratory for Electric Drive Applications and Research (LEDAR), which has been recognized as the most advanced research facility of its kind in a Canadian university.

Dr. Wu has worked closely with Canadian companies and assisted them in achieving technical and commercial success through research and innovation. He has authored/coauthored more than 200 technical reports. Some of his inventions and patents have been adopted by industry and implemented in the production line, resulting in significant economic benefits.

Dr. Wu received the Gold Medal of the Governor General of Canada in 1993, the Premier's Research Excellence Award in 2001, the NSERC Synergy Award for Innovation in 2002, the Ryerson Distinguished Scholar Award in 2003, the Ryerson FEAS Research Excellence Award in 2007, and the Ryerson YSGS Outstanding Contribution to Graduate Education Award and the Professional Engineers Ontario (PEO) Engineering Excellence Medal in 2014. He is a fellow of the Royal Society of Canada (RSC), Institute of Electrical and Electronics Engineers (IEEE), Engineering Institute of Canada (EIC), and Canadian Academy of Engineering (CAE). Dr. Wu is a registered professional engineer in the Province of Ontario, Canada.

Navid Zargari received his B.Eng. degree in electrical engineering from Tehran University, Tehran, Iran, in 1987, and the M.A.Sc. and Ph.D. degrees in electrical and computer engineering from Concordia University, Montreal, Canada, in 1991 and 1995, respectively.

Dr. Zargari has been with Rockwell Automation, Canada since November 1994, first as a senior designer, then as the manager of the Medium Voltage R&D Department, and currently as a product architect. For the past 20 years, he has been involved with simulation, analysis, and design of power converters for medium voltage (MV) AC drives. His field of expertise includes converter thermal analysis, converter topologies, and semiconductor devices including SCR, IGCT, SGCT, and IGBT and their associated circuitry. As a power architect, he is a consultant to a multidisciplinary team of more than 40 electrical, mechanical, application engineers and product managers to design, develop, and bring to market medium voltage drives (2300–11,000 V, up to 10,000 HP). He creates technology road maps for the medium voltage drive and starter products, and provides technical solutions and support to the existing and potential new customers and applications.

Dr. Zargari is also an adjunct professor at Ryerson University and his research interests include power converter topologies, advanced control schemes, power semiconductors, and renewable energy systems. He collaborates closely with LEDAR research team at Ryerson University in developing the leading-edge technologies for high-power MV drives. He has coauthored more than 150 research papers as well as a Wiley-IEEE Press book, *Power Conversion and Control of Wind Energy Systems*. He holds more than 40 US granted/pending patents in the Power Converters and Medium Voltage Applications, and received a First Prize Paper Award from the *IEEE Transactions on Industry Applications*. Dr. Zargari was awarded a very prestigious Premier's Catalyst Award – Innovator of the Year in 2009 with a funding of 200,000 CAD for research in wind energy technology and commercialization. Dr. Zargari is a fellow of the Institute of Electrical and Electronics Engineers (IEEE) and a registered professional engineer in the Province of Ontario, Canada.

PREFACE

The high-power multilevel converters are well known in industry and academia as the preferred choices for efficient power conversion. Over the past decade, several power converters are developed and commercialized in the form of standard and customized products that power a wide range of industrial applications. The standard power converters require either a step-up transformer or devices in series to reach the high-voltage operation. The former solution is costly and increases the size and volume of the converter system. The latter solution causes unequal voltage stress on the semiconductor devices. Currently, the modular approach is employed in the designing of the power converters for the high-voltage, high-power applications. This approach is cost-effective and uses low-voltage, low-cost semiconductor technology to build the high-power converters. There are few developments namely; cascaded H-bridge, H-bridge neutral-point clamped, and modular multilevel converter (MMC).

Among them, the MMC is one of the fast-growing technologies and received wide acceptance from both industry and academia. Modular multilevel converters feature the highest degree of modularity and the redundancy due to the large number of submodules they have, as well as the lowest harmonic content due to the large number of output voltage levels they produce. The larger number of submodules substantially increases the converter controller complexity, but each submodule offers a simple structure, reducing the manufacturing costs. Currently, the modular multilevel converters are used in applications like variable speed motor drives, grid integration of off-shore wind farms, HVDC transmission systems, railway traction power supplies, and power quality improvement. Over the past decade, a numerous books on conventional high-power converters focusing on specific areas of control, modeling and analysis, and applications have been published. However, books with a comprehensive analysis, control, and modeling of MMC technologies for a wide range of applications seems not available. This book is intended to fill this gap.

This book covers a wide range of topics from electrical engineering that involves power electronics, electrical machines, control theory, and power systems. The contents of this book include an overview of high-power converters, reference frame theory, modeling and operation of conventional and advanced modular multilevel converters, pulse width modulation schemes, classical and advanced control methods, modeling of AC drives, advanced drive control schemes, transformer-less motor drives, low/zero-speed operation, modeling and control of two-terminal and multi-terminal HVDC systems, active and reactive power control, DC-side fault protection, power quality problems, reactive power, unbalance compensation, modeling and control of static synchronous compensator (STATCOM), and unified power quality conditioner (UPQC). The book reflects the latest developments in the field, and it is a valuable reference book for academic researchers, practicing engineers, and other professionals. Also, this book can be used as a textbook for graduate-level courses.

ORGANIZATION OF BOOK CONTENTS

This book contains three parts with nine chapters. The overview and content of each chapter are described below:

Part I is composed of four chapters related to the analysis, modeling, and control of the conventional MMC. Chapter 1 provides an overview of high-power converters such as voltage source converters, current source converters, matrix converters, and modular multilevel converters, and their technical challenges and applications. Chapter 2 introduces the fundamentals of the MMC, including the submodule (SM) configurations and their operation. Also, the implementation and comprehensive analysis of pulse width modulation schemes (PWM) for an MMC are presented. Chapter 3 presents the classical control approach for the MMC, including submodule capacitor voltage, output current, and circulating current control schemes. Chapter 4 discusses the direct and indirect model predictive control (MPC) schemes for an MMC, where the important issues such as discrete-time modeling, extrapolation techniques, cost function formulation, selection of weight factors, and computational complexity are addressed.

Part II contains three chapters focused on the analysis, modeling, and control of advanced modular multilevel converters. Chapter 5 presents the circuit configuration, control and low/zero-frequency operation of the passive cross-connected modular multilevel converter (PC-MMC). Chapter 6 discusses the configuration, operation, and control of the active cross-connected modular multilevel converter (AC-MMC). Also, the comprehensive analysis of submodule capacitor voltage ripple and circulating currents are presented. Chapter 7 discusses the configuration, operation, and control of the star-channel modular multilevel converter. Also, the delta-channel modular multilevel converter is introduced in this chapter.

Part III contains two chapters focused on the applications of modular multilevel converter including the medium-voltage motor drives, HVDC transmission systems, and power quality improvement. Chapter 8 presents the system configurations, drive control schemes, low/zero-speed operation, common-mode voltage issues, and transformer-less MMC drive. Chapter 9 emphasizes the operation and control of two-terminal and multi-terminal MMC-HVDC systems, DC-side fault protection, static synchronous compensator, and unified power quality conditioner.

SALIENT FEATURES

In summary, this book is a unique and comprehensive work dealing with the technical challenges, modeling, and control aspects of the conventional and advanced modular multilevel converters in a wide range of applications including the medium-voltage motor drives, HVDC transmission systems, and power quality technologies. We developed 179 figures, 26 tables, and 37 case studies in the main body of the book to transfer our knowledge to the readers effectively. Furthermore, this book has the following features:

- Reflects the latest developments in the high-power converters.
- Presents the comprehensive analysis on a variety of modular multilevel converters.
- Presents the detailed modeling, pulse width modulation schemes, classical and advanced control methods for various modular multilevel converters.
- Discusses the technical challenges associated with the modular multilevel converters in the field of medium-voltage motor drives, HVDC transmission systems, and power quality technologies.
- Discusses the important concepts with the simulations.
- Serves as a valuable reference for academic researchers, practicing engineers, and other professionals.
- Provides adequate technical background for graduate level teaching.

SIXING DU
APPARAO DEKKA
BIN WU
NAVID ZARGARI

Toronto, Canada

ACKNOWLEDGMENTS

This book has been written as a result of many years of our research on modular multilevel converters. The manuscript took almost 2 years to complete. We would like to thank our colleagues and friends who have supported and helped us in this endeavor. The references section of each chapter also acknowledges the research works of various scholars who have been passionate about this research area.

We are also very thankful to the reviewers for painstakingly reviewing our book proposal and final manuscript and providing us constructive comments to improve the quality and readability of the book. Our special thanks to the Wiley/IEEE Press editor Mary Hatcher, editorial assistant Victoria Bradshaw, and production team members Danielle Lacourciere, Vishnu Narayanan, and Anand Amarnath for their help in the preparation of this manuscript. We express our intense gratitude to our families and friends who supported and inspired us in every possible way to see the completion of this work.

SIXING DU
APPARAO DEKKA
BIN WU
NAVID ZARGARI

Part I of this book is developed as an outgrowth of my Ph.D. dissertation entitled "High-Power Modular Multilevel Converters: Modeling, Modulation, and Control." I am thankful to Dr. Dewei Xu at Ryerson University, and Dr. Aleksandar Prodic at the University of Toronto for their encouragement in transforming my Ph.D. dissertation into a book. I express my sincere gratitude to my former Ph.D. dissertation supervisor, Dr. Bin Wu, for his guidance and encouragement in writing this manuscript. Also, I am thankful to Dr. Navid Zargari at Rockwell Automation, Canada for his timely feedback and suggestions on our research. This book would not have been envisaged without their help. I wish to thank Drs. Venkata Yaramasu, Marcelo Perez, and Ricardo Lizana for the collaborative works on MMCs and MPC.

APPARAO DEKKA

ACRONYMS

2L	Two-level (converter)
3L	Three-level (converter)
4L	Four-level (converter)
5L	Five-level (converter)
3ϕ	Three-phase (generator/grid)
AC	Alternating current
AC-MMC	Active cross-connected modular multilevel converter
ANPC	Active neutral-point clamped
ABB	ASEA brown boveri
APOD	Alternate phase-opposition-disposition
BM	Blocking mode
BTB	Back-to-back
CCV	Cycloconverter
CD	Double clamp
CEPRI	China electric power research institute
CH	Cascaded half-bridge
CHB	Cascaded H-bridge
CMV	Common-mode voltage
CNPC	Cascaded neutral-point clamped
CSC	Current source converter
CSI	Current source inverter
CSR	Current source rectifier
CT	Continuous time

Modular Multilevel Converters: Analysis, Control, and Applications, First Edition.
Sixing Du, Apparao Dekka, Bin Wu, and Navid Zargari.

DC	Direct current
DFOC	Direct field-oriented control
DMC	Direct matrix converter
DMPC	Direct model predictive control
DSP	Digital signal processor
DT	Discrete time
DTC	Direct torque control
FB	Full-bridge
FC	Flying capacitor
FPGA	Field-programmable gate array
FOC	Field-oriented control
GCT	Gate-commutated thyristor
GTO	Gate turn-off thyristor
HB	Half-bridge
HNPC	H-bridge neutral-point clamped
HV	High-voltage
HVAC	High-voltage alternating current
HVDC	High-voltage direct current
IEGT	Injection-enhanced gate transistor
IFOC	Indirect field-oriented control
IGBT	Insulated-gate bipolar transistor
IGCT	Integrated gate-commutated thyristor
IMC	Indirect matrix converter
IMPC	Indirect model predictive control
LC	Load-commutated
LCC	Line-commutated converter
LCS	Load commutation switch
LPF	Low-pass filter
LSC	Level-shifting carrier
LV	Low-voltage
MB	Main breaker
MC	Matrix converter
MMC	Modular multilevel converter
MMMC	Multi-modular matrix converter

MPC	Model predictive control
MV	Medium-voltage
NASA	National aeronautics and space administration
NLM	Nearest level modulation
NNPC	Nested neutral-point clamped
NPC	Neutral-point clamped
PCC	Point of common coupling
PC-MMC	Passive cross-connected modular multilevel converter
PD	Phase-disposition
PF	Power factor
PI	Proportional and integral
PLL	Phase locked loop
POD	Phase-opposition-disposition
PSC	Phase-shifting carrier
PTG	Pole-to-ground
PTP	Pole-to-pole
PWM	Pulse width modulation
RES	Resonant
RMS	Root mean square
SAM	Sampled average modulation
SCIM	Squirrel cage induction motor
SCR	Silicon-controlled rectifier
SGCT	Symmetric gate-commutated thyristor
SHE	Selective harmonic elimination
SM	Submodule
STATCOM	Static synchronous compensator
SVM	Space vector modulation
THD	Total harmonic distortion
UFD	Ultra-fast disconnector
UPQC	Unified power quality conditioner
VOC	Voltage-oriented control
VSC	Voltage source converter
VSI	Voltage source inverter
VSR	Voltage source rectifier

SYMBOLS

Notation

P, Q, S	Active, reactive, and apparent power
R, L, C	Resistance, inductance, and capacitance
v, i, ψ	Voltage, current, and flux linkage (Peak values)
V, I, Ψ	Voltage, current, and flux linkage (RMS values)
x	Phase variable $\in \{a, b, c\}$
y	Arm variable $\in \{u, l\}$

Superscripts of the Variable x

$(x)^n$	Normalized variable
$(x)^m$	Measured variable
$(x)^p$	Predicted/previous instant variable
$\widehat{(x)}$	Estimated variable
$(x)^*$	Reference variable
$\widehat{(x)^*}$	Extrapolated reference variable
$\overline{x}$	Complementary/DC value
$\vec{x}$	Space phasor
$(x)^+$	Positive sequence component
$(x)^-$	Negative sequence component
$\lvert x \rvert$	Magnitude of space phasor

Subscripts of the Variable x

$\vec{()}_r$	Reference space vector
$(x)_{abc}$	Stationary (abc) frame variables

Modular Multilevel Converters: Analysis, Control, and Applications, First Edition.
Sixing Du, Apparao Dekka, Bin Wu, and Navid Zargari.

$(x)_B$	Base value
$(x)_{dq0}$	Synchronous ($dq0$) frame variables
$(x)_{dc}$	DC-system variable
$(x)_g$	Grid-related variable
$(x)_i$	Inverter-related variable
$(x)_r$	Rectifier/rotor-related variable
$(x)_{\alpha\beta0}$	Stationary ($\alpha\beta0$) frame variables

Significant Symbols

a_m	Coefficient of extrapolated variable
a^*, b^*, c^*	Phase modulation signals
a_l^*, b_l^*, c_l^*	Lower arm modulation signals
a_u^*, b_u^*, c_u^*	Upper arm modulation signals
a_y^*, b_y^*, c_y^*	Modulation signals of yth-arm
A_d	Direction of the arm current
AI_h	Actual submodule capacitor voltage index number
C	Submodule capacitance
C_{dc}	DC-bus or DC-link capacitor
C_f	AC filter or commutation capacitor
C_M	Motor leakage capacitance
C_T	Transformer leakage capacitance
C_x	Cross-connected capacitor in PC-MMC
$C_a, C_b, C_c,$	Cross-connected capacitors in PC-MMC
C_{c1}, C_{c2}	Flying capacitors
D_{xy}	Number of inserted submodules in present cycle
D_{xu}, D_{xl}	Number of inserted submodules in the arm
D_{xy}^p	Number of inserted submodules in previous cycle
e_{max}, e_{min}	Maximum and minimum value of submodule energy
E	Battery voltage
E_{xyh}	Submodule energy
ΔE_{xyh}	Submodule energy fluctuation
f_1	AC grid-I frequency
f_2	AC grid-II frequency
f_3	AC grid-III frequency
f_4	AC grid-IV frequency
f_α, f_β	Stationary-$\alpha\beta$ frame signals
f_a, f_b, f_c	Stationary-abc frame signals

f_c	Carrier frequency
f_d, f_q	Synchronous-dq frame signals
f_g	Grid frequency
f_h	High frequency
F_m	Peak value of space phasor
f_o	Inverter output frequency
f_r	Resonant frequency
f_s	Sampling/stator frequency
f_{sr}	Rated stator frequency
f_{sw}	Converter switching frequency
$f_{_{SM}}$	Submodule switching frequency
g	Cost function of model predictive control
g_c	Cost function of submodule capacitor voltage control
g_{dc}	Cost function of DC-bus current control
g_o	Cost function of output current control
g_{op}	Optimal cost function value
g_{t1}, g_{t2}	Gating signals of the voltage levels
g_{y1}, g_{y2}	Gating signals of the switching vectors
g_{y3}, g_{y4}	Gating signals of the switching vectors
g_z	Cost function of AC circulating current control
$G_{1,xyh}$	Gating signals of converter station-I
$G_{2,xyh}$	Gating signals of converter station-II
$G_{3,xyh}$	Gating signals of converter station-III
$G_{4,xyh}$	Gating signals of converter station-IV
$G_{_{LC}}$	Transfer function
G_{xyh}	Submodule gating signals of the arm
G_{xuh}, G_{xlh}	Gating signals of the upper and lower arm submodules
h	Harmonic order
$\mathbf{i_1}$	Peak value of AC grid-I current
$\mathbf{i_2}$	Peak value of AC grid-II current
$\mathbf{i_3}$	Peak value of AC grid-III current
$\mathbf{i_4}$	Peak value of AC grid-IV current
$\mathbf{i_g}$	Peak value of grid current
$\mathbf{i_i}$	Peak value of inverter output current
$\mathbf{i_m}$	Peak value of magnetizing current
$\mathbf{i_o}$	Peak value of load current
$\mathbf{i_r}$	Peak value of rotor current
$\mathbf{i_s}$	Peak value of stator current

$i_{\alpha g}, i_{\beta g}$	Grid currents in $\alpha\beta$-frame
$i_{\alpha i}, i_{\beta i}$	Inverter output currents in $\alpha\beta$-frame
$i_{\alpha z}, i_{\beta z}$	AC circulating currents in $\alpha\beta$-frame
i_{ag}, i_{bg}, i_{cg}	Grid currents in abc-frame
i_{ai}, i_{bi}, i_{ci}	Inverter output currents in abc-frame
i_{al}, i_{bl}, i_{cl}	Converter lower arm currents in abc-frame
i_{ao}, i_{bo}, i_{co}	Load currents in abc-frame
i_{as}, i_{bs}, i_{cs}	Stator currents in abc-frame
i_{au}, i_{bu}, i_{cu}	Converter upper arm currents in abc-frame
i_{aw}, i_{bw}, i_{cw}	Cross-connected arm currents in abc-frame
i_{az}, i_{bz}, i_{cz}	AC circulating currents in abc-frame
$i_{al}^{m}, i_{bl}^{m}, i_{cl}^{m}$	Outer lower sub-arm currents in abc-frame
$i_{au}^{m}, i_{bu}^{m}, i_{cu}^{m}$	Outer upper sub-arm currents in abc-frame
$i_{am}^{l}, i_{bm}^{l}, i_{cm}^{l}$	Lower star-channel currents in abc-frame
$i_{am}^{u}, i_{bm}^{u}, i_{cm}^{u}$	Upper star-channel currents in abc-frame
i_{d1}, i_{q1}	AC grid-I currents in dq-frame
i_{d2}, i_{q2}	AC grid-II currents in dq-frame
i_{d3}, i_{q3}	AC grid-III currents in dq-frame
i_{d4}, i_{q4}	AC grid-IV currents in dq-frame
i_{dg}, i_{qg}	Grid currents in dq-frame
i_{di}, i_{qi}	Inverter output currents in dq-frame
i_{do}, i_{qo}	Load currents in dq-frame
i_{ds}, i_{qs}	Stator currents in dq-frame
i_{dz}, i_{qz}	AC circulating currents in dq-frame
$\Delta i_{di}, \Delta i_{qi}$	Inverter output current error in dq-frame
$\Delta i_{dz}, \Delta i_{qz}$	AC circulating current error in dq-frame
i_C	Submodule capacitor current
i_{dc}	DC-bus current
i_{jx}	Current direction
i_{xd}	DC component in the arm current
i_{xh}	High-frequency component in the arm current
i_{xi}	Inverter output current of the phase-x
i_{xl}	Lower arm current of the phase-x
i_{xo}	Passive load current of the phase-x
i_{xs}	Stator current of the phase-x
i_{xse}	Series compensator current of the phase-x
i_{xsh}	Shunt compensator current of the phase-x

i_{xu}	Upper arm current of the phase-x
i_{xw}	Cross-connected arm current of the phase-x
i_{xy}	Arm current of the phase-x
i_{xz}	AC circulating current of the phase-x
i_{xzf}	AC+DC circulating current components in the arm
i_{xd}^{f}	Feed-forward DC current component
i_{xh}^{f}	Feed-forward high-frequency current component
i_{xz}^{f}	Feed-forward circulating current component
i_{xm}^{u}, i_{xm}^{l}	Upper and lower star-channel current
i_{Cyh}^{x}	Submodule capacitor current
I	Peak value of output current
I_{B}	Base value of output current
I_{dc}	Peak value of DC-bus current
I_{h}	Peak value of high-frequency current
I_{n}	Peak value of negative sequence current
I_{p}	Peak value of positive sequence current
I_{s}	RMS value of stator current
I_{xz}	Peak value of AC circulating current
j	Switching state
j_{op}	Optimal switching state
J	Moment of inertia
k	Present sampling instant in discrete time
$k-1$	Past sampling instant in discrete time
$k+1$	Future sampling instant in discrete time
k_{c}	Current limiting factor
k_{h}	Coefficient of resonant controller
k_{id}	Integral gain of d-axis current PI controller
k_{ii}	Integral gain of current PI controller
k_{iq}	Integral gain of q-axis current PI controller
k_{iv}	Integral gain of voltage PI controller
k_{pd}	Proportional gain of d-axis current PI controller
k_{pi}	Proportional gain of current PI controller
k_{pq}	Proportional gain of q-axis current PI controller
k_{pv}	Proportional gain of voltage PI controller
K_{dc}	Slope of the voltage droop
K_{T}	Motor torque constant

l	Lower arm
L_{1t}	AC grid+transformer leakage inductance
L_d	Delta-channel inductor
L_g	AC grid inductance
L_{lr}	Rotor leakage inductance
L_{ls}	Stator leakage inductance
L_m	Motor magnetizing inductance
L_r	Rotor self-inductance
L_s	Stator self-inductance
L_{se}	Filter inductance of series compensator
L_{sh}	Filter inductance of shunt compensator
m	Number of voltage levels
m_a	Amplitude modulation index
m_f	Frequency modulation index
m_{xuh}, m_{xlh}	Normalized upper and lower arm submodule voltages
m_{xch}	Normalized cross-connected arm reference voltage
m_{xu}^n, m_{xl}^n	Normalized upper and lower arm voltage levels
m_c^x	Normalized cross-connected arm voltage
m_l^x	Normalized lower arm voltage
m_{l1}^x	Normalized sub-arm III voltage
m_{l2}^x	Normalized sub-arm IV voltage
m_u^x	Normalized upper arm voltage
m_{u1}^x	Normalized sub-arm I voltage
m_{u2}^x	Normalized sub-arm II voltage
n	Order of extrapolation in discrete time
n	Neutral point of the load
nu, nl	Neutral point of star-channel arms
N	Number of submodules per arm
N_c	Number of submodules in the star-channel
N_p	Predictive horizon
N_r	Reference rotor speed
o	DC-bus mid-point
p_{eg1}^x	Submodule power in the sub-arm
p_{eg}^x, p_{rs}^x	Upper and lower arm powers in PC-MMC
p_{xyh}	Submodule instantaneous power
$P_{o,3}$	Initial value of active power
P_1	Active power of converter station-I

P_2	Active power of converter station-II
P_3	Active power of converter station-III
P_4	Active power of converter station-IV
P	Number of pole pairs
P_B	Base active power
P_{dc}	DC-bus power
P_g	Grid active power
P_s	Stator active power
Q_1	Reactive power of converter station-I
Q_2	Reactive power of converter station-II
Q_3	Reactive power of converter station-III
Q_4	Reactive power of converter station-IV
Q_g	Grid reactive power
r	Magnitude of polar coordinate system
r, L	Resistance and inductance of the arm
r_o, L_o	Resistance and inductance of the load
r_{dc}, L_{dc}	Resistance and inductance of the DC-line
R_r	Rotor winding resistance
R_s	Stator winding resistance
R_{xyh}	Submodule gating signals on the rectifier
s	Laplace domain
S_h	Wave shape of high-frequency signal
S_{hw}	RMS value of high-frequency voltage
S_l	Lower arm switching signals
S_o	Load apparent power
S_p	Wave shape of cross-connected arm voltage
S_u	Upper arm switching signals
S_{xy}	Arm voltage level
S_{xyh}	Submodule switching state in the present cycle
S^p_{xyh}	Submodule switching state in the previous cycle
t	Time
t_1	Sampling instant
T	Triangular carrier signals
T_e	Motor electromagnetic torque
T_L	Motor load torque
T_s	Sampling time
T_{xy}	Dwell time of the arm

T_{x1}, T_{x1}	Dwell time of the voltage levels
$T_{y0},\ T_{y1},\ T_{y2},\ T_{y3}, T_{y4}$	Dwell time of the switching vectors
$\mathbf{T}_{abc \Rightarrow \alpha\beta}$	abc-frame to $\alpha\beta$-frame transformation matrix
$\mathbf{T}_{\alpha\beta \Rightarrow abc}$	$\alpha\beta$-frame to abc-frame transformation matrix
$\mathbf{T}_{abc \Rightarrow dq}$	abc-frame to dq-frame transformation matrix
$\mathbf{T}_{dq \Rightarrow abc}$	dq-frame to abc-frame transformation matrix
$\mathbf{T}_{abc \Rightarrow dq0}$	abc-frame to $dq0$-frame transformation matrix
u	Upper arm
v_{a1}, v_{b1}, v_{c1}	AC grid-I voltages in abc-frame
v_{a2}, v_{b2}, v_{c2}	AC grid-II voltages in abc-frame
v_{a3}, v_{b3}, v_{c3}	AC grid-III voltages in abc-frame
v_{a4}, v_{b4}, v_{c4}	AC grid-IV voltages in abc-frame
$\mathbf{v_g}$	Peak value of grid voltage
$\mathbf{v_i}$	Peak value of inverter output voltage
$\mathbf{v_r}$	Peak value of rotor voltage
$\mathbf{v_s}$	Peak value of stator voltage
v_{ab}	Voltage between phase-a and phase-b
v_{abn}	n^{th}- harmonic component of line-to-line voltage
v_{ag}, v_{bg}, v_{cg}	Grid voltages in abc-frame
v_{ai}, v_{bi}, v_{ci}	Inverter output voltages in abc-frame
v_{an}, v_{bn}, v_{cn}	Motor phase-to-neutral voltages in abc-frame
v_{ao}, v_{bo}, v_{co}	Motor phase-to-DC-bus mid-point voltages in abc-frame
v_{ap}, v_{bp}, v_{cp}	High-frequency pulse voltages in abc-frame
v_{ar}, v_{br}, v_{cr}	Rectifier voltages in abc-frame
v_{as}, v_{bs}, v_{cs}	Stator voltages in abc-frame
v_{az}, v_{bz}, v_{cz}	Voltage components of AC circulating current in abc-frame
v_C	Submodule capacitor voltage
v_{CD}	Output voltage of double clamp SM
v_{Cg}	Average DC voltage of star-channel MMC
v_{Cg}	Average overall DC voltage of MMC
v_{CH}	Output voltage of cascaded half-bridge SM
v_{cmi}	Inverter common-mode voltage
v_{cmr}	Rectifier common-mode voltage
v_{cmv}	Drive common-mode voltage
v_{Cu}, v_{Cl}	Average value of arm capacitor voltage
v_{Cx}	Average value of leg voltage

Δv_{Cx}	Average leg voltage error
v_d	DC component in the arm voltage
v_{d1}, v_{q1}	AC grid-I voltages in dq-frame
v_{d2}, v_{q2}	AC grid-II voltages in dq-frame
v_{d3}, v_{q3}	AC grid-III voltages in dq-frame
v_{d4}, v_{q4}	AC grid-IV voltages in dq-frame
v_{di}, v_{qi}	Inverter output voltage in dq-frame
v_{dg}, v_{qg}	Grid voltages in dq-frame
v_{dr}, v_{qr}	Rectifier voltages in dq-frame
v_{ds}, v_{qs}	Stator voltages in dq-frame
v_{dz}, v_{qz}	Voltage components of AC circulating current in dq-frame
v_F	Output voltage of full-bridge SM
v_{FC}	Output voltage of flying capacitor SM
v_H	Output voltage of half-bridge SM
v_{Lev}	Number of arm voltage levels
v_n	Harmonic components in the arm voltage
v_{ng}	Voltage between motor neutral and grid ground
v_{no}	Voltage between motor neutral and DC-bus mid-point
v_{og}	Voltage between grid ground and DC-bus mid-point
v_{sw}	Switching voltage ripple
v_{xf}	Fundamental-frequency component in the arm voltage
v_{xh}	High-frequency component in the arm voltage
v_{xi}	Inverter output voltage
v_{xp}	High-frequency pulse voltage in an AC-MMC
v_{xs}	Stator voltage
v_{xu}, v_{xl}	Upper and lower arm voltages
v_{xz}	Voltage component corresponding to AC circulating current
$\vec{v}_r$	Voltage space vector
$\vec{v}_{ru}, \vec{v}_{rl}$	Upper and lower arm voltage space vectors
$\vec{v}_{ry}$	Voltage space vector of yth-arm
$\vec{v}_{ryo}$	Offset of voltage space vector
$\vec{v}_{ryt}$	Two-level voltage space vector
$\vec{v}_{y1}, \vec{v}_{y2}$	Switching vectors of SVM
$\vec{v}_{y3}, \vec{v}_{y4}$	Switching vectors of SVM
$\Delta v^*_{1,xyh}$	Compensating voltage command of converter station-I
$\Delta v^*_{2,xyh}$	Compensating voltage command of converter station-II
$\Delta v^*_{3,xyh}$	Compensating voltage command of converter station-III

$\Delta v^*_{4,xyh}$	Compensating voltage command of converter station-IV
v^*_{dtx}	Reference value of leg voltage controller
v^*_{xuh}, v^*_{xlh}	Reference voltage of upper and lower arm submodules
Δv^*_{xh}	Compensating high-frequency voltage
Δv^*_{xyh}	Compensating submodule voltage command
v^a_{nuo}, v^b_{nuo},	Voltage between star-channel neutral point and
v^c_{nuo}	DC-bus mid-point in *abc*-frame
v^n_{rxi}	Normalized reference inverter output voltage
v^n_{xi}	Normalized inverter output voltage
v^n_{xu}, v^n_{xl}	Normalized upper and lower arm voltages
v^x_{Cyh}	Submodule capacitor voltage
v^x_{Cuh}, v^x_{Clh}	Submodule capacitor voltage in the upper and lower arms
v^x_{eg}	Upper arm voltage in PC-MMC
v^x_{em}	Sub-arm I voltage in PC-MMC
v^x_{Fyh}	Full-bridge capacitor voltage in star-channel
v^x_{mg}	Sub-arm II voltage in PC-MMC
v^x_{mnu}	Voltage between star-channel neutral point and arm mid-point
v^x_{nuo}	Voltage between star-channel neutral point and DC-bus mid-point
v^x_{mw}	Voltage across cross-connected arm in an AC-MMC
v^x_{rs}	Lower arm voltage in PC-MMC
v^x_{rw}	Sub-arm III voltage in PC-MMC
v^x_{Syh}	Submodule output voltage
v^x_{ws}	Sub-arm IV voltage in PC-MMC
Δv^x_{Cyh}	Submodule capacitor voltage fluctuations
VI_h	Submodule capacitor voltage index number
V_1	RMS value of AC grid-I voltage
V_2	RMS value of AC grid-II voltage
V_3	RMS value of AC grid-III voltage
V_4	RMS value of AC grid-IV voltage
V_B	Base voltage
V_{dc}	DC-bus or DC-link voltage
ΔV_{dc}	DC-bus voltage error
V_{dcl}	Lower limit of DC-bus voltage
V_{dcm}	Maximum value of DC-bus voltage
V_{dcn}	Minimum value of DC-bus voltage
V_{dcu}	Upper limit of DC-bus voltage

V_f	Peak value of fundamental frequency voltage
V_g	RMS value of grid voltage
V_h	Peak value of high-frequency voltage
V_n	Peak value of negative sequence voltage
V_o	Output RMS line-to-line voltage
V_p	Peak value of positive sequence voltage
V_s	RMS value of stator voltage
V_{dc}^o	Initial value of DC-bus voltage
V_{dc}^+	Positive pole voltage
V_{dc}^-	Negative pole voltage
V_{x1}, V_{x2}	Nearest output voltage levels
V_{xk}	Nearest k^{th}-voltage level
V_{xl1}, V_{xl2}	Lower arm nearest voltage levels
V_{xu1}, V_{xu2}	Upper arm nearest voltage levels
V_{xuk}, V_{xlk}	Upper and lower arm k^{th}-voltage level
V_C^*	Reference submodule capacitor voltage
$V_{1,C}^*$	Reference submodule capacitor voltage in station-I
$V_{2,C}^*$	Reference submodule capacitor voltage in station-II
$V_{3,C}^*$	Reference submodule capacitor voltage in station-III
$V_{4,C}^*$	Reference submodule capacitor voltage in station-IV
ω_o	Inverter output angular frequency
ω_s	Stator angular frequency
ω	Angular frequency of arbitrary reference frame
ω_r	Rotor electrical angular sppeed
ω_{sl}	Angular slip frequency
ω_h	Angular high frequency
ω_g	Angular grid frequency
ω_f	Feed-forward angular grid frequency
ω_{mr}	Motor mechanical speed
ω_2	Angular frequency of AC grid-II
z	Discrete time sampling delay
Z_B	Base impedance
τ_r	Rotor electrical time constant
θ_1	AC grid-I angle
θ_2	AC grid-II angle
θ_3	AC grid-III angle
θ_4	AC grid-IV angle

θ	Phase angle in the polar coordinate system
θ	Phase angle between the space phasor and reference axis
θ_f	Rotor flux vector position
θ_g	Grid angle
θ_o	Initial phase angle of the space phasor
θ_r	Rotor position
θ_{sl}	Slip angle
θ_x	Phase angle difference between three-phase signals
λ_c	Weighting factor of submodule capacitor voltage control
λ_{dc}	Weighting factor of DC-bus current control
λ_o	Weighting factor of output current control
λ_z	Weighting factor of AC circulating current control
ψ_r	Peak value of rotor flux
ψ_{dr}, ψ_{qr}	Rotor flux components in dq-frame
$\boldsymbol{\psi}_\mathbf{r}$	RMS value of rotor flux vector
$\boldsymbol{\psi}_\mathbf{s}$	RMS value of stator flux vector
φ	Load power factor angle
φ_g	Grid power factor angle
ϕ	Phase angle between d-axis and dq-space phasor
ϕ_c	Phase angle between the carrier signals
ϕ_{ci}	Angle between the upper and lower arm carrier signals
δ	Phase angle between positive and negative sequence voltages
α	Phase angle between positive sequence voltage and current
β	Phase angle between negative sequence voltage and current
Φ_{dc}, Γ_{dc}	DC-bus current constants in discrete time
Φ_o, Γ_o	Output current constants in discrete time
Φ_z, Γ_z	AC circulating current constants in discrete time

Vectors

$\mathbf{f_{abc}}$	Stationary-abc frame signals = $[f_a\ f_b\ f_c]^T$
$\mathbf{f_{dq}}$	Synchronous-dq frame signals = $[f_d\ f_q]^T$
$\mathbf{f}_{\alpha\beta}$	Stationary-$\alpha\beta$ frame signals = $[f_\alpha\ f_\beta]^T$
$\mathbf{i_{1,abc}}$	AC grid-I currents in abc-frame = $[i_{a1}\ i_{b1}\ i_{c1}]^T$
$\mathbf{i_{2,abc}}$	AC grid-II currents in abc-frame = $[i_{a2}\ i_{b2}\ i_{c2}]^T$
$\mathbf{i_{3,abc}}$	AC grid-III currents in abc-frame = $[i_{a3}\ i_{b3}\ i_{c3}]^T$
$\mathbf{i_{4,abc}}$	AC grid-IV currents in abc-frame = $[i_{a4}\ i_{b4}\ i_{c4}]^T$
$\mathbf{i_{a,ul}}$	Phase-a arm currents = $[i_{au}\ i_{al}]^T$

$\mathbf{i_{g,abc}}$	Grid currents in *abc*-frame = $[i_{ag}\ i_{bg}\ i_{cg}]^T$
$\mathbf{i_{i,abc}}$	Inverter output currents in *abc*-frame = $[i_{ai}\ i_{bi}\ i_{ci}]^T$
$\mathbf{i_{o,abc}}$	Load currents in *abc*-frame = $[i_{ao}\ i_{bo}\ i_{co}]^T$
$\mathbf{i_{s,abc}}$	Stator currents in *abc*-frame = $[i_{as}\ i_{bs}\ i_{cs}]^T$
$\mathbf{i_{xi}}$	Inverter output currents = $[i_{ai}\ i_{bi}\ i_{ci}]^T$
$\mathbf{i_{xl}}$	Lower arm currents = $[i_{al}\ i_{bl}\ i_{cl}]^T$
$\mathbf{i_{xs}}$	Stator currents = $[i_{as}\ i_{bs}\ i_{cs}]^T$
$\mathbf{i_{xu}}$	Upper arm currents = $[i_{au}\ i_{bu}\ i_{cu}]^T$
$\mathbf{i_{xz}}$	AC circulating currents = $[i_{az}\ i_{bz}\ i_{cz}]^T$
$\mathbf{i_{1,dq}}$	AC grid-I currents in *dq*-frame = $[i_{d1}\ i_{q1}]^T$
$\mathbf{i_{2,dq}}$	AC grid-II currents in *dq*-frame = $[i_{d2}\ i_{q2}]^T$
$\mathbf{i_{3,dq}}$	AC grid-III currents in *dq*-frame = $[i_{d3}\ i_{q3}]^T$
$\mathbf{i_{4,dq}}$	AC grid-IV currents in *dq*-frame = $[i_{d4}\ i_{q4}]^T$
$\mathbf{i_{i,dq}}$	Inverter output currents in *dq*-frame = $[i_{di}\ i_{qi}]^T$
$\mathbf{i_{g,dq}}$	Grid currents in *dq*-frame = $[i_{dg}\ i_{qg}]^T$
$\mathbf{i_{s,dq}}$	Stator currents in *dq*-frame = $[i_{ds}\ i_{qs}]^T$
$\mathbf{i^+_{i,dq0}}$	Positive sequence inverter currents = $[i^+_{di}\ i^+_{qi}\ i^+_{0i}]^T$
$\mathbf{i^-_{i,dq0}}$	Negative sequence inverter currents = $[i^-_{di}\ i^-_{qi}\ i^-_{0i}]^T$
$\mathbf{v_{1,abc}}$	AC grid-I voltages in *abc*-frame = $[v_{a1}\ v_{b1}\ v_{c1}]^T$
$\mathbf{v_{2,abc}}$	AC grid-II voltages in *abc*-frame = $[v_{a2}\ v_{b2}\ v_{c2}]^T$
$\mathbf{v_{3,abc}}$	AC grid-III voltages in *abc*-frame = $[v_{a3}\ v_{b3}\ v_{c3}]^T$
$\mathbf{v_{4,abc}}$	AC grid-IV voltages in *abc*-frame = $[v_{a4}\ v_{b4}\ v_{c4}]^T$
$\mathbf{v_{a,ul}}$	Phase-a arm voltages = $[v_{au}\ v_{al}]^T$
$\mathbf{v_{g,abc}}$	Grid voltages in *abc*-frame = $[v_{ag}\ v_{bg}\ v_{cg}]^T$
$\mathbf{v_{i,abc}}$	Inverter output voltages in *abc*-frame = $[v_{ai}\ v_{bi}\ v_{ci}]^T$
$\mathbf{v_{r,abc}}$	Rectifier output voltages in *abc*-frame = $[v_{ar}\ v_{br}\ v_{cr}]^T$
$\mathbf{v_{s,abc}}$	Stator voltages in *abc*-frame = $[v_{as}\ v_{bs}\ v_{cs}]^T$
$\mathbf{v_{xl}}$	Lower arm voltages = $[v_{al}\ v_{bl}\ v_{cl}]^T$
$\mathbf{v_{xu}}$	Upper arm voltages = $[v_{au}\ v_{bu}\ v_{cu}]^T$
$\mathbf{v_{xz}}$	Voltage components of AC circulating current = $[v_{az}\ v_{bz}\ v_{cz}]^T$

$\mathbf{v_{z,abc}}$	Voltage components of AC circulating current controller $= [v_{az}\ v_{bz}\ v_{cz}]^T$
$\mathbf{v_{1,dq}}$	AC grid-I voltages in dq-frame $= [v_{d1}\ v_{q1}]^T$
$\mathbf{v_{2,dq}}$	AC grid-II voltages in dq-frame $= [v_{d2}\ v_{q2}]^T$
$\mathbf{v_{3,dq}}$	AC grid-III voltages in dq-frame $= [v_{d3}\ v_{q3}]^T$
$\mathbf{v_{4,dq}}$	AC grid-IV voltages in dq-frame $= [v_{d4}\ v_{q4}]^T$
$\mathbf{v_{g,dq}}$	Grid voltages in dq-frame $= [v_{dg}\ v_{qg}]^T$
$\mathbf{v^+_{g,dq0}}$	Positive sequence grid voltages $= [v^+_{dg}\ v^+_{qg}\ v^+_{0g}]^T$
$\mathbf{v^-_{g,dq0}}$	Negative sequence grid voltages $= [v^-_{dg}\ v^-_{qg}\ v^-_{0g}]^T$

ABOUT THE COMPANION WEBSITE

This book is accompanied by a companion website:

www.wiley.com/go/Du/ModularMultilevelConverters

The website includes:

- Demo project
- Video

PART I

GENERAL ASPECTS OF CONVENTIONAL MMC

CHAPTER 1

REVIEW OF HIGH-POWER CONVERTERS

1.1 INTRODUCTION

Power converters are well known in industries and the academia as one of the preferred choices for high-efficiency, electric power conversion systems. For the past few years, several power converters were developed and commercialized in the form of standard and customized products. These converters power a wide range of industrial applications such as pumps, compressors, fans, mills, conveyors, transportation (e.g., electric vehicles and, railway traction), manufacturing, petrochemical, high-voltage direct current (HVDC) transmission, reactive power compensation, and wind energy conversion systems [1–3].

This chapter deals with state-of-the-art and recent advancements in high-power converter topologies. The classification of high-power converters is presented in Section 1.2. The first category of power converters uses an intermediate DC-link to achieve the power conversion from fixed AC to variable AC supply. Based on the type of intermediate DC-link, power converters are categorized into voltage source converters (VSC) and current source converters (CSC). A brief description of VSC and CSC topologies is presented in Sections 1.3 and 1.4, respectively. An alternative approach is the direct power conversion from fixed AC to variable AC without

Modular Multilevel Converters: Analysis, Control, and Applications, First Edition.
Sixing Du, Apparao Dekka, Bin Wu, and Navid Zargari.

any intermediate DC-link. Cycloconverters (CCV) and matrix converters (MC) belong to this category. The circuit configuration and operation of these topologies are presented in Section 1.5.

Voltage source converters have the highest market share and are widely used in industries. Several VSCs were developed in the past years. Among them, modular multilevel converter (MMC) is a highly regarded topology, which is the main focus of this book. The modular multilevel converter has several features such as modular construction, high reliability, and cost-effectiveness because it uses low-voltage, insulated gate bipolar transistor (LV-IGBT) technology to reach high-voltage (HV) operation. MMC is directly connected to medium-voltage (MV) and high-voltage (HV) grids without any step-up transformer. These features are potentially useful for a wide range of applications. This topology is commercially implemented in medium-voltage motor drives, HVDC transmission, multi-terminal HVDC systems, offshore wind farms, and static synchronous compensators (STATCOM). MMC also requires a control scheme with a more complex structure and function to manage several control objectives simultaneously. Furthermore, it requires a high-performance control platform to process a large number of gating signals. The applications and technical challenges of MMC are presented in Section 1.6. The summary of this chapter is discussed in Section 1.7.

1.2 OVERVIEW OF HIGH-POWER CONVERTERS

Most power converter topologies are available in the form of standard products for an operating voltage of 2.3–13.8 kV only. These power converter topologies require either a step-up transformer or high-voltage semiconductor devices for a high-voltage operation. The former solution is costly and increases the size and volume of the converter system. The new solution does not increase the size and volume of the converter. The main limitation of this solution is the availability of HV semiconductor devices. The available semiconductor devices and their voltage and current ratings are shown in Table 1.1. Insulated gate bipolar transistor (IGBT) devices are widely used in VSCs and matrix converters. These devices are available with a maximum voltage 6.5 kV and a maximum current of 2.4 kA. The integrated gate commutated thyristor (IGCT) is another important device technology used in voltage source converters. Other variants of the IGCT technology include asymmetrical, reverse conducting, and symmetrical types. Symmetrical gate commutated thyristors (SGCT) are used in the CSCs, which are available for a maximum voltage of 10 kV and a maximum current of 5 kA [3, 4]. These MV rated devices cannot block the higher operating voltage.

Alternatively, the medium-voltage devices are connected in series to increase the operating voltage of power converters. The series connected devices, and their gate drivers may not exhibit similar static and dynamic performance. Also, these devices may not equally share the total voltage during blocking mode or switching transients. This approach requires a voltage equalization circuit to achieve the equal voltage sharing during blocking mode. These additional circuits increase the power losses in

Table 1.1 Market overview of power semiconductor devices

Parameter	Power Diode	Thyristor	GTO	GCT/IGCT/ SGCT	LV-IGBT/ HV-IGBT
Voltage	8.5 kV@1.2 kA	12 kV@1.5 kA	6 kV@6 kA	10 kV@1.7 kA	6.5 kV@0.75 kA
Current	9.6 kA@1.8 kV	5 kA@0.4 kV	6 kA@6 kV	5 kA@4.5 kV	2.4 kA@1.7 kV

power converters. The series connection of devices does not improve output voltage and current waveform quality.

A modular approach has been developed to overcome the limitations of semiconductor device ratings and their series connection to achieve higher operating voltage. In a modular approach, the identical low-voltage submodules are connected in cascade to reach the higher operating voltage. With the modular approach, the scaling of voltage and power rating of power converters becomes easy, the complexity of assembly and maintenance of power converters is reduced, and power converters can be operated at reduced capacity during submodule failure. Hence, modular power converters become popular in both industries and the academia for high-voltage, high-power applications.

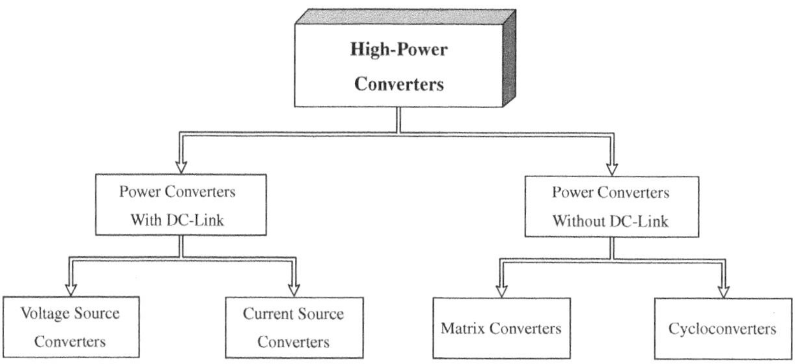

Figure 1.1 Classification of high-power converters.

High-power converter topologies are categorized into two groups as shown in Figure 1.1. The first group consists of power converters with an intermediate DC-link (DC energy storage component). This group of converters performs the power conversion in two stages: AC–DC and DC–AC between the AC grid and a machine. This group is further categorized into VSCs and CSCs, depending on the nature of the DC energy storage component [5–9]. VSCs normally employ DC capacitors as DC energy storage component, whereas CSCs employ DC inductors in the DC-bus. VSCs are referred to as voltage source rectifiers (VSR) in the AC–DC conversion stage, and as voltage source inverters (VSI) in the DC–AC conversion stage. Similarly, CSCs are referred to as current source rectifier (CSR) and current source inverter (CSI), based on their functionality.

The second group consists of power converters without an intermediate DC-link (DC energy storage component). This group of converters performs a direct AC–AC conversion between the AC grid and a machine. CCVs and matrix converters belong to this group. The CCV is a widely used topology in high-power applications, which use an array of thyristors for the direct connection of the AC grid to a machine. CCV allows efficient power flow in both directions. However, CCV has limited dynamic performance, operating voltage, and frequency range, and low power factor at low motor speed [3, 5]. The matrix converter is a new topology capable of the direct AC–AC conversion without DC energy storage element. The main features of matrix converters are sinusoidal input or output waveforms, a controllable input power factor, a wide range of output voltage and frequency control, and light-weight design unlike that of CCV [3, 10]. The power converters shown in Figure 1.1, are briefly discussed in the following sections.

1.3 VOLTAGE SOURCE CONVERTERS

VSCs are a matured technology that uses a DC capacitor as a storage element in the DC-link. Over the past decade, VSCs exhibited higher market penetration and more evident development compared to CSCs and matrix converters. The most prominent developments in VSCs are shown in Figure 1.2. The two-level converter is a standard VSC; that is limited to low-voltage (LV) and low-power applications. For high-power applications, two-level converters require either device in parallel to carry high-current at low-voltage operation or devices in series to reach medium-voltage operation with low current-carrying capacity. The parallel and series connections of devices do not introduce any additional benefits in the improvement of power quality (reduction of voltage and current harmonic distortion) and $\frac{dv}{dt}$ reduction. Power losses and voltage blocking are uneven in series connected devices. Thus, two-level converters are not very popular for high-power applications [7–12].

New converter topologies with low-cost semiconductor technology, also known as multilevel converters, are developed for high-power applications. Multilevel converters have significant advantages compared to high-power, two-level converters. These advantages include low $\frac{dv}{dt}$ and voltage harmonic distortion, near-sinusoidal currents, smaller size of input and output filters (if necessary), high-efficiency, low common-mode voltage, and possible fault-tolerant operation in certain cases. Multilevel converters are composed of an array of low-voltage or medium-voltage semiconductor devices and DC capacitive voltage sources. Multilevel converters generate a stepped voltage waveform with variable magnitude and frequency with a proper arrangement of devices, capacitive voltage sources, and control methods [7–12].

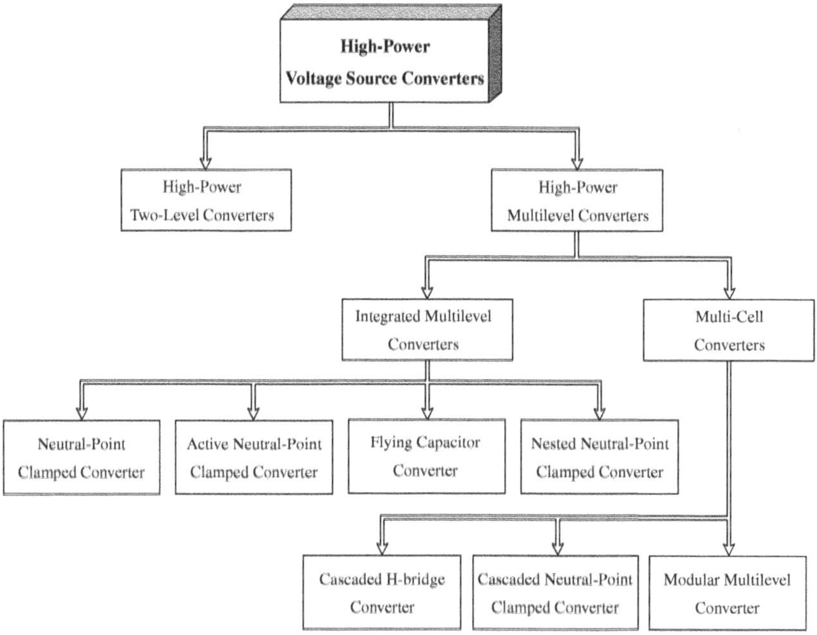

Figure 1.2 Classification of voltage source converters.

Several multilevel converter topologies were established in the past years, and very few succeeded in the industry as shown in Figure 1.2. Neutral-point clamped converter (NPC), active neutral-point clamped (ANPC), flying capacitor (FC), and nested neutral-point clamped (NNPC) converters can handle only a voltage of 2.3–4.16 kV. These converters require significant modifications to increase the operating voltage and a number of voltage levels, which is not cost-effective. These converters must be shut down during faults and other failures, leading to a significant loss of production in industrial processes.

Cascaded converter topologies, also referred to as multi-cell converters, are developed to address the above problems. Among them, the cascaded H-bridge (CHB) and cascaded neutral-point clamped (CNPC) converters are quite popular. These topologies are constructed with a cascade connection of low-power submodules with an isolated DC source in each phase. These topologies have a modular construction, can achieve operating voltages higher than 6 kV up to 13.8 kV, and operate with reduced capacity during faults. The number of submodules in each phase can be increased to achieve higher operating voltage. However, CHB and CNPC requires a phase-shifting transformer with multiple secondary windings to generate isolated DC sources. The transformer increases the overall size and cost of the converter system [7–12]. The details and commercial products based on the above converter topologies are presented in the following subsections.

Recently, the MMC has been developed to address the problems associated with the current cascaded converter topologies. MMC preserves the features of cascaded converters and do not require isolated DC sources. Hence, the phase-shifting transformer can be eliminated, which significantly minimizes the cost and size of the converter system. The unique features of MMC are that, it can reach an operating voltage up to 400 kV without a transformer. This most prominent feature is exploited and successfully applied in various industrial applications. This book mainly focused on the control and applications of MMCs, and are separately discussed in Section 1.6.

1.3.1 Neutral-Point Clamped Converter

Three-level neutral-point clamped (3L-NPC) converters are basically composed of two traditional two-level VSCs stacked one over the other with some minor modifications as shown in Figure 1.3. The negative bar of the upper converter and the positive bar of the lower converter are joined to form a new output phase, whereas the original output terminals are connected via two clamping diodes to form a neutral point "o" [9, 11].

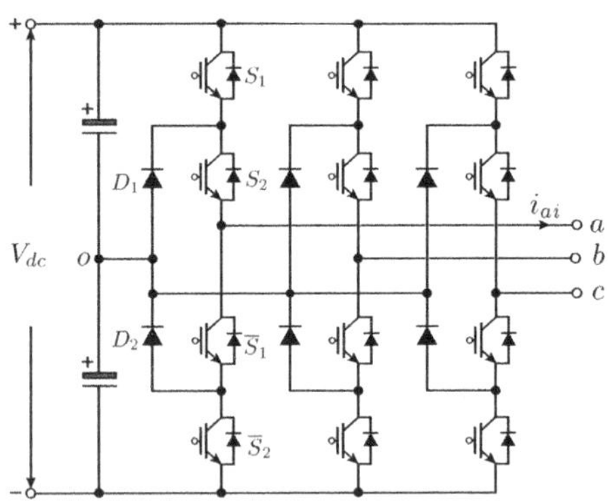

Figure 1.3 Three-level neutral-point clamped converter.

The neutral point divides the total DC-link voltage into two halves. Each power device has to block only half of the total DC-link voltage. With the same semiconductor technology, the operating voltage and power rating of the 3L-NPC converter can be doubled compared to that of a two-level converter. The neutral point of 3L-NPC enables the generation of zero voltage level along with $-\frac{V_{dc}}{2}$ and $+\frac{V_{dc}}{2}$. This process results in low $\frac{dv}{dt}$ and harmonic content in the output voltage and current waveforms.

One of the major problems in 3L-NPC is the neutral-point voltage balancing and low-frequency ripple at higher modulation indices [13, 14]. The neutral-point voltage oscillations increase the voltage stress on devices and generate a sixth-order harmonics in the output voltage. Traditionally, the DC-link capacitance is significantly increased to minimize the voltage oscillations. On the other hand, several modulation schemes have been presented to address the issue of neutral-point voltage balancing in the literature [15, 16]. Among them, the carrier based PWM schemes are simple and easy to implement. In this approach, the zero sequence voltage included in the reference voltage. The injected zero sequence voltage does not affect the output line voltages, and it is only influence the device switching states and neutral point potential [15, 16].

Alternatively, the nearest three vectors and virtual vector based space vector modulation (SVM) schemes are presented for the 3L-NPC [15, 16]. In these approaches, redundant switching vectors are employed to balance the neutral-point potential. The zero sequence voltage and neutral-point current are important variables used in the selection of redundant switching vectors [15, 16].

The NPC topology with a higher number of voltage levels is required for high-power applications. Hence, the NPC topology requires additional power devices, clamping diodes, and DC capacitors to increase the number of output voltage levels. However, the NPC with a higher number of voltage levels is preferable because of the uneven distribution of power losses between the outer and inner devices. Also, the series connection of clamping diodes is required to block the total DC-link voltage. The series connection of devices introduces additional conduction losses and produces reverse recovery currents during the commutation that affects the switching losses of other devices. Furthermore, the balancing of DC-link capacitor voltages is difficult to achieve over the entire operating range [9, 11].

Table 1.2 Selection of devices in 3L-NPC

System Voltage	Device Rating
2.3 kV	3.3 kV
3.3 kV	4.5 kV
4.16 kV	6.5 kV

Among the high-power converters shown in Figure 1.2, the 3L-NPC is the most widely used in all industrial applications in the range of 2.3–6.6 kV. The rating of semiconductor devices in NPC is selected based on the system voltage as shown in Table 1.2. A 3L-NPC converter requires 3.3 kV devices to handle a voltage of 2.3 kV [17]. With the current semiconductor technology, 3L-NPC allows an output voltage up to 4.16 kV without connecting the devices in series. Today, several manufacturers offer 3L-NPC with different type of devices and system voltages as indicated in Table 1.3. Asea Brown Boveri (ABB) offers 3.3 kV converters with 4.5 kV IGCT, Converteam offers 3.3 kV converters with 4.5 kV press-pack IGBT technology, and Siemens offers a range of 2.3–6.6 kV converters with 3.3 kV and 6.5 kV IGBT technology.

Table 1.3 Market overview of VSCs for medium-voltage drives

Topology	Manufacturer	Product Model	Power (MVA)	Voltage (kV)	Semiconductor
3L-NPC	ABB	ACS 1000	0.3–5	2.3–4.16	IGCT
		ACS 6000	3.0–27	2.3–3.3	IGCT
	Converteam	VDM 7000	7–9.5	3.3	GTO/MV-IGBT
		MV 7000	0.4–40	1.25–6.6	IGCT/HV-IGBT
	Siemens	Sinamics GM150	0.6–10.1	2.3–6.6	MV-IGBT
		SM 150	5–28	3.3	IGCT
	TMEIC-GE	Dura-Bilt5i MV	0.3–2.4	4.16	IGBT
	Ingedrive	Ingedrive MV100	0.8–15	3.15–4.16	HV-IGBT
		Ingedrive MV500	6–36	3.15–3.3	IGCT
3L-ANPC	ABB	PCS 8000	6–100	6–220	IGCT
4L-FC	Alstom	VDM 6000	0.3–8	2.3–4.2	MV-IGBT
CHB	Siemens	Perfect Harmony	0.3–30	2.3–13.8	LV-IGBT
	Hitachi	Hivectol-HVI-E	0.31–10	2.3–11	IGBT
	LS Ind. Systems	LS MV Drive	0.2–3	3.3–6.6	IGBT
	Schneider	HARSVERT-A	0.25–6.25	3.3–11	LV-IGBT
	Rongxin Power	MVC	0.25–10	10	IGBT
	Rockwell Automation	PF6000	0.2–5.6	3–11	IGBT
5L-HNPC	ABB	ACS 5000	1.7–27	6–6.9	IGCT
	TMEIC-GE	TMdrive-XL85	30–120	7.2	GCT
MMC	Siemens	Sinamics SM120	6–13.7	3.3–7.2	IGBT
	Benshaw	M2L 3000	0.224–7.466	2.3–6.6	IGBT

1.3.2 Active Neutral-Point Clamped Converter

One of the drawbacks of the 3L-NPC is the unequal loss distribution between the inner and outer semiconductor devices in each leg. Thus, the semiconductor devices are cooled with separate heat sinks and cooling systems, resulting in the asymmetrical distribution of junction temperatures between the semiconductor devices. This process affects the cooling system design and limits the output power and switching frequency of the converter. This drawback is overcome by replacing the clamping diodes with active switches as illustrated in Figure 1.4. The corresponding circuit configuration is called active neutral-point clamped converter (ANPC) [18, 19]. The active switches provide a controllable path for the neutral-point current, which allows the symmetrical loss distribution among the devices and helps to balance the neutral-point voltage [3, 8, 10].

The three-level ANPC is commercially available in the form of a back-to-back configuration by ABB as shown in Table 1.3. This configuration covers a power range of 20–200 MVA and can be connected to a 200 kV grid through a transformer. The ANPC concept is extended to a five-level hybrid multilevel converter, which combines a 3L-ANPC leg with a three-level flying capacitor cell. This circuit configuration effectively increases the number of output voltage levels and is referred to as a five-level ANPC [20]. The five-level ANPC is commercially introduced in a back-to-back configuration for an operating voltage of 6–6.9 kV and 0.4–1 MVA power capacity with a maximum output frequency of 75 Hz.

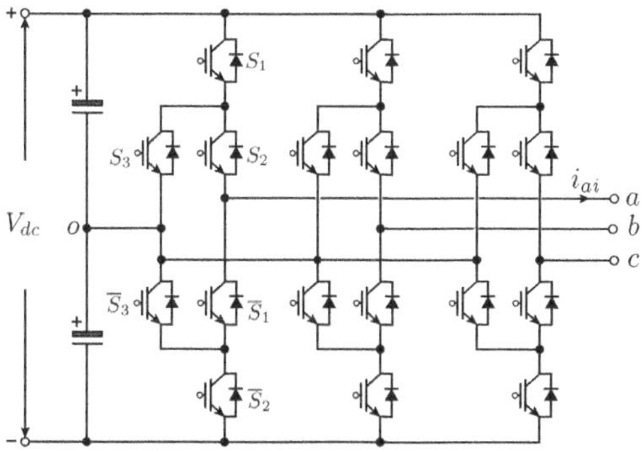

Figure 1.4 Three-level active neutral-point clamped converter.

1.3.3 Flying Capacitor Converter

A flying capacitor (FC) converter is quite similar to the NPC converter, provided that the clamping diodes in NPC are replaced with flying capacitors as shown in Figure 1.5. In FC converter, each pair of semiconductor devices with a flying capacitor forms a power cell. The number of output voltage levels can be increased by adding the power cells to a converter. Thus, it is considered as a modular converter [9].

The circuit configuration of four-level FC (4L-FC) topology is shown in Figure 1.5. The 4L-FC topology has two flying capacitors C_{c1} and C_{c2} with rated voltages of $\frac{V_{dc}}{3}$ and $\frac{2\,V_{dc}}{3}$, respectively, and six semiconductor devices with a voltage rating of $\frac{V_{dc}}{3}$. The flying capacitors have zero initial voltage and must be regulated at their rated value to generate a four-level voltage waveform at the output. Hence, the FC converter requires an additional pre-charging circuit during the start-up process and balancing control during steady-state operation. The natural voltage balancing of flying capacitors can be obtained by using the phase-shifted carrier modulation (PSC-PWM) scheme with high carrier frequency [21]. Also, the PSC-PWM scheme ensures symmetrical loss distribution among the devices, which makes this topol-

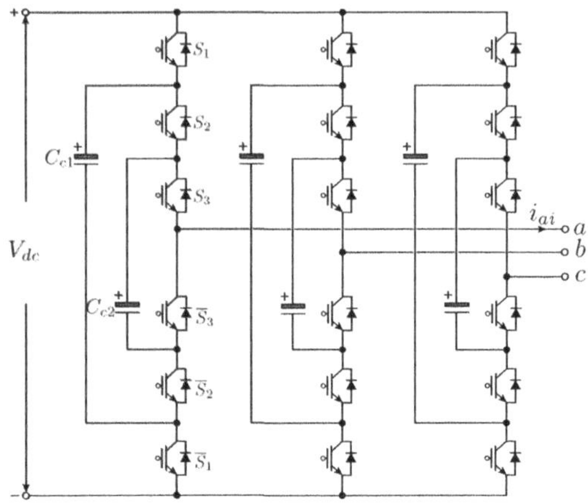

Figure 1.5 Four-level flying capacitor converter.

ogy more attractive for the high-speed drives or applications with low current ripple requirements.

The capacitor voltage ripple is one of the major drawbacks in the flying capacitor converter. The magnitude of the voltage ripple proportionally increases with the magnitude of the load current and is inversely proportional to the switching frequency. To maintain the voltage ripple within the permissible limits, either large-size flying capacitors or a higher switching frequency is required. The larger size of the flying capacitors significantly increases the cost and size of the converter. Also, the converter operation at higher switching frequency increases the switching losses and effects the converter efficiency. These issues limit the application of flying capacitor converters to the medium switching frequency applications only. The 4L-FC topology is commercially available for traction applications and pumps in the water industry. The 4L-FC converter is equipped with a medium voltage IGBT technology and available for an operating voltage of 2.3–4.2 kV and 0.3–8 MVA power capacity as shown in Table 1.3.

1.3.4 Nested Neutral-Point Clamped Converter

Nested neutral-point clamped (NNPC) converter is a recent development in multilevel converters. NNPC combines the structure of an NPC converter and FC converter as shown in Figure 1.6. The flying capacitor in NNPC is divided into two equal parts, and its mid-point is clamped with the diodes. The flying capacitor voltage is maintained at one-third of the total DC-bus voltage such that the NNPC generates a four-level voltage waveform at the output. Also, the switching devices block identical voltages during reverse blocking mode [22]. The nested neutral-point clamped converter requires fewer semiconductor devices and passive components compared to that of other four-level multilevel converter topologies [22, 23].

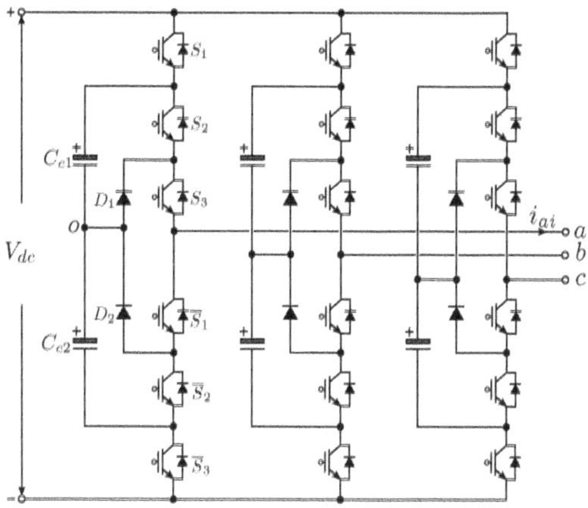

Figure 1.6 Four-level nested neutral-point clamped converter.

In NNPC converter, the load current charges and discharges the flying capacitors, causing voltage ripple. The voltage ripple in the flying capacitors is very high during the low-speed operation, resulting in a poor quality of voltage and current waveforms at the output. Also, the voltage ripple increases the voltage stress on devices and capacitors, which leads to a possible device failure and affects the reliability of the converter. Similar to the FC converter, the NNPC converter requires large-sized capacitors or injection of common-mode voltage with modulation signals to minimize the voltage ripple. The voltage ripple minimization and capacitor voltage balancing at a low-speed operation are the current research focus.

A four-level NNPC converter can operate at a voltage range of 2.4–7.2 kV without devices in series. The other variations of nested neutral-point clamped converter are five-level NNPC and cascaded NNPC converter topologies, which are designed to handle higher operating voltages. Cascaded NNPC converter topology can handle a voltage of 10 kV and requires a phase-shifting transformer to generate the isolated DC sources [24, 25].

1.3.5 Cascaded H-bridge Converter

Multilevel converters such as NPC, ANPC, FC, and NNPC topologies can handle a voltage up to 6.6 kV without devices in series. These topologies are not cost-effective for an operating voltage greater than 6.6 kV. Cascaded converter topologies can handle voltages greater than 6.6 kV (up to 13.8 kV) using low-cost, and low-voltage IGBT technology. The most popular cascaded converter topology is the CHB converter, which consists of several three-level H-bridge power modules in cascade as shown in Figure 1.7. Each H-bridge power module requires an isolated DC source, which is generated using a phase-shifting transformer with multiple secondary windings and a diode bridge rectifier. In regenerative applications, the three-phase, two-level VSC is used instead of a diode bridge rectifier [8, 12, 26]. The transformer and rectifier configuration forms a multi-pulse rectifier on an AC-grid side, which improves the power quality and eliminates the dominant harmonic components from the AC grid voltage.

The series connection of H-bridge power modules naturally increases the converter voltage, and consequently the power rating. The output voltage waveform with a higher number of voltage levels and lowest harmonic distortion can be obtained. The number of voltage levels in the output voltage is $2N + 1$, where N represents the number of H-bridge power modules per phase. However, the CHB topology requires a large number of isolated DC sources to power each H-bridge, which limits the number of H-bridge power modules in each phase. Also, the CHB topology requires a complicated phase-shifting transformer, which increases the volume, size, and cost of the overall converter system.

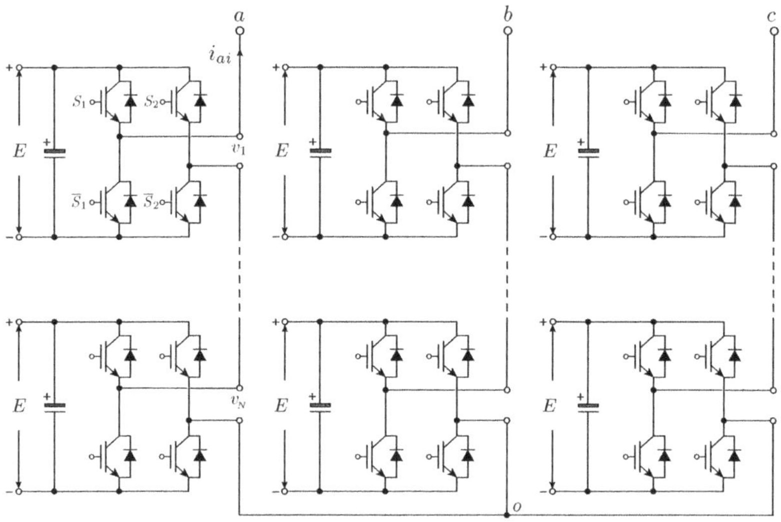

Figure 1.7 Cascaded H-bridge converter.

The CHB topology has a modular construction and allows the fault-tolerant operation with redundant power modules [27]. The CHB topology is commercially available in different numbers of voltage levels with a maximum operating voltage of 13.8 kV as shown in Table 1.3. `Siemens` developed a CHB topology with 17 voltage levels to handle a system voltage of 13.8 kV. This topology uses low-voltage IGBT technology that is available in the market under the trade name `Perfect Harmony`. In CHB topology, each H-bridge DC source voltage is maintained at an identical value. This topology is referred to as a symmetrical CHB. It is also possible to maintain a certain DC voltage ratio between the H-bridge power modules. Hence, an output voltage waveform with the higher number of voltage levels can be obtained without increasing the number of H-bridge power modules in each phase. This topology is referred to as an asymmetrical CHB.

The asymmetrical CHB concept is introduced for magnetic resonance imaging applications with the powers of two voltage ratios ($1 : 2 : \cdots : 2^{N-1}$). This ratio allows the generation of seven-level voltage waveform with two H-bridge power modules per phase. Another important voltage ratio called powers of three ($1 : 3 : \cdots : 3^{N-1}$) was introduced, which eliminates all redundant switching states and maximizes the number of voltage levels at the output. This converter is referred to as a "trinary hybrid multilevel converter." Aside from the exponential increase in the number of voltage levels, the asymmetrical topology minimizes the switching losses and improves the efficiency [7, 10]. However, it requires devices with different voltage ratings to build the H-bridge modules, and different thermal designs for each power module. Hence, the modularity feature no longer exists in the asymmetrical CHB.

1.3.6 Cascaded Neutral-Point Clamped Converter

Asymmetrical CHB converter generates an output voltage with a higher number of levels using a smaller number of lower H-bridge power modules in each phase. This topology loses the modularity feature and requires the devices with different voltage ratings and thermal designs. These drawbacks can overcome by replacing the three-level H-bridge power modules with other multilevel converters. One of the most popular topologies is the cascaded neutral-point clamped converter (CNPC), which is shown in Figure 1.8. This topology uses two three-level neutral-point clamped converter legs to form an H-bridge structure in each phase, which is referred to as an HNPC.

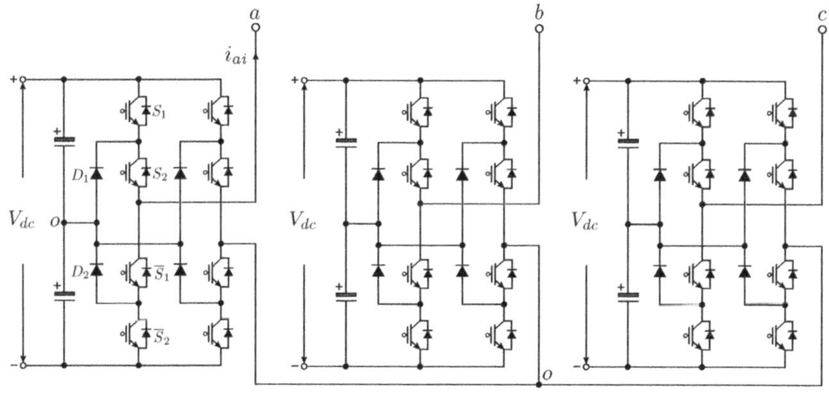

Figure 1.8 Cascaded neutral-point clamped converter.

Each HNPC power module generates a five-level voltage waveform at the output [3, 10, 28]. The HNPC power module uses identical voltage rated devices and capacitors, unlike asymmetrical CHB converter. Due to the control complexity, the number of HNPC modules in the cascaded neutral-point clamped converter is limited to one. Also, the CNPC converter requires a phase-shifting transformer with a rectifier unit to generate the isolated DC source for each HNPC module. Thus, the maximum operating voltage of the cascaded neutral-point clamped converter is limited to 6.6–7.2 kV without devices in series. The five-level CNPC converter is commercially available with a 36-pulse rectifier system featuring IGCT devices for compressors and conveyor applications. Several other topologies were developed with a maximum operating voltage of 7.2 kV, featuring MV-IGBTs, IEGT, and GCT devices as presented in Table 1.3.

1.4 CURRENT SOURCE CONVERTERS

The current source converter (CSC) technology is well suited for high-power applications. The main features of CSC include a simple converter structure, low switch count, low switching $\frac{dv}{dt}$, reliable, and over current and short-circuit protection, unlike VSCs. Despite all of these advantages, it has a limited dynamic performance because of the large DC inductor used in the DC-link [3, 5]. The CSCs are classified into load-commutated current source converter (LC-CSC) and pulse width modulated current source converter (PWM-CSC) as shown in Figure 1.9. These topologies are briefly discussed in the following subsections.

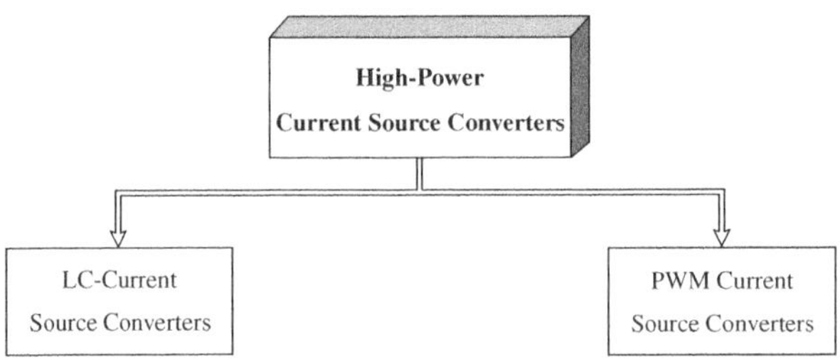

Figure 1.9 Classification of current source converter.

1.4.1 Load-Commutated Current Source Converter

The load-commutated current source converter (LC-CSC) is one of the earliest topologies used in the high-power variable speed drives and HVDC systems. The thyristor devices are employed in the LC-CSC topology as shown in Figure 1.10. The phase-controlled thyristor rectifier is used to adjust the magnitude of DC-bus current i_{dc}. The DC inductor smoothened the DC-link current and fed to the LC-CSC. The thyristor devices in LC-CSC does not have the self-turn-off capability. These devices are naturally commutated by the load voltage with a leading power factor. Therefore, the ideal load for the LC-CSC is a synchronous motor that operates at a leading power factor. The leading power factor operation can be easily achieved by adjusting the rotor field current of a synchronous motor [5]. On the other hand, the rectifier input current is highly distorted. Hence, the LC-CSC fed drives are equipped with harmonic filters to reduce the line current harmonic distortion. These filters can also serve as a power factor compensator.

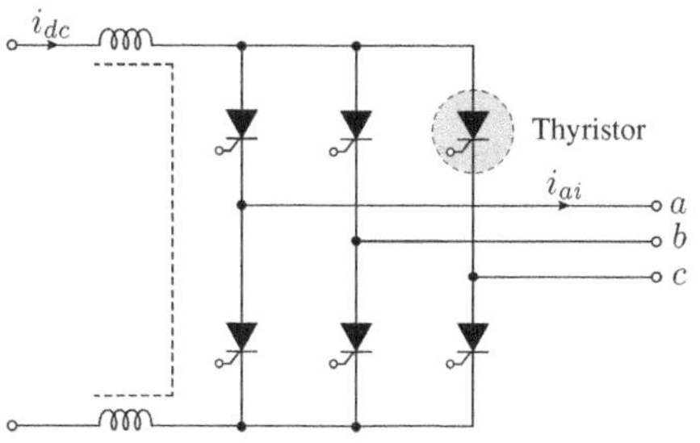

Figure 1.10 Load-commutated current source converter.

The thyristor is a natural-commutated device that has low power losses and high-efficiency. Therefore, LC-CSCs are suitable for very large electric drives with tens of megawatt power capacity. The use of thyristor leads to a lower manufacturing cost and high reliability compared to that of other motor drives based on IGBT and IGCT devices. A typical example of LC-CSC drive is a 100 MW synchronous motor drive installed in NASA's wind tunnel facility [5]. Several commercial products based on the LC-CSCs were developed by the leading industrial manufacturers, such as `Siemens`, `ABB`, and `Alstom` as indicated in Table 1.4. These products have a maximum power rating greater than 70 MW with a maximum operating voltage greater than 10 kV.

Table 1.4 Market overview of CSCs for MV drives

Topology	Manufacturer	Product Model	Power (MVA)	Voltage (kV)	Semiconductor
LC-CSC	Siemens	Sinamics GL150	2.8–75	2.3–12	Thyristor
	ABB	MEGADRIVE-LCI	2.0–72	2.1–10	
	Alstom	ALSPA SD7000	2.0–100	1.0–10	
PWM-CSC	Rockwell Automation	Power Flex 7000	0.15–25.4	2.4–6.6	SGCT

1.4.2 PWM Current Source Converter

Unlike LC-CSC, the pulse width modulated current source converters (PWM-CSC) use force-commutated devices such as GTO and GCT. Therefore, the induction motors and synchronous motors can be employed in PWM-CSC fed drives. The PWM-CSC is also used in the grid integration of offshore wind farms and HVDC systems. PWM-CSC can be controlled by employing a pulse width modulation scheme such as selective harmonic elimination (SHE), trapezoidal-PWM, and current space vector modulation [5, 28–30]. The typical configuration of SGCT based PWM-CSC is shown in Figure 1.11. The PWM-CSC requires a three-phase capacitor (C_f) at its output to assist the commutation of SGCT devices. The capacitor provides a current path for the energy trapped in the leakage inductance of a motor when SGCT is turned off. Otherwise, a high-voltage spike would be induced and damage the switching devices. The capacitor C_f also serves as a harmonic filter, thus improves the motor current and voltage waveform quality [28].

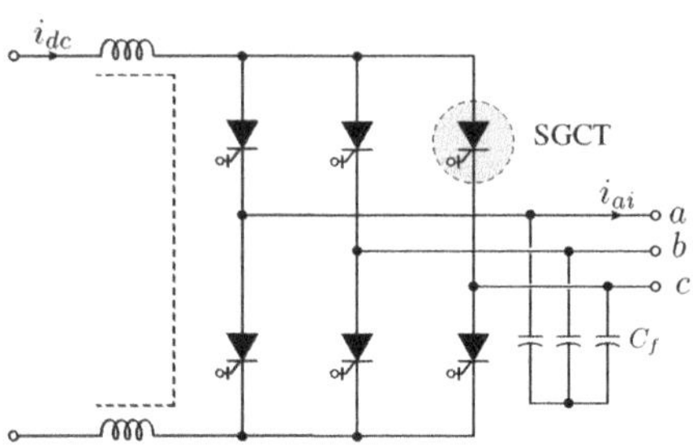

Figure 1.11 Pulse width modulated current source converter.

PWM-CSC generates a common-mode voltage across the motor terminals, causing premature failure of motor winding insulation and motor bearings. PWM-CSC fed drive requires an isolation transformer to block the common-mode voltages [31]. The transformer leakage inductance also serves as a harmonic filter and reduces the line current harmonic distortion. This transformer can be arranged with multiple phase-shifted secondary windings for harmonic cancellation in multi-pulse rectifiers [5].

The modern PWM-CSC fed drives consist of an integrated common-mode DC inductor in place of a conventional DC inductor. The integrated common-mode DC inductor consists of both differential and common-mode inductance, and effectively blocks the common-mode voltage. The integrated inductor eliminates the isolation transformer and significantly reduces the overall drive cost, size, and weight [31]. Modern PWM-CSC fed drive with an integrated common-mode DC inductor is referred to as a transformer-less drive. The transformer-less drives are developed by `Rockwell Automation` and are commercially available under the trade name `Power Flex 7000` as shown in Table 1.4. `Power Flex 7000` is available with a targeted voltage range of 2.4–6.6 kV and power capacity of 0.15–25.4 MVA.

1.5 MATRIX CONVERTERS

The matrix converter (MC) is a force-commutated converter that belongs to the direct AC–AC power conversion family. Unlike CCVs, the matrix converter (MC) employs fully controllable IGBT devices, and are controlled with PWM schemes. The conventional matrix converter has three inputs and three outputs only. However, it can be extended to any number of inputs and outputs, depending on the application. The main features of MC are four-quadrant operation and fast dynamic response because of the single-stage power conversion. It provides a sinusoidal input and output waveforms with controllable input power factor. The output frequency can be adjusted to either higher or lower than that of the input frequency [32–34]. Despite all of these advantages, the major issues in matrix converters are commutation problems, switching losses, and leakage currents because of the common-mode voltage [35–37]. The matrix converters are classified into direct matrix converter (DMC), indirect matrix converter (IMC), and multi-modular matrix converter (MMMC) as shown in Figure 1.12. These converters are discussed in the following subsections.

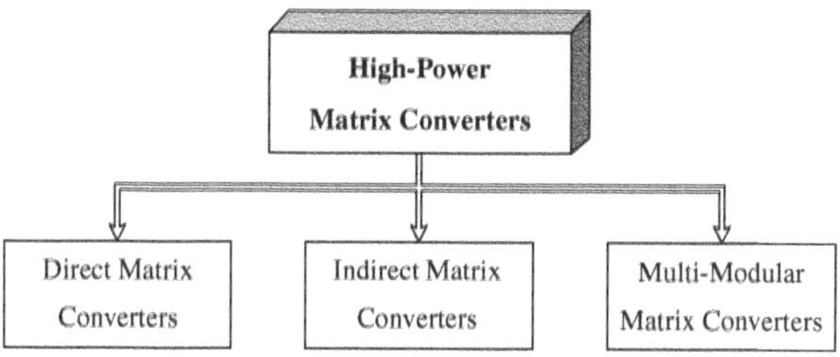

Figure 1.12 Classification of matrix power converters.

1.5.1 Direct Matrix Converter

Figure 1.13 shows the configuration of a three input and three output direct matrix converter (DMC), where input phases (a_g, b_g, c_g) are directly connected to the output phases (a_i, b_i, c_i) through bidirectional switches and an input filter. The bidirectional switches should carry the current and block the voltage in both the directions [38]. These switches are implemented by an anti-series connection of two IGBT devices (with a common collector or emitter) as shown in Figure 1.13. The input filter (C_f) is designed to attenuate the higher-order harmonic components in the input current and to reduce the input voltage harmonic distortion.

Direct matrix converters have several issues such as the maximum voltage transfer ratio from the input to the output; that is limited to 0.5 in the linear modulation region. Also, there is limited control over the input power factor. Several switching methods are presented in the literature to increase the voltage transfer ratio. The switching method based on the three line–line input voltages produce a 0.75, and two line–line input voltages produce a 0.866 transfer ratio with input power factor

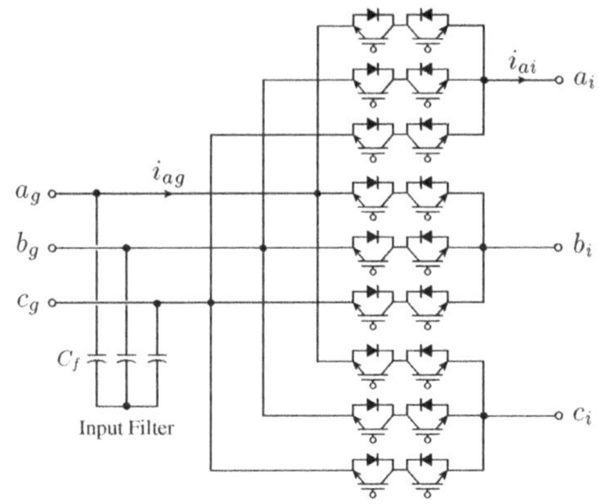

Figure 1.13 Configuration of direct matrix converter.

control in the entire operating range [39]. However, these methods cause a low-frequency distortion in the input currents and the output voltages.

Space vector modulation (SVM) approach is exploited and successfully used in the control of the matrix converters. This approach has full control over the instantaneous value of output voltages and the input power factor and provides a voltage ratio of 0.866. The SVM approach is further improved to achieve a voltage ratio of 1.155 at the cost of high-switching frequency. The other major issue is the simultaneous commutation of bidirectional switches without generating an over current or over voltage spikes during the interruption of inductive load current. This problem is solved with the multi-step commutation techniques, which allows the safe operation of devices [38].

The direct matrix converters are commercially used in the low-voltage motor drive products of `FRENIC-MC` series by `Fuji` Electric Systems and, `AC7` and `U1000` Matrix drive by `Yaskawa`. These products are available in 30 kW at 230 V input and 45 kW at 480 V input options.

1.5.2 Indirect Matrix Converter

Unlike direct matrix converters, the indirect matrix converters (IMC) consist of a rectifier stage and an inverter stage. The rectifier stage is constructed with either unidirectional or bidirectional switches, and an inverter stage is constructed using a conventional two-level voltage source converter (2L-VSC) as shown in Figure 1.14. Therefore, the conventional modulation schemes such as carrier and space vector modulation schemes can be independently applied to the rectifier and inverter stages. The main features of indirect matrix converter include the elimination of commutation problem presented in the DMC, and less number of devices is required compared to that of the DMC [37, 40]. However, the voltage transfer ratio remains the same as that of the DMC. The Z-source network is employed in the IMC to increase the voltage transfer ratio.

The performance of the IMC can be improved by replacing the 2L-VSC with a multilevel VSC technology, such as NPC and FC converters. These topologies are referred to as multilevel matrix converters. The multilevel VSC generates a stepped voltage waveform at the motor terminals and reduces the voltage harmonic distortion and switching $\frac{dv}{dt}$.

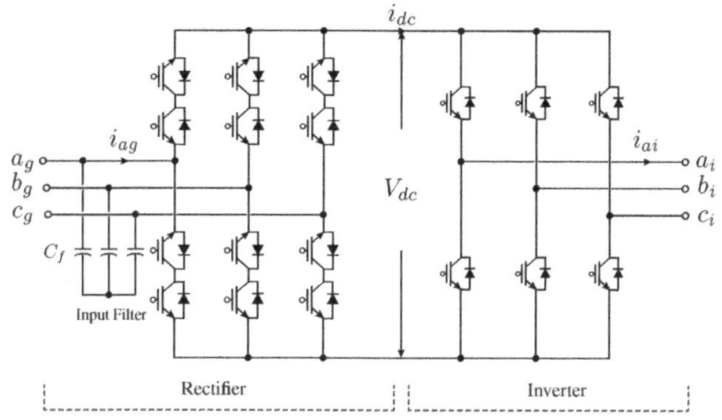

Figure 1.14 Configuration of indirect matrix converter.

1.5.3 Multi-Modular Matrix Converter

The direct and indirect matrix converters are available for low-voltage, and low-power applications only. These topologies are difficult to apply for the medium- and high-voltage, and high-power applications because of the limited voltage rating of semiconductor devices. The modular multilevel converter technology is employed to increase the voltage and power rating of the matrix converters. This new converter technology is referred to as an multi-modular matrix converter (MMMC) or cascaded matrix converter as shown in Figure 1.15.

The multi-modular matrix converter consists of identical power modules in each phase [41, 42]. These power modules are simple three input and two output direct matrix converters (3 x 2 DMC) as shown in Figure 1.16. The three-phase input of each power module is connected to the AC grid through a phase-shifting transformer. This phase-shifting transformer provides an isolation and improves the input and output power quality. However, the transformer increases the weight and volume of the converter. The output of each 3 x 2 DMC module is connected in cascade to increase the number of voltage levels. Several power modules such as regenerative H-bridge and non-regenerative H-bridge modules are studied for MMMC [43]. These power modules are built with low-voltage IGBT technology.

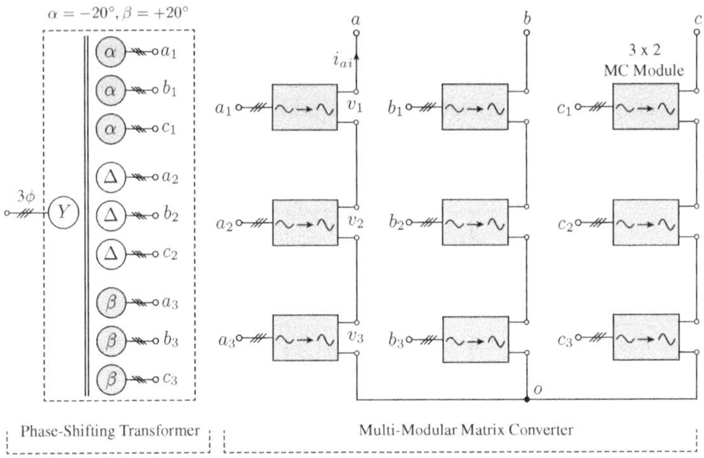

Figure 1.15 Configuration of multi-modular matrix converter.

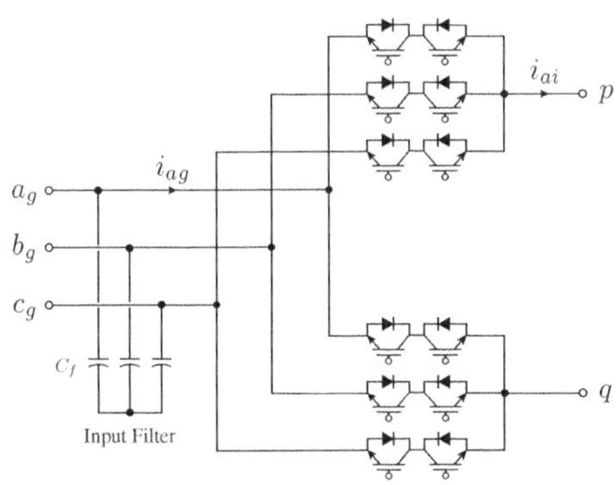

Figure 1.16 Configuration of 3 x 2 matrix converter.

Multi-modular matrix converters are commercially available for the medium-voltage motor drives, and their specifications are shown in Table 1.5. The commercial product `FSDrive MX1S` is designed by `Yaskawa` with a targeted voltage of 3 kV with a power rating of 0.2–3 MVA, and 6.6 kV with a power rating of 0.4–6 MVA. This product consists of 3 x 2 DMC modules in each phase (three modules per phase for 3 kV and six modules per phase for 6.6 kV drive).

Table 1.5 Market overview of MCs for MV drives

Topology	Manufacturer	Product Model	Power (MVA)	Voltage (kV)	Semiconductor
MMMC	Yaskawa	FSDrive MX1S	0.2–6	3–6.6	IGBT

1.6 MODULAR MULTILEVEL CONVERTERS

Cascaded converter topologies have a modular construction, voltage and power scalability with a cascade connection of power modules, fault-tolerant operation, and high reliability and use a low-cost, low-voltage semiconductor technology, unlike other multilevel converters. These topologies can achieve higher operating voltages without the series connection of semiconductor devices. However, the cascaded converter topologies require a large number of isolated DC sources and complicated phase-shifting transformer with multiple secondary windings. The phase-shifting transformer improves the power quality on the AC grid side, but increases the overall size and cost of the converter.

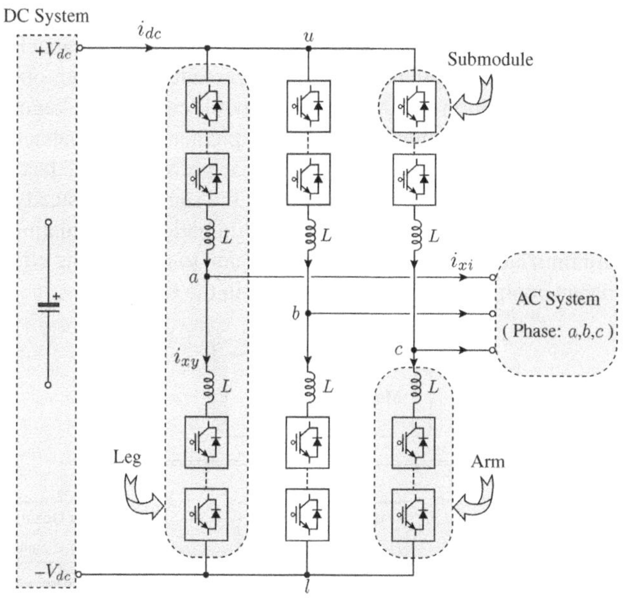

Figure 1.17 Configuration of modular multilevel converter.

The MMC is one the recent developments in the multi-cell converter family [44, 45]. The MMC preserves the features of cascaded converter topologies and does not require isolated DC sources and complicated phase-shifting transformer. Therefore, MMCs can reach any operating voltage from the medium (2.3–13.8 kV) to high-voltage (33–400 kV) and power rating of 0.226–1000 MW. These features have attracted researchers from both industry and the academia and developed several commercial products for a wide range of high-power applications, such as medium-voltage motor drives, HVDC transmission systems, multi-terminal HVDC systems, offshore wind farms, and static synchronous compensators (STATCOM). The MMC also requires a complex control structure to meet several control objectives [44, 46,

47]. The applications and technical challenges associated with the operation and control of MMC are presented in the following subsections.

1.6.1 Converter Technology

The circuit configuration of the MMC is shown in Figure 1.17. The voltage and power capacity of MMC can be naturally increased by adding the submodules in each arm. Similar to other cascaded converter topologies, any power converter circuit such as full-bridge or H-bridge, flying capacitor, and neutral-point clamped converters can be utilized as a submodule (SM) in MMC [48, 49]. The three-phase MMC can be extended to multi-phase applications because of its modular construction. Several submodule topologies have been developed in the past few years to meet the application requirements. The details of the most widely used submodule configurations are presented in the Chapter 2. These submodules have floating capacitors and are pre-charged through the DC system, unlike other cascaded converter topologies. For the safe and reliable operation of MMC, the submodule capacitor voltage in each arm must be maintained at their nominal value. The implementation of capacitors voltage control methods is discussed in the Chapter 3. Each arm of MMC has an inductor to limit the inrush and circulating currents during the start-up and steady-state operation of MMC [50, 51]. However, the complete elimination of circulating currents is necessary to minimize the converter power losses and to improve the efficiency. The circulating current control methods are presented in the Chapter 3.

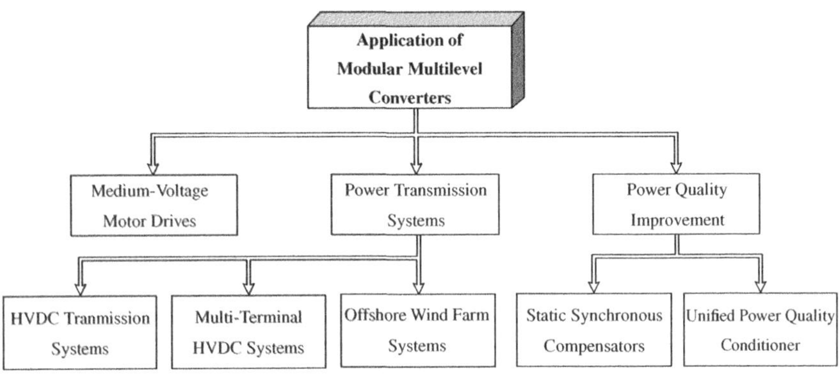

Figure 1.18 Applications of modular multilevel converter.

1.6.2 Applications

Modular multilevel converters are commercialized in the form of various standard and customized products for high-power applications. MMC is used in several industrial applications such as motor drives, power transmission systems, and power quality improvement as shown in Figure 1.18. In this book, the most popular applications such as medium-voltage motor drives, HVDC transmission systems, multi-terminal HVDC systems, offshore wind farms, static synchronous compensator (STATCOM), and unified power quality conditioner (UPQC) are discussed in detail [50, 51]. The summary of commercially available MMC-based products is presented as well.

(1) Medium-Voltage Motor Drives: One of the applications of MMC is the medium-voltage motor drive system. The MMC offers a simple voltage scalability by connecting the SMs in series, operating with standard motors without output filters, and offers a simple construction with less engineering effort, unlike conventional multilevel converters.

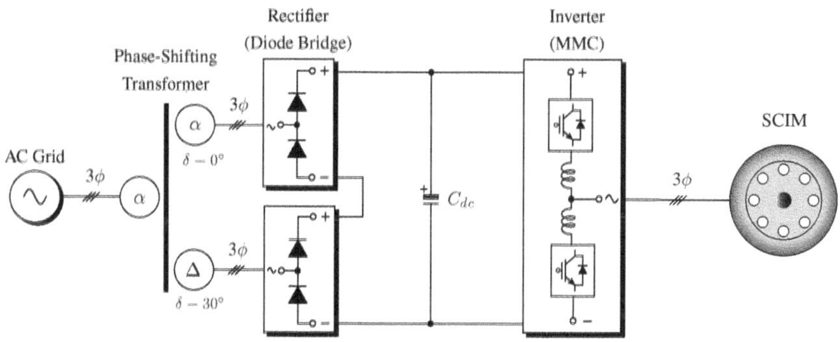

Figure 1.19 MMC-based motor drive with passive rectifier.

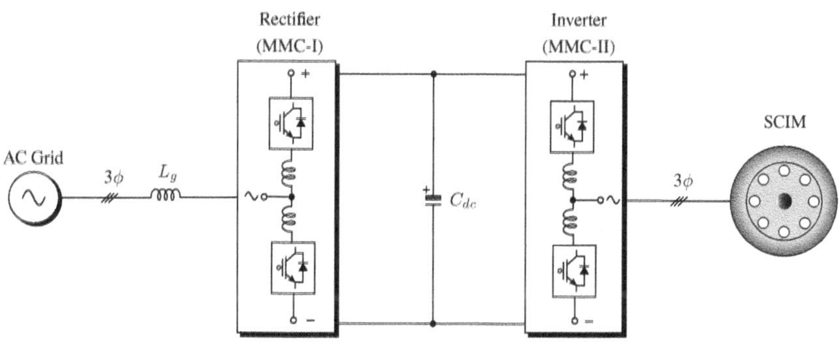

Figure 1.20 MMC-based motor drive with active rectifier.

The typical configuration of MMC-based medium-voltage motor drive developed by the Benshaw is shown in Figure 1.19. The motor drive uses a 12-pulse diode rectifier on the AC grid side and an MMC with a cascade half-bridge submodule on the motor side. The 12–pulse rectifier eliminates the lower-order harmonics in the line current and improves power quality on the AC grid side. Also, the transformer on the AC grid side blocks the common-mode current from entering into the system. On the other hand, the MMC consists of six submodules in each arm and generates a phase-voltage with 13 levels at the motor terminals. Hence, the output filters are not required.

For regenerative applications, the motor drive systems employ MMCs on both the AC grid and motor side as shown in Figure 1.20. The MMC-I is controlled with a voltage-oriented control (VOC) approach, and generates a sinusoidal current on the AC grid side and improves the grid power factor. The MMC-II is controlled with a field-oriented control (FOC) to meet the motor requirements such as speed and torque. An in-depth analysis and control of MMC-fed motor drive systems are presented in the Chapter 8. Currently, the MMC-based motor drives are commercially available with a maximum voltage of 7.2 kV and power rating of 13.7 MVA by `Siemens` as indicated in Table 1.6.

Table 1.6 Market overview of MMC-based motor drive systems

Project Name	Power (MVA)	Voltage (kV)	Developed by
Sinamics SM120 MV Drive	6–13.7	3.3–7.2	Siemens
M2L 3000 Series MV Drive	0.224–7.466	2.3–6.6	Benshaw

(2) HVDC Transmission Systems: HVDC systems are employed to interconnect two asynchronous AC systems over a long distance using a back-to-back converter configuration. The HVDC system can be implemented with either CSCs or VSCs. The CSC-based HVDC system works effectively with silicon-controlled rectifier (SCR) devices and can handle higher powers and voltage ratings than that of the IGBT-based VSC-HVDC system. However, the VSC overcomes the important disadvantages of CSCs such as the reduction of harmonic distortion on both the AC and DC side. Thus, the size of output filters can be reduced. Also, the active and reactive powers can be independently controlled, and ability to operate with weak AC grids.

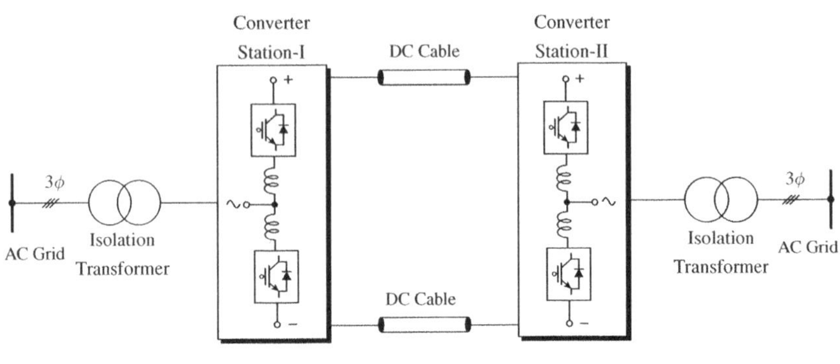

Figure 1.21 MMC-based HVDC system.

Table 1.7 Market overview of MMC-based HVDC systems

Project Name	Power (MW)	Voltage (±kV)	DC Cable (km)	Company	HVDC Tech.
Trans Bay	400	200	85	Siemens	HVDC Plus
INELFE	2 x 1000	320	65	Siemens	HVDC Plus
Sydvastlanken	2 x 720	300	260	Alstom	HVDC MaxSine
Piemonte-Savoia	2 x 600	320	190	Alstom	HVDC MaxSine
Dalian City Feed	1000	320	43	C-EPRI	HVDC Flexible
Xiamen Island HVDC	1000	320	10.7	C-EPRI	HVDC Flexible

A typical MMC-based VSC-HVDC transmission system is shown in Figure 1.21, where the MMC is employed in converter station-I and station-II. The AC system of converter station-I and II is connected to the AC grid through an isolation transformer, and their DC systems are inter-connected through a DC cable several kilometers long. The MMC provides a high-quality output voltage and current waveforms and eliminates the need for harmonic filters in the DC and AC system side.

(a) (b) (c)

Figure 1.22 MMC-based HVDC plus technology : (a) converter leg, (b) converter arm, and (c) submodule (Courtesy: Siemens).

Currently, four types of MMC-based HVDC technologies were commercialized as follows: `HVDC Plus`, `HVDC Light Gen.4`, `HVDC Flexible`, and `HVDC MaxSine`, which can reach up to a voltage ±320 kV [52]. The list of MMC-based HVDC projects is summarized in Table 1.7. These technologies adopt different type of submodules and arm structure to configure the MMC. San Francisco's `Trans Bay` project is the first MMC-based HVDC system installed by `Siemens` in 2010. This project supplies a power 400 MW at a voltage rating of ±200 kV through a sub-sea cable 85 km long. In this project, the half-bridge submodules are connected in cascade to form an MMC. In `HVDC MaxSine` technology, the full-bridge submodules along with IGBT devices are connected in series to form the MMC. These variations of HVDC technologies and possible submodule configurations are discussed in the Chapters 2 and 9.

Commercially available MMC-based `Siemens` HVDC technology is presented in Figure 1.22. MMC consists of 200 half-bridge submodules in each arm to handle a voltage of ±200 kV. The structure of half-bridge submodule is shown in Figure 1.22, which includes the protection, cooling, sensor boards, local controller, and communication circuits to the central controller. Currently, the MMC handles a power of 1000 MW at a voltage of ±320 kV.

(3) Multi-Terminal HVDC Systems: A typical configuration of MMC-based multi-terminal HVDC system is shown in Figure 1.23, which allows the interconnection of different HVDC systems to form an HVDC grid with high controllability, efficiency, and reliability unlike that of the AC grid. The DC system of MMCs are interconnected through the DC cable, and the AC system is connected to the AC-grid through an isolation transformer. The power converters are used to control the active and reactive power flow between the asynchronous AC grids.

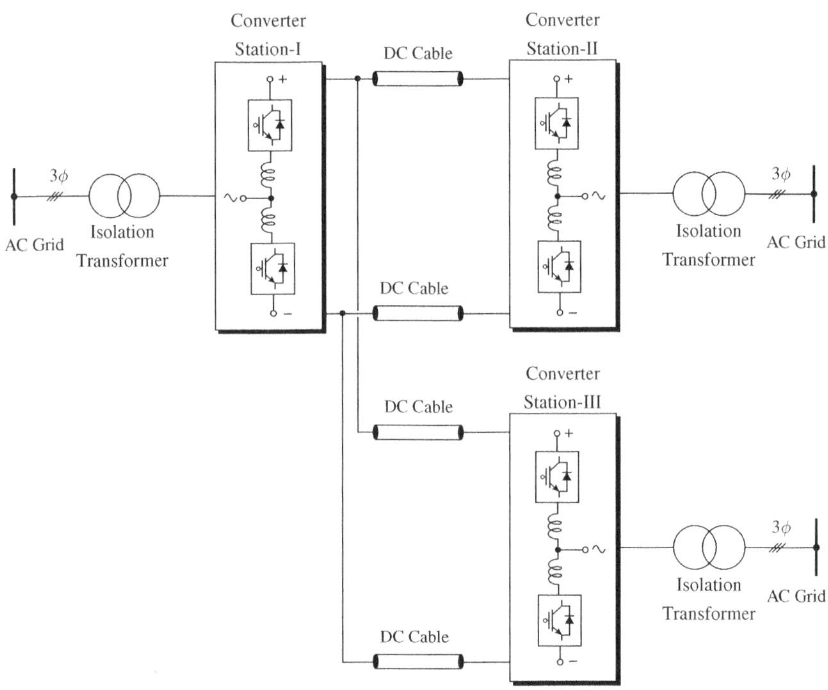

Figure 1.23 Multi-terminal MMC-HVDC system.

Multi-terminal HVDC grids are most vulnerable to DC-side faults and asymmetrical AC faults. The application of MMCs for multi-terminal HVDC system needs further evaluation of fast protection systems and robust power management schemes to withstand the asymmetrical AC faults, which is an on-going research topic. The China Electric Power Research Institute (CEPRI) installed an MMC-based multi-terminal HVDC system in Zhoushan to distribute the power among multiple regions as indicated in Table 1.8. In this project, the `HVDC Flexible` technology is employed to handle a voltage ±200 kV.

Table 1.8 Market overview of multi-terminal MMC-based HVDC systems

Project Name	Power (MW)	Voltage (±kV)	DC Cable (km)	Company	HVDC Tech.
Zhoushan	400/300/100 /100/100	200	134	CEPRI	HVDC Flexible

(4) Offshore Wind Farm: Typically, offshore wind farms are located in the sea, which is far from the utility grids. These offshore wind farms are connected to the AC grid for the subsequent distribution and consumption of generated power. The MMC-HVDC system is a suitable candidate to transfer the power over a long distance. The connection of an offshore wind farm with the AC grid through a submarine cable and MMC-HVDC network is shown in Figure 1.24. The MMC-HVDC provides a black start operation, requires less space and compact structure, and does not require output filters and reactive power support compared to that of the line-commutated converter (LCC) based HVDC system [53].

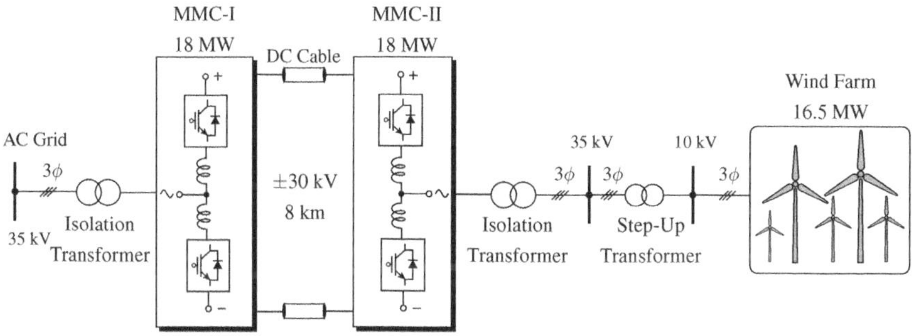

Figure 1.24 MMC-HVDC with submarine cable for offshore wind farm.

The MMC-based HVDC systems are used to interconnect offshore wind farms with the AC grid as shown in Table 1.9. These projects are mainly installed in the European region with a maximum voltage of ±320 kV and power rating of 900 MW.

Table 1.9 Market overview of MMC-HVDC-based offshore wind farms

Project Name	Power (MW)	Voltage (±kV)	DC Cable (km)	Company	HVDC Tech.
Borwin-2	800	300	200	Siemens	HVDC Plus
Sylwin-1	864	320	205	Siemens	HVDC Plus
Dolwin-2	900	320	135	ABB	HVDC Light Gen. 4
Dolwin-3	900	320	160	Alstom	HVDC MaxSine
Nanhui	18	30	8.4	C-EPRI	HVDC Flexible

(5) Static Synchronous Compensator: Another potential application of MMC is to operate as a static synchronous compensator (STATCOM) to compensate both harmonics and reactive power demanded by the distorted and unbalanced loads. MMC-based STATCOM has modularity and voltage scalability features, and allows the direct connection (without transformer) to the medium- or high-voltage power system networks as shown in Figure 1.25.

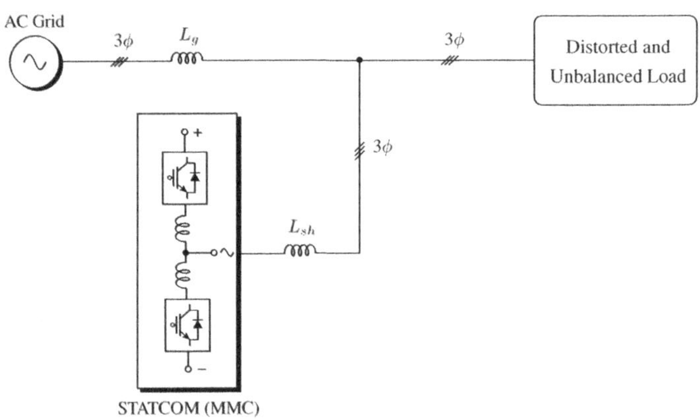

Figure 1.25 MMC-based STATCOM system.

Table 1.10 Market overview of MMC-based STATCOM

Project Name	Power (±MVAr)	Voltage (kV)	Company	STATCOM Tech.
Kikiwa SVC Plus	2 x 35	220/11	Siemens	SVC Plus M
Mocuba Substation	35	33	Siemens	SVC Plus M
Inter Island Link pole 3	50	220	Siemens	SVC Plus C
Rio Branco SVC Plus	55	230	Siemens	SVC Plus C
Offshore Greater Gabbard	50	13.9	Siemens	SVC Plus L

In this application, the half-bridge and full-bridge submodules are preferred for MMCs. The MMC-based STATCOM technologies were developed by `Siemens` to handle a voltage of 33 kV without a transformer and 220 kV with a transformer as shown in Table. 1.10. These technologies are installed in offshore wind farms, substations and arc furnace industries to supply the reactive power.

(6) Unified Power Quality Conditioner: The unified power quality conditioner (UPQC) is another important and widely used technology in the power quality improvement. The UPQC system has two stages of compensation with the series and parallel connection of the power converters. The UPQC can be realized using a back-to-back connection of MMC as shown in Figure 1.26.

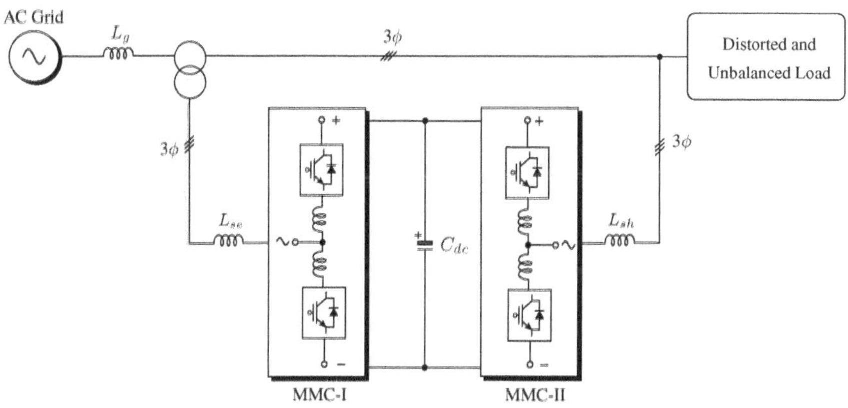

Figure 1.26 MMC-based UPQC system.

The MMC-I is connected to the AC grid through a transformer. It provides the series compensation, where it handles the voltage sag and voltage swell problems of the AC grid. The MMC-II provides a shunt compensation, which is directly connected to the AC grid without any transformer. Eliminating the transformer in shunt compensation is one of the advantages of MMC [54, 55]. The shunt compensator handles the reactive power and harmonic compensation of the distorted load. High-quality output waveforms and the fast current control of MMCs helps to improve the active filtering capability of the UPQC system. The control and power quality improvement using UPQC is discussed in Chapter 9.

1.6.3 Technical Challenges

In this section, the technical challenges associated with the operation and control of MMCs, including the design constraints, submodule capacitor pre-charging process, submodule capacitor voltage control, submodule capacitor voltage ripple, circulating current, and fault-tolerant issues are presented [50, 51]. These technical challenges and corresponding control methods are presented in the subsequent chapters of this book.

(1) Design Constraints: Arm inductance (L) and submodule capacitor (C) are the only passive components used in MMCs. Arm inductance filters the switching frequency harmonics in the arm current and limits the DC short-circuit current. Therefore, sizing of the arm inductor depends on the arm current ripple and short-circuit current. The suppression of undesirable low-frequency currents needs to be considered during the design of an arm inductor. Arm inductor size can be reduced by using an integrated arm inductor, where the upper and lower arm inductors in each leg are wounded on the same core. The submodule capacitor is sized based on the tradeoff between the size or cost and capacitor voltage ripple. It is designed to provide a permissible peak-to-peak ripple at twice the fundamental frequency.

(2) Submodule Capacitor Pre-Charging Process: The submodules in each arm of MMC has floating capacitors with zero initial voltage. These submodule capacitors must be charged to its nominal voltage level before starting the normal operation. However, the charging process of submodule capacitors during startup and after a fault can lead to a large inrush current because of the small equivalent impedance of the converter. The pre-charging of submodule capacitors without inrush current is one of the major challenges in MMC.

(3) Submodule Capacitor Voltage Control: The submodule capacitors voltage must be regulated at the given reference voltage value to produce a multilevel stepped waveform at the output of MMC. The MMC has several submodules in each arm and controlling these submodules is one of the challenging tasks. The capacitor voltage control is usually separated into three stages named as, leg voltage control, voltage balance among the arms, and voltage balance among the submodules within the arm. The first stage can be integrated into the control scheme. In this stage, the distribution of energy among converter legs is controlled by changing the DC-link current reference. The voltage unbalance among the arms causes a circulating current. Therefore, the circulating current control is commonly employed to ensure a voltage balance among the arms. Voltage balance among the submodules within the arm depends on the direction of the arm current and capacitor voltage error. It can be implemented at either the control stage or modulation stage. Implementing the submodule capacitor voltage control requires a significant number of sensors to measure the submodule capacitor voltages and arm currents.

(4) Submodule Capacitor Voltage Ripple: The interaction between the arm currents and voltage causes a voltage ripple in submodule capacitors. The ripple in submodule capacitors is dominated by the fundamental and second-order harmonic components only. The magnitude of voltage ripple is inversely proportional to the fundamental frequency. Therefore, the voltage ripple is severe in variable speed motor drive systems rather than HVDC applications. In motor drive applications, each submodule requires a large capacitance value to suppress the voltage ripple at zero and low speed of operation. Thus, it increases the converter cost. Hence, the suppression of voltage ripple is a trade-off between the performance/efficiency/reliability and the cost/size of the MMC.

(5) Circulating Currents: Circulating currents originate from the voltage difference between the arms in each converter leg. These currents mainly contain negative sequence components, which flows internally among the converter legs only. The circulating currents do not affect the AC output voltages and currents. However, it increases the peak and RMS value of the arm current, which consequently increases converter power losses and the ripple in submodule capacitor voltage. Therefore, the circulating currents must be suppressed for a reliable and efficient operation of the MMC.

(6) Fault Tolerance: The modular multilevel converter is always designed with the redundant submodules to continue their operation during the faults. Typically, the submodules are designed with a bypass switch connected across the AC output terminals. During the fault condition, the bypass switch is used to disconnect the faulty submodule and insert one of the redundant submodules in the arm. Thus, the effect of faults on the MMC operation can be significantly minimized. However, detecting of faults and inserting redundant submodules without inrush current is a great challenge.

The DC-link short-circuit fault is another major issue in the MMC-based HVDC systems. The DC circuit breaker is commonly employed to protect the system during the DC faults. On the other hand, the submodules with DC fault-blocking capability can be employed in the MMC-HVDC system. During the fault condition, the submodules are controlled to generate the negative voltage level at the AC output terminal, which blocks the fault current flowing through the devices.

1.7 SUMMARY

The rapid industrialization in various industrial sectors such as manufacturing, petrochemical, and mining sectors increases their power demand and rate of production. To address these issues, several high-power converters were developed and commercialized in the form of standard and customized products. In this Chapter, a comprehensive overview of several high-power converters including VSCs, CSCs, and matrix converters, and their limitations for high-voltage, high-power applications are presented. Among them, VSCs achieved a major market share and had noticeable developments in the past decade. The technical challenges, limitations, and voltage and power capacity of the most popular and commercially available VSCs are presented. With the present semiconductor technology, VSCs can handle an operating voltage up to 6.6 kV without devices in series. For the higher operating voltage, some of the topologies are not cost-effective. Another few topologies require isolated DC sources and a complicated phase-shift transformer, which increases the cost and volume of the converter.

The most recent developments in the voltage source converter is a modular multilevel converter. This topology has a modular construction; voltage and power rating of the converter can be increased by adding the submodules in series. Also, the phases-shifting transformer is not required, which significantly minimizes the cost of the converter. These features led to the commercial success of MMC, and a wide range of products are available in the market. The applications and market overview of the modular multilevel converter are presented. These applications are further analyzed in the subsequent chapters of this book. Despite all its advantages, MMC requires a complex control system to meet multiple control objectives. The technical challenges, which are associated with the control of MMC, are also presented. An in-depth analysis of these challenges is discussed in the following Chapters of this book.

REFERENCES

1. H. Stemmler, "High-power industrial drives," *Proc. IEEE*, vol. 82, no. 8, pp. 1266–1286, Aug 1994.
2. L. M. Tolbert, F. Z. Peng, and T. G. Habetler, "Multilevel converters for large electric drives," *IEEE Trans. Ind. Appl.*, vol. 35, no. 1, pp. 36–44, Jan 1999.
3. S. Kouro, J. Rodriguez, B. Wu, S. Bernet, and M. Perez, "Powering the future of industry: High-power adjustable speed drive topologies," *IEEE Ind. Appl. Mag.*, vol. 18, no. 4, pp. 26–39, Jul 2012.
4. L. G. Franquelo, J. Rodriguez, J. I. Leon, S. Kouro, R. Portillo, and M. A. M. Prats, "The age of multilevel converters arrives," *IEEE Ind. Electron. Mag.*, vol. 2, no. 2, pp. 28–39, Jun 2008.
5. B. Wu, J. Pontt, J. Rodriguez, S. Bernet, and S. Kouro, "Current-source converter and cycloconverter topologies for industrial medium-voltage drives," *IEEE Trans. Ind. Electron.*, vol. 55, no. 7, pp. 2786–2797, Jul 2008.
6. J. Rodriguez, J.-S. Lai, and F. Z. Peng, "Multilevel inverters: a survey of topologies, controls, and applications," *IEEE Trans. Ind. Electron.*, vol. 49, no. 4, pp. 724–738, Aug 2002.
7. H. Abu-Rub, J. Holtz, J. Rodriguez, and G. Baoming, "Medium-voltage multilevel converters: State of the art, challenges, and requirements in industrial applications," *IEEE Trans. Ind. Electron.*, vol. 57, no. 8, pp. 2581–2596, Aug 2010.
8. J. Rodriguez, L. G. Franquelo, S. Kouro, J. I. Leon, R. C. Portillo, M. Á. M. Prats, and M. A. Perez, "Multilevel converters: An enabling technology for high-power applications," *Proc. IEEE*, vol. 97, no. 11, pp. 1786–1817, Nov 2009.
9. J. Rodriguez, S. Bernet, B. Wu, J. Pontt, and S. Kouro, "Multilevel voltage-source-converter topologies for industrial medium-voltage drives," *IEEE Trans. Ind. Electron.*, vol. 54, no. 6, pp. 2930–2945, Dec 2007.
10. S. Kouro, M. Malinowski, K. Gopakumar, J. Pou, L. G. Franquelo, B. Wu, J. Rodriguez, M. A. Perez, and J. I. Leon, "Recent advances and industrial applications of multilevel converters," *IEEE Trans. Ind. Electron.*, vol. 57, no. 8, pp. 2553–2580, Aug 2010.
11. J. Rodriguez, S. Bernet, P. Steimer, and I. Lizama, "A survey on neutral-point-clamped inverters," *IEEE Trans. Ind. Electron.*, vol. 57, no. 7, pp. 2219–2230, Jul 2010.
12. M. Malinowski, K. Gopakumar, J. Rodriguez, and M. Pèrez, "A survey on cascaded multilevel inverters," *IEEE Trans. Ind. Electron.*, vol. 57, no. 7, pp. 2197–2206, Jul 2010.
13. N. Celanovic and D. Boroyevich, "A comprehensive study of neutral-point voltage balancing problem in three-level neutral-point-clamped voltage source pwm inverters," *IEEE Trans. Power Electron.*, vol. 15, no. 2, pp. 242–249, Mar 2000.
14. C. Wang and Y. Li, "Analysis and calculation of zero-sequence voltage considering neutral-point potential balancing in three-level npc converters," *IEEE Trans. Ind. Electron.*, vol. 57, no. 7, pp. 2262–2271, Jul 2010.
15. S. Busquets-Monge, J. Bordonau, D. Boroyevich, and S. Somavilla, "The nearest three virtual space vector pwm - a modulation for the comprehensive neutral-point balancing in the three-level npc inverter," *IEEE Power Electr. Letters*, vol. 2, no. 1, pp. 11–15, Mar 2004.
16. H. du Toit Mouton, "Natural balancing of three-level neutral-point-clamped pwm inverters," *IEEE Trans. Ind. Electron.*, vol. 49, no. 5, pp. 1017–1025, Oct 2002.

17. J. A. Sayago, T. Bruckner, and S. Bernet, "How to select the system voltage of mv drives-a comparison of semiconductor expenses," *IEEE Trans. Ind. Electron.*, vol. 55, no. 9, pp. 3381–3390, Sep 2008.

18. T. Bruckner, S. Bernet, and H. Guldner, "The active npc converter and its loss-balancing control," *IEEE Trans. Ind. Electron.*, vol. 52, no. 3, pp. 855–868, Jun 2005.

19. P. Barbosa, P. Steimer, L. Meysenc, M. Winkelnkemper, J. Steinke, and N. Celanovic, "Active neutral-point-clamped multilevel converters," in *2005 IEEE 36th Power Electronics Specialists Conference*, Jun 2005, pp. 2296–2301.

20. F. Kieferndorf, M. Basler, L. A. Serpa, J. H. Fabian, A. Coccia, and G. A. Scheuer, "A new medium voltage drive system based on anpc-5l technology," in *2010 IEEE International Conference on Industrial Technology*, Mar 2010, pp. 643–649.

21. B. P. McGrath and D. G. Holmes, "Natural capacitor voltage balancing for a flying capacitor converter induction motor drive," *IEEE Trans. Power Electron.*, vol. 24, no. 6, pp. 1554–1561, Jun 2009.

22. M. Narimani, B. Wu, Z. Cheng, and N. R. Zargari, "A new nested neutral point-clamped (nnpc) converter for medium-voltage (mv) power conversion," *IEEE Trans. Power Electron.*, vol. 29, no. 12, pp. 6375–6382, Dec 2014.

23. M. Narimani, B. Wu, Z. Cheng, and N. R. Zargari, "A novel and simple single-phase modulator for the nested neutral-point clamped (nnpc) converter," *IEEE Trans. Power Electron.*, vol. 30, no. 8, pp. 4069–4078, Aug 2015.

24. M. Narimani, B. Wu, and N. R. Zargari, "A novel five-level voltage source inverter with sinusoidal pulse width modulator for medium-voltage applications," *IEEE Trans. Power Electron.*, vol. 31, no. 3, pp. 1959–1967, Mar 2016.

25. M. Narimani, B. Wu, K. Tian, Z. Cheng, and N. R. Zargari, "A new h-bridge nnpc converter for 10 kV class motor drives," in *IECON 2014–40th Annual Conference of the IEEE Industrial Electronics Society*, Oct 2014, pp. 1067–1071.

26. J.-S. Lai and F. Z. Peng, "Multilevel converters-a new breed of power converters," *IEEE Trans. Ind. Appl.*, vol. 32, no. 3, pp. 509–517, May 1996.

27. J. Rodriguez, P. W. Hammond, J. Pontt, R. Musalem, P. Lezana, and M. J. Escobar, "Operation of a medium-voltage drive under faulty conditions," *IEEE Trans. Ind. Electron.*, vol. 52, no. 4, pp. 1080–1085, Aug 2005.

28. B. Wu, *High-Power Converters and AC Drives*. New york: Wiley-IEEE Press, Mar 2006.

29. K. P. Phillips, "Current-source converter for ac motor drives," *IEEE Trans. Ind. Appl.*, vol. IA-8, no. 6, pp. 679–683, Nov 1972.

30. J. Dai, Y. Lang, B. Wu, D. Xu, and N. R. Zargari, "A multisampling svm scheme for current source converters with superior harmonic performance," *IEEE Trans. Power Electron.*, vol. 24, no. 11, pp. 2436–2445, Nov 2009.

31. N. Zhu, D. Xu, B. Wu, N. R. Zargari, M. Kazerani, and F. Liu, "Common-mode voltage reduction methods for current-source converters in medium-voltage drives," *IEEE Trans. Power Electron.*, vol. 28, no. 2, pp. 995–1006, Feb 2013.

32. P. W. Wheeler, J. Rodriguez, J. C. Clare, L. Empringham, and A. Weinstein, "Matrix converters: a technology review," *IEEE Trans. Ind. Electron.*, vol. 49, no. 2, pp. 276–288, Apr 2002.

33. J. W. Kolar, T. Friedli, J. Rodriguez, and P. W. Wheeler, "Review of three-phase pwm ac–ac converter topologies," *IEEE Trans. Ind. Electron.*, vol. 58, no. 11, pp. 4988–5006, Nov 2011.

34. L. Empringham, J. W. Kolar, J. Rodriguez, P. W. Wheeler, and J. C. Clare, "Technological issues and industrial application of matrix converters: A review," *IEEE Trans. Ind. Electron.*, vol. 60, no. 10, pp. 4260–4271, Oct 2013.

35. H. N. Nguyen and H. H. Lee, "A modulation scheme for matrix converters with perfect zero common-mode voltage," *IEEE Trans. Power Electron.*, vol. 31, no. 8, pp. 5411–5422, Aug 2016.

36. H. M. Nguyen, H. H. Lee, and T. W. Chun, "Input power factor compensation algorithms using a new direct-svm method for matrix converter," *IEEE Trans. Ind. Electron.*, vol. 58, no. 1, pp. 232–243, Jan 2011.

37. M. Chai, D. Xiao, R. Dutta, and J. E. Fletcher, "Space vector pwm techniques for three-to-five-phase indirect matrix converter in the overmodulation region," *IEEE Trans. Ind. Electron.*, vol. 63, no. 1, pp. 550–561, Jan 2016.

38. R. K. Gupta, K. K. Mohapatra, A. Somani, and N. Mohan, "Direct-matrix-converter-based drive for a three-phase open-end-winding ac machine with advanced features," *IEEE Trans. Ind. Electron.*, vol. 57, no. 12, pp. 4032–4042, Dec 2010.

39. P. Szczesniak, *Three-Phase AC–AC Power Converters Based on Matrix Converter Topology*. Springer, 2013.

40. T. D. Nguyen and H. H. Lee, "Development of a three-to-five-phase indirect matrix converter with carrier-based pwm based on space-vector modulation analysis," *IEEE Trans. Ind. Electron.*, vol. 63, no. 1, pp. 13–24, Jan 2016.

41. J. Wang, B. Wu, D. Xu, and N. R. Zargari, "Multimodular matrix converters with sinusoidal input and output waveforms," *IEEE Trans. Ind. Electron.*, vol. 59, no. 1, pp. 17–26, Jan 2012.

42. J. Wang, B. Wu, D. Xu, and N. R. Zargari, "Indirect space-vector-based modulation techniques for high-power multimodular matrix converters," *IEEE Trans. Ind. Electron.*, vol. 60, no. 8, pp. 3060–3071, Aug 2013.

43. J. Wang, B. Wu, D. Xu, and N. R. Zargari, "Phase-shifting-transformer-fed multimodular matrix converter operated by a new modulation strategy," *IEEE Trans. Ind. Electron.*, vol. 60, no. 10, pp. 4329–4338, Oct 2013.

44. M. Glinka and R. Marquardt, "A new ac/ac multilevel converter family," *IEEE Trans. Ind. Electron.*, vol. 52, no. 3, pp. 662–669, 2005.

45. A. Lesnicar and R. Marquardt, "An innovative modular multilevel converter topology suitable for a wide power range," in *Power Tech Conference Proceedings, 2003 IEEE Bologna*, vol. 3, Jun 2003, pp. 23–26.

46. H. Akagi, "Classification, terminology, and application of the modular multilevel cascade converter (mmcc)," *IEEE Trans. Power Electron.*, vol. 26, no. 11, pp. 3119–3130, Nov 2011.

47. S. Rohner, S. Bernet, M. Hiller, and R. Sommer, "Modulation, losses, and semiconductor requirements of modular multilevel converters," *IEEE Trans. Ind. Electron.*, vol. 57, no. 8, pp. 2633–2642, 2010.

48. V. Dargahi, A. Sadigh, M. Abarzadeh, S. Eskandari, and K. Corzine, "A new family of modular multilevel converter based on modified flying-capacitor multicell converters," *IEEE Trans. Power Electron.*, vol. 30, no. 1, pp. 138–147, Jan 2015.

49. A. Nami, J. Liang, F. Dijkhuizen, and G. Demetriades, "Modular multilevel converters for hvdc applications: Review on converter cells and functionalities," *IEEE Trans. Power Electron.*, vol. 30, no. 1, pp. 18–36, Jan 2015.

50. S. Debnath, J. Qin, B. Bahrani, M. Saeedifard, and P. Barbosa, "Operation, control, and applications of the modular multilevel converter: A review," *IEEE Trans. Power Electron.*, vol. 30, no. 1, pp. 37–53, Jan 2015.

51. M. Perez, S. Bernet, J. Rodriguez, S. Kouro, and R. Lizana, "Circuit topologies, modeling, control schemes, and applications of modular multilevel converters," *IEEE Trans. Power Electron.*, vol. 30, no. 1, pp. 4–17, Jan 2015.

52. H. J. Knaak, "Modular multilevel converters and hvdc/facts: A success story," in *Power Electronics and Applications (EPE 2011), Proceedings of the 2011-14th European Conference on*, Aug 2011, pp. 1–6.

53. J. Glasdam, J. Hjerrild, L. H. Kocewiak, and C. L. Bak, "Review on multi-level voltage source converter based hvdc technologies for grid connection of large offshore wind farms," in *Power System Technology (POWERCON), 2012 IEEE International Conference on*, Oct 2012, pp. 1–6.

54. V. Khadkikar, "Enhancing electric power quality using upqc: A comprehensive overview," *IEEE Trans. Power Electron.*, vol. 27, no. 5, pp. 2284–2297, May 2012.

55. Z. Liu, J. Lu, Z. Ou, M. Ma, C. Yuan, X. Xiao, and H. Wang, "The start control strategy design of unified power quality conditioner based on modular multilevel converter," in *Electric Machines Drives Conference (IEMDC), 2013 IEEE International*, May 2013, pp. 933–937.

CHAPTER 2

FUNDAMENTALS OF MODULAR MULTILEVEL CONVERTER

2.1 INTRODUCTION

The evolution of high-power converters was presented in Chapter 1. It was noted that the integrated multilevel converters can be directly connected to the medium-voltage (2.3–4.16 kV) power systems. Moreover, these topologies require a higher number of output voltage levels to operate at the higher operating voltage, which subsequently increases the complexity of control and the assembly of the converter. Also, to achieve the higher operating voltages, they require a step-up transformer, which increases the overall system cost and size [1, 2]. On the other hand, multi-cell converters are modular and are easier to scale up to higher voltages. Among them, the CHB and CNPC topologies require an isolated DC sources, and their operating voltage is limited to 6.6–13.8 kV.

The modular multilevel converter (MMC) is one of the promising topologies in multi-cell converter family. The MMC eliminates the need for an isolated DC sources and the transformer for the high-voltage operation. It employs a cascade connection of submodules to reach the desired system voltage while producing a high-quality multilevel output voltage waveform. The submodule is a building block of the MMC and can be configured in various forms by using IGBT devices and DC

Modular Multilevel Converters: Analysis, Control, and Applications, First Edition.
Sixing Du, Apparao Dekka, Bin Wu, and Navid Zargari.

capacitors. The number of submodules used in the MMC varies with the application, operating voltage, and the rating of IGBT devices [3]. For example, the medium-voltage motor drives with 3.3–13.8 kV operating voltage require 5–20 SM/arm [4, 5], the high-voltage direct current (HVDC) transmission system needs 200–400 SM/arm to reach $\pm$ 320 kV (DC) voltage [6, 7], and the static synchronous compensators employ 15–200 SM/arm to reach operating voltages of 13.9–220 kV [8].

This chapter deals with the general aspects of the modular multilevel converter, including topologies, operation, and pulse width modulation (PWM) schemes. The circuit configuration of the modular multilevel converter with different submodule structures and, their functionality and features are discussed in Section 2.2. The operation of MMC with half-bridge (HB) submodule is also presented. The implementation and performance of various PWM schemes including sine-triangle, staircase, sampled average, and space vector modulation (SVM) schemes are presented in Section 2.3. Finally, the summary of this chapter is discussed in Section 2.4.

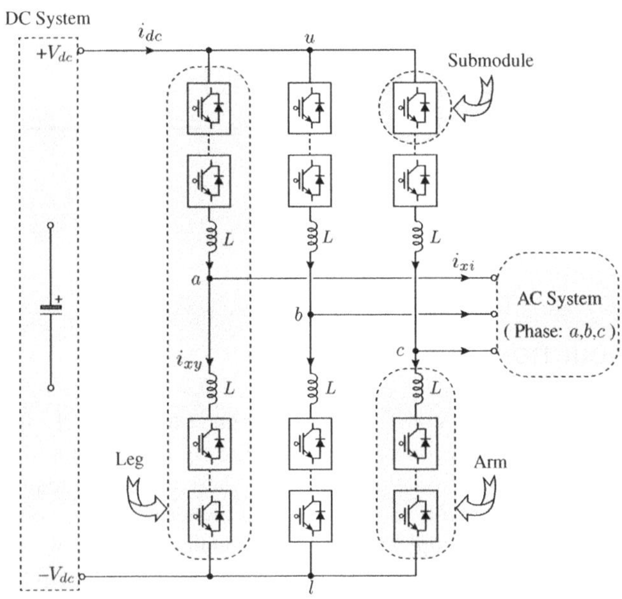

Figure 2.1 Configuration of modular multilevel converter.

2.2 MODULAR MULTILEVEL CONVERTER

The modular multilevel converter (MMC) topology is a promising solution for high-voltage, high-power industrial applications. The details of the modular multilevel converter including the converter configuration, the choice of submodules, the principle of operation, and their features are presented in this section.

2.2.1 Converter Configuration

The generalized circuit configuration of a three-phase MMC is shown in Figure 2.1. The DC system of an MMC is often referred to as a DC-bus or DC-link, connected to the positive and negative bars of the converter legs. The three-phase AC system is connected to the mid-point of each leg (a, b, c). Each leg of the MMC is divided into two arms. The arms connected to the positive bar are referred to as the upper arms (u), and the arms connected to the negative bar are referred to as the lower arms (l). Each arm has a group of submodules and an inductor (L). The arm inductor is connected in series with each group of submodules to limit the current due to the instantaneous voltage difference between the arms. The main features of MMC are as follows [9]:

- It has a modular construction with an ability to scale the voltage and power rating.
- It can generate an output voltage and current waveforms with reduced $\frac{dv}{dt}$ and ripple. Typically, an output voltage with a very low total harmonic distortion (THD) can be obtained.
- MMC can produce the output voltage waveform with very large number of voltage levels. Hence, it is possible to operate the submodules with a very low switching frequency.
- It can employ redundant submodules in each arm to achieve a fault-tolerant operation.

2.2.2 Configuration of Submodules

A submodule (SM) is a simple DC–AC power conversion circuit. Usually, low-voltage IGBT devices (1200–1700 V) and DC capacitors are used to configure a submodule. The commonly used submodule configurations in a modular multilevel converter are [10, 11]:

- Half-bridge (HB) submodule
- Full-bridge (FB) submodule
- Flying capacitor (FC) submodule
- Cascaded half-bridge (CH) submodule
- Double clamp (CD) submodule

The above submodules will be presented and analyzed in detail in the following subsections.

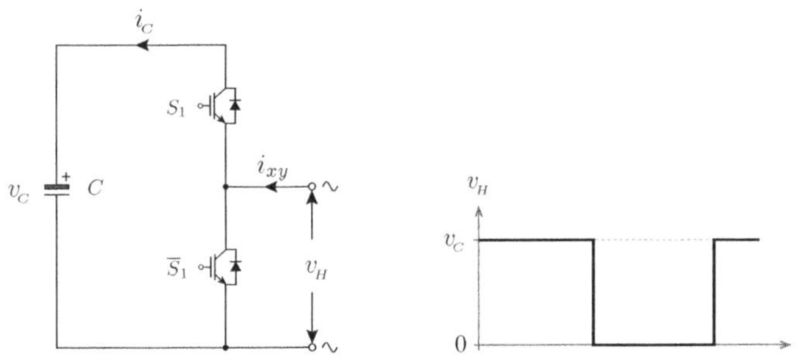

Figure 2.2 Half-bridge submodule and output voltage waveform.

(1) Half-bridge Submodule: The half-bridge (HB) submodule is often referred to as a chopper cell. The circuit configuration of half-bridge submodule is shown in Figure 2.2. It is composed of two IGBT devices with antiparallel diodes (S_1 and $\overline{S}_1$) and one DC capacitor (C).

The two IGBT devices are operated in a complementary manner to regulate the DC capacitor voltage at a value of v_C. The DC capacitor voltage is given by

$$v_C = \frac{1}{C}\int_{0+}^{t} i_C(\tau)d\tau \tag{2.1}$$

The DC capacitor current in terms of AC current (i_{xy}) and the switching state of top-device S_1 is given by

$$i_C = S_1\, i_{xy} \tag{2.2}$$

Depending on the state of top switch S_1, the DC capacitor current is equal to either the AC current or zero. The possible states of S_1, and their effect on the DC capacitor voltage for different directions of AC current are shown in Table 2.1.

Table 2.1 Switching states of HB SM

State	S_1	v_H	$i_{xy} > 0$	$i_{xy} \leq 0$
②	1	v_C	$v_C \uparrow$	$v_C \downarrow$
①	0	0	$v_C \approx$	$v_C \approx$

$\approx$ = No change, $\uparrow$ = Increasing, $\downarrow$ = Decreasing

The AC output voltage of half-bridge submodule has two voltage levels, "0" and "v_C" as shown in Figure 2.2. When the top switch is "ON," the AC output voltage is equal to v_C. In this mode, the DC capacitor voltage increases for the positive direction of the current and decreases in the negative direction of the current. When the top switch is "OFF," the AC output voltage is equal to "0." In this mode, the DC capacitor voltage remains constant, irrespective of the current direction. The AC output voltage of half-bridge submodule can be represented in terms of DC capacitor voltage and the switching state of top device S_1 as

$$v_H = S_1 \, v_C \tag{2.3}$$

(2) Full-bridge Submodule: The full-bridge (FB) submodule is also referred to as an H-bridge converter. The circuit configuration of full-bridge submodule is shown in Figure 2.3. It is composed of two half-bridge legs (S_1, $\overline{S}_1$ and S_2, $\overline{S}_2$) with one DC capacitor (C). Each leg consist of two IGBT devices with antiparallel diodes and are operated in a complementary manner.

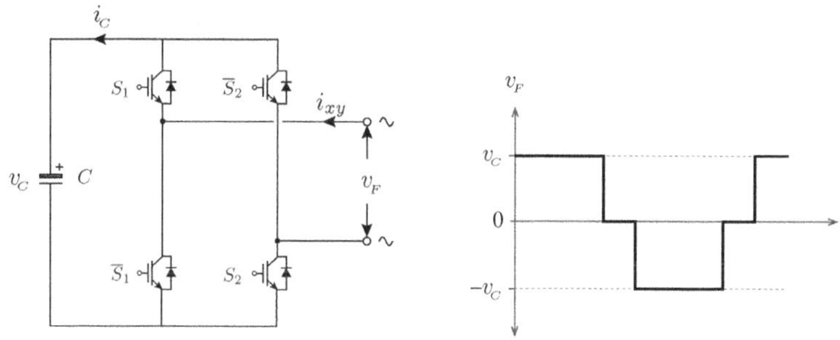

Figure 2.3 Full-bridge submodule and output voltage waveform.

The DC capacitor voltage is regulated by controlling the IGBT devices S_1 and S_2. The expression for the DC capacitor voltage of full-bridge submodule is same as that of the half-bridge submodule given in equation (2.1). The current flowing through the DC capacitor depends on the state of devices S_1 and S_2. The DC capacitor current is given by

$$i_C = (S_1 \, S_2 - \overline{S}_1 \, \overline{S}_2) \, i_{xy} \tag{2.4}$$

All possible combination of switching states for S_1 and S_2 are shown in Table 2.2. The four switching combinations generate three voltage levels, "0," "v_C," and "$-v_C$."

Table 2.2 Switching states of FB SM

State	S_1	S_2	v_F	$i_{xy} > 0$	$i_{xy} \leq 0$
③	1	1	v_C	$v_C \uparrow$	$v_C \downarrow$
②	1	0	0	$v_C \approx$	$v_C \approx$
①	0	1	0	$v_C \approx$	$v_C \approx$
BM	0	0	$-v_C$	$v_C \downarrow$	$v_C \uparrow$

$\approx$ = No change, $\uparrow$ = Increasing, $\downarrow$ = Decreasing, BM = Blocking mode

The AC output voltage of full-bridge submodule is shown in Figure 2.3. When IGBT devices S_1, S_2 are "ON," the AC output voltage is equal to "v_C." In this mode, the DC capacitor voltage increases when the current direction is positive and decreases in the negative direction of the current. The AC output voltage will be "0" for switching states S_1–"ON," S_2–"OFF" and S_1–"OFF," S_2–"ON." In this mode, the DC capacitor voltage remains constant irrespective of the current direction. The full-bridge submodule has two switching states corresponding to the voltage level "0." The additional switching states corresponding to the same voltage level are referred to as redundant switching states. The redundant switching states are used to achieve the symmetrical distribution of power losses between S_1 and S_2 devices.

The full-bridge submodule generates a voltage level "$-v_C$" when the devices S_1, S_2 are "OFF." This switching state is used to limit the current during DC-side faults. The AC output voltage equation, which represents the operation of full-bridge submodule is given by

$$v_F = (S_1\, S_2 - \overline{S}_1\, \overline{S}_2)\, v_C \tag{2.5}$$

(3) Flying Capacitor Submodule: Figure 2.4(a) shows the configuration of flying capacitor (FC) submodule. It is composed of four IGBT devices with antiparallel diodes (S_1, $\overline{S}_1$ and S_2, $\overline{S}_2$) and two DC capacitors (C_1 and C_2). The voltage of DC capacitors C_1 and C_2 is given by

$$\begin{aligned} v_{C1} &= \frac{1}{C1}\int_{0+}^{t} i_{C1}(\tau)d\tau \\ v_{C2} &= \frac{1}{C2}\int_{0+}^{t} i_{C2}(\tau)d\tau \end{aligned} \tag{2.6}$$

The DC capacitors current is given in terms of the device switching states and the AC current as

$$\begin{aligned} i_{C1} &= S_1\, i_{xy} \\ i_{C2} &= (S_2 - S_1)\, i_{xy} \end{aligned} \tag{2.7}$$

Table 2.3 Switching states of FC SM

State	S_1	S_2	v_{FC}	$i_{xy} > 0$	$i_{xy} \leq 0$
④	1	1	v_{C1}	$v_{C1}\uparrow$,$v_{C2}\approx$	$v_{C1}\downarrow$,$v_{C2}\approx$
③	0	1	v_{C2}	$v_{C1}\approx$,$v_{C2}\uparrow$	$v_{C1}\approx$,$v_{C2}\downarrow$
②	1	0	$v_{C1} - v_{C2}$	$v_{C1}\uparrow$,$v_{C2}\downarrow$	$v_{C1}\downarrow$,$v_{C2}\uparrow$
①	0	0	0	$v_{C1}\approx$,$v_{C2}\approx$	$v_{C1}\approx$,$v_{C2}\approx$

$\approx$ = No change, $\uparrow$ = Increasing, $\downarrow$ = Decreasing

The device switching states (S_1 and S_2) of flying capacitor submodule are shown in Table 2.3. There are a total four switching combinations, which generates three voltage levels of "0," $v_{C1} - v_{C2}$ or v_{C2}, and v_{C1} as shown in Figure 2.4(a). The DC capacitor voltage v_{C1} is regulated at twice the DC capacitor voltage v_{C2}, resulting in symmetrical steps in the output voltage and equal voltage across IGBT devices during the reverse blocking mode. When the devices S_1, S_2 are "ON," then the AC

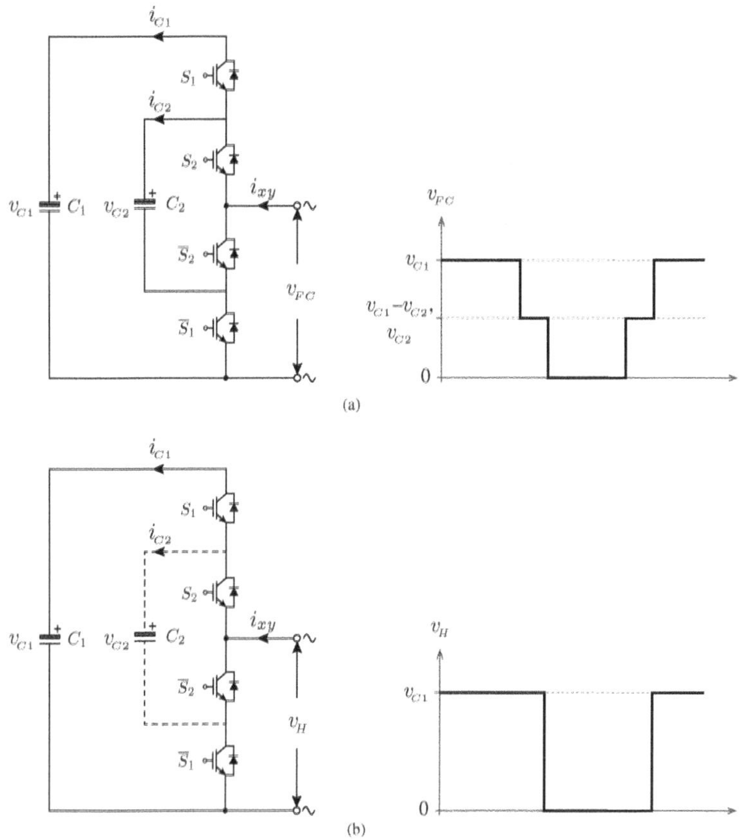

Figure 2.4 Flying capacitor submodule and output voltage waveform: (a) three-level operation, and (b) two-level operation.

output voltage is equal to "v_{C1}." In this mode, the DC capacitor voltage v_{C1} increases for the positive direction of the current and decreases for the negative direction of the current. The DC capacitor voltage v_{C2} remains constant.

The AC output voltage will be "$v_{C1} - v_{C2}$" for a switching state S_1–"ON," S_2–"OFF." In this mode, the DC capacitor voltage v_{C1} increases and v_{C2} decreases for the positive direction of the current, and vice versa. Similarly, for a switching state S_1–"OFF," S_2–"ON," the AC output voltage is equal to v_{C2}. In this mode, the DC capacitor voltage v_{C2} will change, while the DC capacitor voltage v_{C1} remains constant. When the switching state S_1, S_2 are "OFF," the AC voltage level is "0." Among them, the states ② and ③ generate identical voltage levels at the output. The switching states corresponding to these states are used to control the DC capacitors voltage. The AC output voltage equation, which represents the operation of the flying capacitor submodule is given by

$$v_{FC} = S_1\, v_{C1} + (S_2 - S_1)\, v_{C2} \tag{2.8}$$

The flying capacitor submodule has a dual functionality feature, and it can be operated as a half-bridge submodule as well [12]. Figure 2.4(b) shows the equivalent operation of half-bridge submodule with a flying capacitor submodule. The switching combinations corresponding to the states ① and ④ are used to generate the voltage levels "0" and "v_{C1}," respectively, at the output. The DC capacitor C_2 is completely bypassed in this mode of operation. The output voltage of half-bridge submodule is given by

$$v_H = S_1\, S_2\, v_{C1} \tag{2.9}$$

(4) Cascaded Half-bridge Submodule: The circuit configuration of cascaded half-bridge (CH) submodule is shown in Figure 2.5. It is composed of two identical half-bridge submodules connected in series.

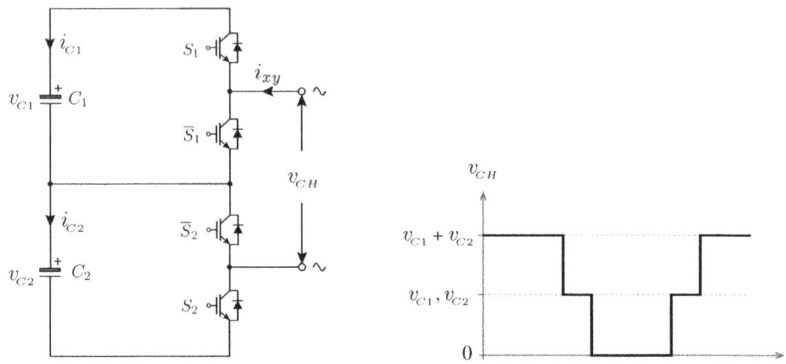

Figure 2.5 Cascaded half-bridge submodule and output voltage waveform.

The cascaded half-bridge submodule has two DC capacitors (C_1 and C_2) with equal nominal voltages. The DC capacitors voltage of C_1 and C_2 can be obtained from equation (2.6). The current flowing through these capacitors are given by

$$\begin{aligned} i_{C1} &= S_1\, i_{xy} \\ i_{C2} &= S_2\, i_{xy} \end{aligned} \tag{2.10}$$

The possible combination of switching states for S_1 and S_2 are shown in Table 2.4. The cascaded half-bridge submodule has four switching combinations and generates a three voltage states "$v_{C1} + v_{C2}$," "v_{C1} and v_{C2}," and "0." The AC output voltage waveform of cascaded half-bridge submodule is shown in Figure 2.5. When the devices S_1, S_2 are "ON," the output voltage is equal to "v_{C1}+v_{C2}." In this state, the DC capacitors C_1 and C_2 voltage increases or decreases based on the direction of the current.

Table 2.4 Switching states of CH SM

State	S_1	S_2	v_{CH}	$i_{xy} > 0$	$i_{xy} \leq 0$
④	1	1	$v_{C1} + v_{C2}$	$v_{C1}\uparrow, v_{C2}\uparrow$	$v_{C1}\downarrow, v_{C2}\downarrow$
③	1	0	v_{C1}	$v_{C1}\uparrow, v_{C2}\approx$	$v_{C1}\downarrow, v_{C2}\approx$
②	0	1	v_{C2}	$v_{C1}\approx, v_{C2}\uparrow$	$v_{C1}\approx, v_{C2}\downarrow$
①	0	0	0	$v_{C1}\approx, v_{C2}\approx$	$v_{C1}\approx, v_{C2}\approx$

$\approx$ = No change, $\uparrow$ = Increasing, $\downarrow$ = Decreasing

For switching states ② and ③, one of the half-bridge submodules is bypassed; i.e., the output voltage is equal to either "v_{C1}" or "v_{C2}." In this mode, the DC capacitor voltage corresponding to the inserted half-bridge submodule increases for the positive direction of the current and vice versa. The DC capacitor voltage of bypassed half-bridge submodule remains constant. The cascaded half-bridge submodule generates "0" voltage level at the output when the devices S_1, S_2 are "OFF." The DC capacitors voltage "v_{C1}" and "v_{C2}" remains constant irrespective of the current direction. The AC output voltage of cascaded half-bridge submodule is mathematically represented as

$$v_{CH} = S_1\, v_{C1} + S_2\, v_{C2} \tag{2.11}$$

(5) Double Clamp Submodule: Figure 2.6 shows the configuration of the double clamp (CD) submodule. It is composed of two identical half-bridge submodules, two blocking diodes (D_{B1} and D_{B2}), and one IGBT device (S_B) [13].

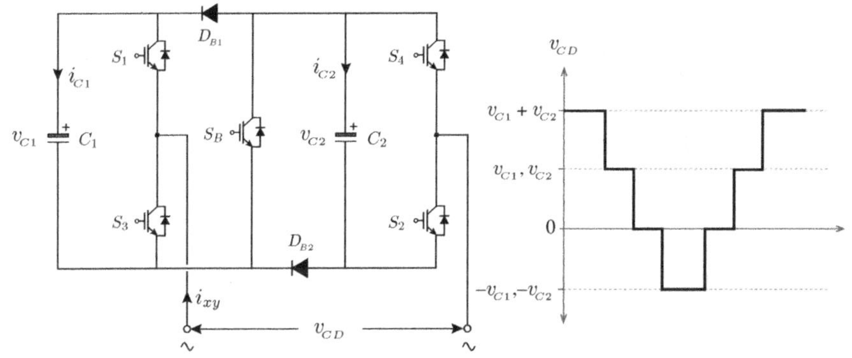

Figure 2.6 Double clamp submodule and output voltage waveform.

Each half-bridge submodule has one DC capacitor whose voltage can be obtained from the equation (2.6). The current flowing through the DC capacitors is given by

$$\begin{aligned} i_{C1} &= S_1\, i_{xy} + (1 - S_B)\, i_{xy} \\ i_{C2} &= S_2\, i_{xy} + (1 - S_B)\, i_{xy} \end{aligned} \tag{2.12}$$

Table 2.5 Switching states of CD SM

State	S_1	S_2	S_3	S_4	S_B	v_{CD}	$i_{xy} > 0$	$i_{xy} \leq 0$
④	1	1	0	0	1	$v_{C1} + v_{C2}$	$v_{C1}\uparrow, v_{C2}\uparrow$	$v_{C1}\downarrow, v_{C2}\downarrow$
③	1	0	0	1	1	v_{C1}	$v_{C1}\uparrow, v_{C2}\approx$	$v_{C1}\downarrow, v_{C2}\approx$
②	0	1	1	0	1	v_{C2}	$v_{C1}\approx, v_{C2}\uparrow$	$v_{C1}\approx, v_{C2}\downarrow$
①	0	0	1	1	1	0	$v_{C1}\approx, v_{C2}\approx$	$v_{C1}\approx, v_{C2}\approx$
BM	0	0	0	0	0	$v_{C1}+v_{C2}$	$v_{C1}\uparrow, v_{C2}\uparrow$	-
BM	0	0	0	0	0	$-v_{C1}, -v_{C2}$	-	$v_{C1}\uparrow, v_{C2}\uparrow$

$\approx$ = No change, $\uparrow$ = Increasing, $\downarrow$ = Decreasing, BM = Blocking mode

The switching states of double clamp submodule are shown in Table 2.5. The normal operation of double clamp submodule is similar to that of cascaded half-bridge submodule, which can generate three voltage levels "$v_{C1} + v_{C2}$", "v_{C1} and v_{C2}," and "0" with four switching combinations. During normal operation, the IGBT device "S_B" is always "ON." When the devices S_1, S_2 are "ON," the two DC capacitors of half-bridge submodules are connected in series through diodes D_{B1} and D_{B2} and switch S_B. In this state, the AC output voltage is equal to "v_{C1}+v_{C2}." To generate the voltage level "v_{C1}, v_{C2}," one of the half-bridge submodule DC capacitor should be bypassed by turning "OFF" either S_1 or S_2. The voltage level "0" will be generated by turning "OFF" devices S_1 and S_2. The AC output voltage equation, which represents the normal operation of double clamp submodule is given by

$$v_{DC} = S_B\,(S_1\,v_{C1} + S_2\,v_{C2}) \tag{2.13}$$

The double clamp submodule also generates negative voltage levels during the blocking mode (BM) of operation. This mode of operation is used to limit current during DC-side faults [13]. In this mode, all the IGBT devices are turned "OFF" as shown in Table 2.5. The submodule output voltage is equal to "v_{C1}+v_{C2}" for the positive direction of current and "$-v_{C1}$, $-v_{C2}$" for the negative direction of the current.

2.2.3 Comparison of Submodules

In Section 2.2.2, several submodules and their principles of operation are presented. These submodules are further evaluated in terms of rating, operation, control and design complexity, losses, and DC fault-blocking capability. A summary of comparison is presented in Table 2.6 [10, 11]. Among all the submodules, the half-bridge (HB) submodule is the most popular and is widely used in commercial products. Its simple construction (two semiconductor devices and one floating capacitor) results in a simple control and design. During the normal operation, only one device (either S_1 or $\overline{S}_1$) will be in "ON" state. Hence, the half-bridge submodule has low power losses and high efficiency. The output voltage of half-bridge submodule consists of positive levels only, and cannot support the bipolar operation and DC fault blocking.

The full-bridge (FB) submodule consists of four semiconductor devices and one DC capacitor. It has twice the number of semiconductors as that of the half-bridge submodule for the same voltage rating. However, the control and design complexity are similar to that of the half-bridge submodule. In the full-bridge submodule, two devices will carry the current during normal operation, resulting in higher power losses and lower efficiency. The negative voltage levels in full-bridge submodule are used to limit the current during DC-side faults.

Similar to the full-bridge SM, the flying capacitor (FC) submodule has four semiconductor devices with a voltage rating of v_C. However, it requires three DC capacitors (rated at v_C), which is three times the capacitors in the full-bridge submodule. The capacitors have different nominal voltages, resulting in a higher design and control complexity. During the normal operation of flying capacitor submodule, a maximum of two devices carry the current. The efficiency and power losses are comparable to that of the full-bridge submodule. The flying capacitor submodule generates only positive voltage levels and cannot limit the current during the DC-side faults.

Table 2.6 Comparison of MMC submodules

Performance Index	HB-SM	FB-SM	FC-SM	CH-SM	CD-SM
No. of output voltage levels	2	3	3	3	4
Max. blocking voltage of SM	v_C	v_C	$2v_C$	$2v_C$	$2v_C$
Max. No. of DC capacitors normalized to v_C	1	1	3	2	2
No. of devices normalized to v_C	2	4	4	4	7
Max. No. of devices in conduction path	1	2	2	2	3
Power losses	Low	Moderate	Moderate	Moderate	High
Bipolar operation	No	Yes	No	No	Yes
SM design complexity	Low	Low	High	Low	High
SM control complexity	Low	Low	High	Low	Low
DC fault blocking	No	Yes	No	No	Yes

The cascaded half-bridge (CH) submodule is formed using two half-bridge submodules connected in series, resulting in a three-level operation. It has a simple control, low design complexity, low power losses, and high efficiency similar to the half-bridge submodule. Similarly, the cascaded half-bridge submodule cannot block the DC-side faults and is not suitable for the bipolar operation.

The double clamp (CD) submodule is also formed using two half-bridge submodules along with three additional devices. This topology can limit the current during DC-side faults. However, the power losses, efficiency, and design complexity are significantly high due to the additional devices.

2.2.4 Principle of Operation

In this section, the operation of a modular multilevel converter using the popular half-bridge submodule is presented. The structure of three-phase modular multilevel converter is shown in Figure 2.7(a), where the DC system is modeled as a split DC source with a voltage $\frac{V_{dc}}{2}$, and a series inductance L_{dc} with an equivalent resistance r_{dc}. The three-phase AC system is connected to a passive load of an inductance L_o and a resistance r_o.

As the name suggests, the modular multilevel converter uses multiple half-bridge submodules with an identical construction in each arm to generate the multilevel stepped waveform at the output. For ease of understanding, only four half-bridge submodules per arm are considered, and their connection is shown in Figure 2.7(b). Each submodule capacitor voltage is rated for v_C, and their output voltages are named as v_{H1}, v_{H2}, v_{H3} and v_{H4}. The submodule output terminals are connected in series to form an arm voltage.

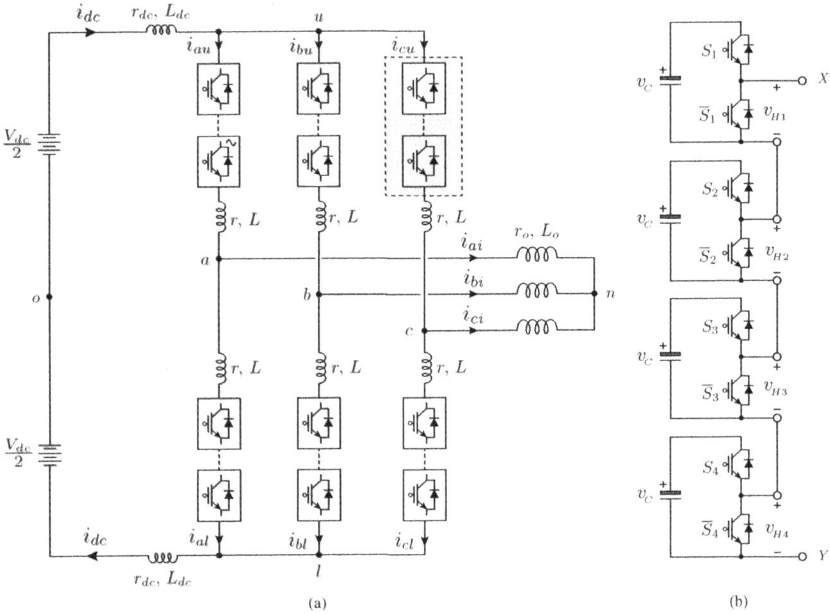

Figure 2.7 Modular multilevel converter and arm configuration: (a) MMC with passive load and (b) connection diagram of HB submodule with in the arm.

The switching states, the output voltage of each submodule, and arm voltage are shown in Table 2.7. With four SMs, the arm voltage has five voltage levels "0, v_C, $2v_C$, $3v_C$, and $4v_C$." The highest voltage level "$4v_C$" is generated by turning "ON" S_1, S_2, S_3, and S_4. Similarly, the lowest voltage level "0" is generated by turn-

Table 2.7 Switching states and voltage levels

S_1	S_2	S_3	S_4	v_{H1}	v_{H2}	v_{H3}	v_{H4}	v_{XY}	**Voltage Level**
0	0	0	0	0	0	0	0	0	0
1	0	0	0	v_C	0	0	0	v_C	
0	1	0	0	0	v_C	0	0	v_C	1
0	0	1	0	0	0	v_C	0	v_C	
0	0	0	1	0	0	0	v_C	v_C	
1	1	0	0	v_C	v_C	0	0	$2v_C$	
1	0	1	0	v_C	0	v_C	0	$2v_C$	
1	0	0	1	v_C	0	0	v_C	$2v_C$	2
0	1	1	0	0	v_C	v_C	0	$2v_C$	
0	1	0	1	0	v_C	0	v_C	$2v_C$	
0	0	1	1	0	0	v_C	v_C	$2v_C$	
1	1	1	0	v_C	v_C	v_C	0	$3v_C$	
1	1	0	1	v_C	v_C	0	v_C	$3v_C$	3
1	0	1	1	v_C	0	v_C	v_C	$3v_C$	
0	1	1	1	0	v_C	v_C	v_C	$3v_C$	
1	1	1	1	v_C	v_C	v_C	v_C	$4v_C$	4

ing "OFF" all the submodule switches. As the Table 2.7 suggests, the intermediate voltage levels can be generated by using multiple switching combinations (redundant switching states). The redundant switching states are often used to control the submodule capacitor voltage in an MMC.

The arm voltage equation, which represents the operation of an arm is given by

$$\begin{aligned} v_{XY} &= v_{H1} + v_{H2} + v_{H3} + v_{H4} \\ &= S_1\, v_C + S_2\, v_C + S_3\, v_C + S_4\, v_C \end{aligned} \tag{2.14}$$

The arm voltage is equal to the summation of submodule output voltages. Each submodule output voltage is equal to the product of submodule capacitor voltage and corresponding submodule switching state. The above philosophy can be easily extended to any number of submodules per arm. The submodules in the upper and lower arms of a modular multilevel converter are controlled to generate the multilevel voltage waveform across the AC system.

2.3 PULSE WIDTH MODULATION SCHEMES

The pulse width modulation (PWM) is widely used for controlling the AC output voltage of a power converter. The targeted (reference) AC output voltage is obtained by varying the duty cycle of switching devices. The PWM schemes are designed to reduce the output voltage harmonic distortion and increase the magnitude of the output voltage at a given switching frequency [14, 15]. The modulation schemes are also designed to achieve other control objectives such as voltage balancing, common-mode voltage reduction, minimization of device switching frequency, reduction of power losses, and minimization of output current ripple. Based on the switching frequency, the PWM schemes for a modular multilevel converter are categorized into high switching frequency, low switching frequency, and fundamental switching frequency modulation schemes as shown in Figure 2.8.

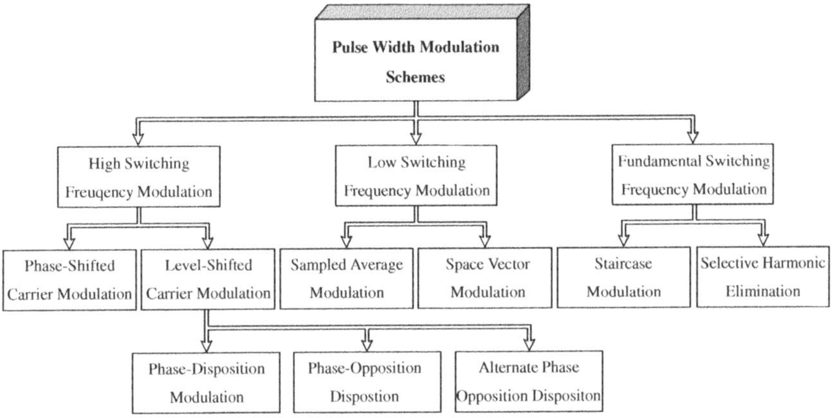

Figure 2.8 Pulse width modulation schemes for MMC.

The carrier modulation scheme is often referred to as a sine-triangular modulation, which is well established for the two-level power converters. In carrier modulation, the modulation signals are compared with a single triangular carrier signal to generate the gating signals for the switching devices of the two-level converter. The carrier modulation schemes are extended to the multilevel converters, where multiple triangular carrier signals are compared with a phase modulation signal to generate the gating signals. This approach is referred to as a multi-carrier modulation scheme.

Depending on the type of carrier arrangement, the multi-carrier modulation schemes are categorized into phase-shifted (PSC-PWM) and level-shifted (LSC-PWM) carrier modulations. These modulation schemes belong to the high switching frequency (f_{sw} >2000 Hz) modulation category. In PSC-PWM, the triangular carrier signals with an identical magnitude are horizontally disposed. In LSC-PWM the triangular carrier signals with an identical magnitude are vertically disposed. Based on the phase relationship between the adjacent carriers, the LSC-PWM is further categorized into phase-disposition (PD-PWM), phase-opposition-disposition (POD-PWM), and alternate phase-opposition-disposition (APOD-PWM) schemes [16]. The carrier modulation schemes are widely applied to control the multilevel power converter due to its simple implementation and ease of extension to higher number of voltage levels. However, these modulation schemes have higher switching losses compared to low and fundamental switching frequency modulation schemes.

The sampled average modulation (SAM) and space vector modulation (SVM) schemes belong to the low switching frequency modulation (100 Hz < f_{sw} < 2000 Hz) category. The sampled average approach directly controls the phase voltages of a modular multilevel converter and allows generating the line-to-line voltage implicitly. Alternatively, the space vector modulation directly controls the line-to-line voltages of a modular multilevel converter and allows generating the phase voltages implicitly [17–19]. However, the three-phase equivalent of sampled average modulation is quite similar to the space vector modulation, except that the distribution of zero vectors at the starting and ending of the switching sequence is not equal. Hence, the sampled average approach generates an output voltage with the highest harmonic distortion compared to the space vector modulation (SVM) approach. The sampled average modulation approach is easy to extend for an MMC with any number of submodules without any modifications in the implementation algorithm. On the other hand, the SVM approach is difficult to implement for an MMC with a large number of submodules ($N > 20$).

Fundamental switching frequency modulation schemes are very popular for the high-power converters such as CHB and MMC. In this approach, the power converters are switched at the fundamental frequency to reduce switching losses. However, the harmonic distortion of the output voltage and current waveforms are significantly increased. The output waveform quality can be improved by increasing the number of submodules (output voltage levels), which is the case for an MMC-based HVDC system. In HVDC applications, the MMC is built with 200–400 submodules in each arm, which generates an output voltage with very low total harmonic distortion (THD < 1%). Therefore, fundamental frequency modulation schemes are suitable for HVDC applications only. The selective harmonic elimination (SHE) and staircase modulation schemes belong to this category [20–22].

The selective harmonic elimination scheme offers high-quality output waveforms and tight control of the output voltage harmonics while maintaining the device switching frequency to a minimum, compared to other modulation schemes. The implementation of SHE scheme mainly involves off-line calculation of switching angles and solving non-linear equations. The number of switching angles drastically increases with the number of voltage levels and it is difficult to obtain the solution for such a large number of switching angles. Therefore, the SHE scheme is not an ideal solution for a modular multilevel converter. On the other hand, the staircase modulation approach is easy to implement and does not involve any complicated calculations like the SHE scheme. However, the staircase modulation scheme requires a higher sampling frequency to generate a high-quality output voltage waveform. The step-by-step design procedure and performance of these modulation schemes are presented in the following subsections.

2.3.1 Phase-Shifted Carrier Modulation

The triangular carrier signals can be arranged either horizontally or vertically within the linear modulation range. The modulation scheme with the horizontal disposition of identical triangular carrier signals is referred to as phase-shifted carrier modulation (PSC-PWM). In PSC-PWM, all triangular signals have the same frequency and peak-to-peak amplitude, but there is a phase-shift between the adjacent triangular signals. This modulation scheme is widely applied to control the multilevel flying capacitor (FC) and cascaded H-bridge (CHB) converters [23, 24]. The PSC-PWM is also attractive for the modular multilevel converter, due to the following distinctive features [25]:

- It provides a natural balancing of submodule capacitors voltage at high switching frequency (carrier frequency).
- The semiconductor stress and the power handled by each submodule are evenly distributed.
- It minimizes the DC-bus current ripple.

The implementation of PSC-PWM scheme for a modular multilevel converter mainly consists of the following steps:

(1) Generation of Modulation Signals: Each leg of the modular multilevel converter is divided into two arms, named as upper and lower arms. The upper and the lower arm submodules are modulated to generate the reference output phase voltage. The modulation signals for the upper and the lower arms are obtained from phase modulation signals. The simplified per-phase equivalent representation of a modular multilevel converter is shown in Figure. 2.9.

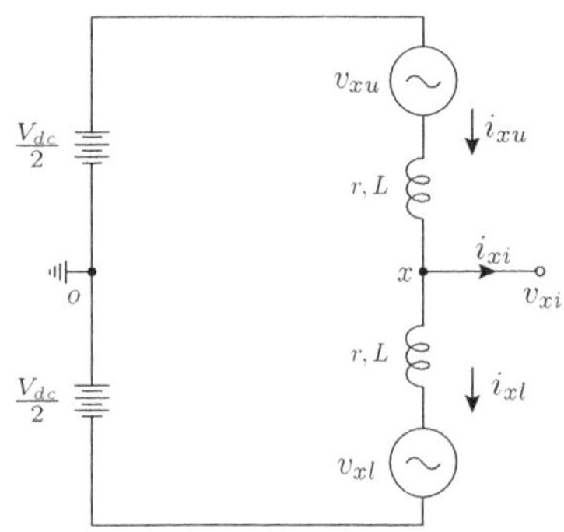

Figure 2.9 Per-phase equivalent circuit of MMC.

The N submodules in the upper and the lower arm are modeled as controlled AC voltage sources. The DC system is modeled as a battery source with a voltage V_{dc} and is divided into two halves. From the equivalent circuit, the arm modulation signals are given by [12, 26]

$$\begin{aligned} v_{xu} &= \frac{V_{dc}}{2} - v_{xi} - L\,\frac{di_{xu}}{dt} - r\,i_{xu} \\ v_{xl} &= \frac{V_{dc}}{2} + v_{xi} - L\,\frac{di_{xl}}{dt} - r\,i_{xl} \end{aligned} \tag{2.15}$$

where $x \in \{a, b, c\}$ represent the phase, v_{xu}, v_{xl} represent the upper and lower arm modulation signals, i_{xu}, i_{xl} represent the upper and lower arm currents, v_{xi} represent the inverter output voltage, and L, r represent the inductance and equivalent resistance of an arm inductor, respectively.

The arm inductor (L) is mainly designed to limit the magnitude of circulating currents flowing through the arms. These circulating currents mainly exist due to the switching mismatch between upper and lower arm submodules, and ripple in submodule floating capacitors voltage. These circulating currents are controlled using a circulating current control. The analysis and design of circulating current control is presented in Chapter 4. To simplify the analysis, the submodule capacitors are replaced with ideal battery sources. The delay in switching between the upper and lower arms is also neglected. Hence, the voltage component v_{xz} corresponding to the circulating currents is neglected. The arm modulation signals at steady state are given by

$$\begin{aligned} v_{xu} &= \frac{V_{dc}}{2} - v_{xi} - v_{xz} \\ v_{xl} &= \frac{V_{dc}}{2} + v_{xi} - v_{xz} \end{aligned} \tag{2.16}$$

where v_{xz} represents the voltage drop across arm inductor corresponding to circulating currents.

Depending on the type of application, the phase modulation signals are generated by using either closed-loop or open-loop control methods. Voltage-oriented control (VOC) or field-oriented control (FOC) method can be used to generate the phase modulation signals. The phase modulation signal is defined as

$$v_{xi} = m_a \times \frac{V_{dc}}{2} \sin(\omega_o t + \theta_x) \tag{2.17}$$

where m_a is the amplitude modulation index of range 0 to 1, ω_o is fundamental angular frequency, and $\theta_x \in \left\{0, -\frac{2\pi}{3}, -\frac{4\pi}{3}\right\}$ represents the phase angle between the three-phase signals in rad.

(2) Generation of Triangular Carrier Signals: Applying the PSC-PWM for a modular multilevel converter with N submodules per arm requires N triangular carrier signals. These N triangular carrier signals are arranged with a phase shift ϕ_c. The phase shift (ϕ_c) between the carrier angles is given by [12, 26]

$$\phi_c = \frac{360^\circ}{N} \tag{2.18}$$

The upper and lower arm triangular signals can be displaced with an interleave angle (ϕ_{ci}), which is given by

$$\phi_{ci} = \frac{360^\circ}{2N} \tag{2.19}$$

With an interleave angle between upper and lower arms, the modular multilevel converter generates an output voltage with $2N + 1$ voltage levels. It is also possible to use similar triangular carrier signals for the upper and the lower arms (i.e., $\phi_{ci} = 0$), resulting in an output voltage with $N + 1$ voltage levels. The reduction in the voltage levels increases the THD of the output voltage.

CASE STUDY 2.1 Analysis of PSC-PWM with an Interleave Angle

Objective: In this case study, the working principle and the performance of a PSC-PWM with an interleave angle are studied through MATLAB simulation.

Parameters: The rated parameters of the MMC are as follows: rated apparent power $S_o = 2$ MVA, output line-to-line voltage $V_o = 6$ kV, rated output frequency $f_o = 60$ Hz, rated power factor (PF) = 0.85 (lag), rated DC-bus voltage $V_{dc} = 10$ kV, number of SMs per arm $N = 4$, and arm inductance $L = 5$ mH.

Analysis:
To study the working principle of the modular multilevel converter, the floating capacitor of each submodule is replaced with an ideal battery source. The DC-bus voltage V_{dc} is equally distributed between the submodules in each arm, and its value is equal to

$$E = \frac{V_{dc}}{N} = \frac{10,000}{4} = 2500 \text{ V} \tag{2.20}$$

The PSC-PWM requires four triangle carrier signals to control the four half-bridge submodules in each arm. These carrier signals are arranged with a horizontal phase-shift of

$$\phi_c = \frac{360^\circ}{N} = \frac{360^\circ}{4} = 90^\circ \tag{2.21}$$

The upper and lower arm carrier signals are arranged with an interleave angle of

$$\phi_{ci} = \frac{360^\circ}{2N} = \frac{360^\circ}{8} = 45^\circ \tag{2.22}$$

to obtain the $2N + 1 = 9$ voltage levels in the phase voltage. Figure 2.10 shows the working principle of phase-a, lower arm with PSC-PWM scheme. The carrier signals T_{l1}, T_{l2}, T_{l3}, and T_{l4} are arranged with an initial angle of 45°, 135°, 225°, and 315°, respectively. The frequency of each carrier signal is set to $f_c = 540$ Hz, and their peak value is $4E$ as shown in Figure 2.10(a).

The phase-a, lower arm reference modulation signal is generated for an amplitude modulation index $m_a = 0.95$, output frequency $f_o = 60$ Hz, and an initial angle $\theta_a = 0^\circ$ as

$$\begin{aligned} v_{au} &= \frac{10,000}{2} - 0.95 \times \frac{10,000}{2} \sin(2\pi \times 60\,t) \\ v_{al} &= \frac{10,000}{2} + 0.95 \times \frac{10,000}{2} \sin(2\pi \times 60\,t) \end{aligned} \tag{2.23}$$

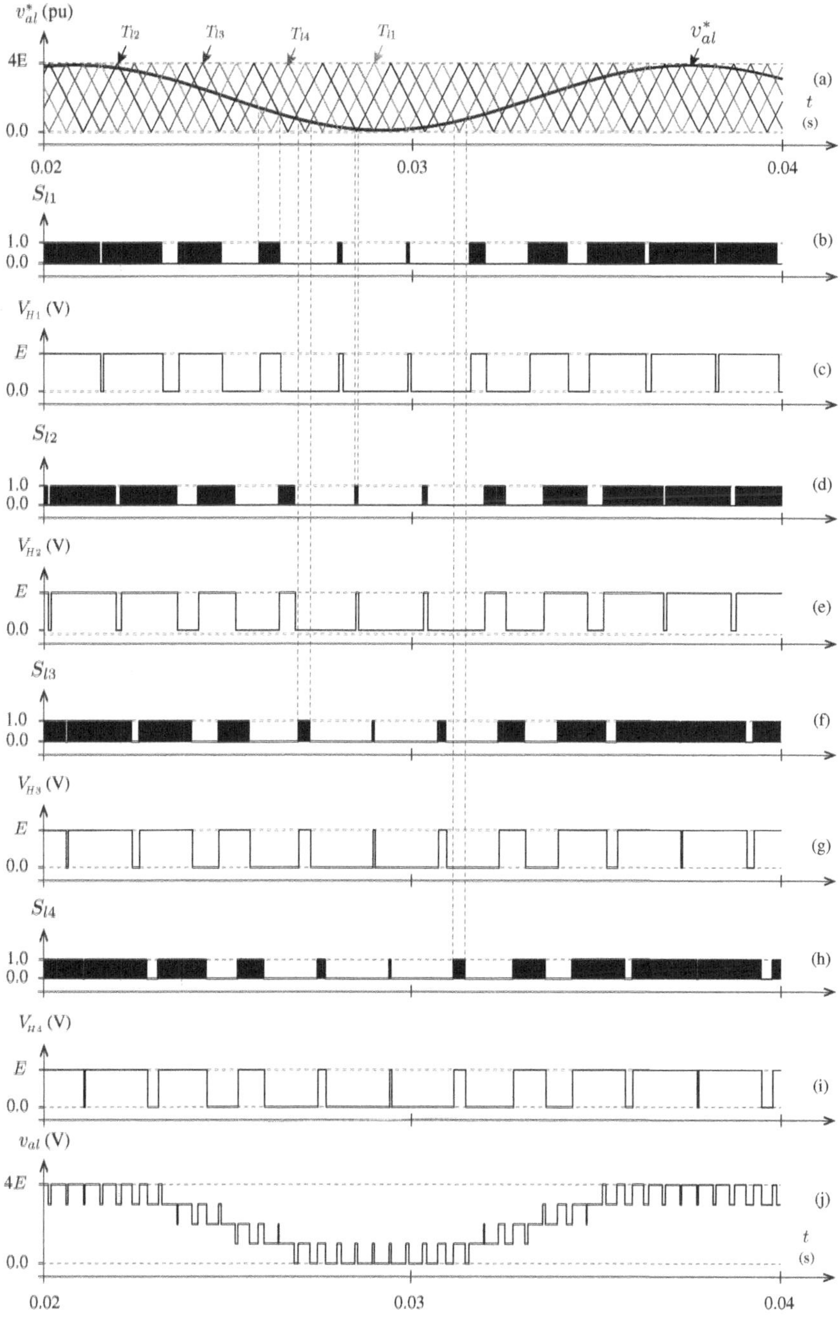

Figure 2.10 Simulation study of PSC-PWM.

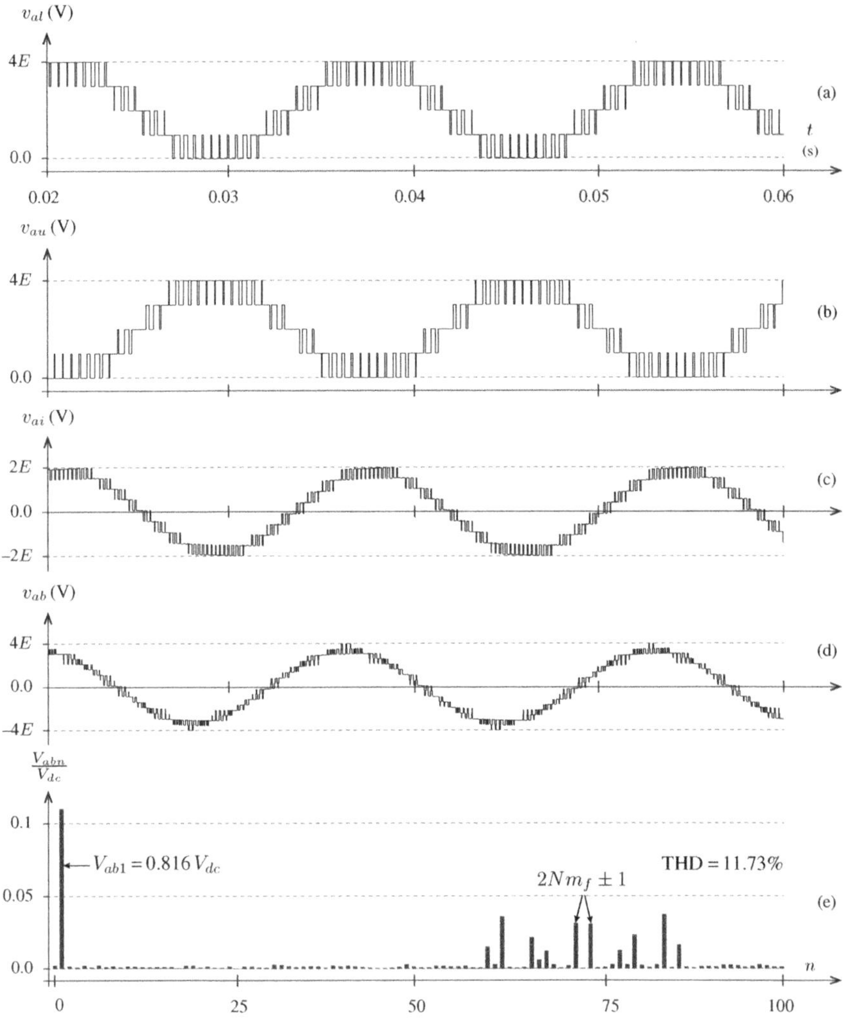

Figure 2.11 Simulation study of PSC-PWM with an interleave angle.

The reference modulation signal is compared with the carrier signals, and the corresponding comparison gives the gating signals S_{l1}, S_{l2}, S_{l3}, and S_{l4}. These gating signals are applied to the top devices of each submodule in the lower arm of phase-a (SM_1, SM_2, SM_3, and SM_4), respectively. The bottom devices are operated in a complementary manner.

The output voltage of each submodule (V_{H1}, V_{H2}, V_{H3}, and V_{H4}) is proportional to their gating signals (S_{l1}, S_{l2}, S_{l3}, and S_{l4}), and their maximum value is equal to E as shown in Figure 2.10(b)–(i). The lower arm voltage is equal to the summation of submodules output voltage. The arm voltage has a total five voltage levels with a voltage step of E. The arm voltage has a maximum value of $4E$ and a minimum value of 0 as shown in Figure 2.10(j). Similarly, the phase-a, upper arm submodules are modulated using four carrier signals T_{u1},

T_{u2}, T_{u3}, and T_{u4} with initial angles of 0°, 90°, 180°, 270°, respectively, and corresponding reference arm modulation signal.

Figure 2.11 shows, the simulated waveforms of the modular multilevel converter with a PSC-PWM scheme. The upper arm output voltage has the same number of voltage levels as that of the lower arm voltage as shown in Figure 2.11(a)–(b). The output phase voltage with respect to the mid-point of DC-bus (v_{ai}) can be obtained from equation (2.16), which is equal to the difference of lower and upper arm voltages as

$$v_{ai} = \frac{v_{al} - v_{au}}{2} \tag{2.24}$$

According to equation (2.24), the output phase voltage has both positive and negative voltage levels with a peak value of $\pm 2E$. The output voltage has a total $2N + 1 = 9$ voltage levels with a step of $\frac{E}{2} = 1250$ V as shown in Figure 2.11(c). The line-to-line voltage (v_{ab}) is measured across phase-a and b, and has a total $4N + 1 = 17$ voltage levels with a step of $\frac{E}{4} = 625$ V as shown in Figure 2.11(d). The line voltage has a THD of 11.73% with a fundamental (60 Hz) voltage component of 0.816 V_{dc} as shown in Figure 2.11(e). Each submodule is switched at a frequency (f_{SM}) equal to the carrier frequency of 540 Hz. The switching frequency of modular multilevel converter is equal to

$$f_{sw} = 2N \times f_c = 8 \times 540 = 4320\,\text{Hz} \tag{2.25}$$

where f_{sw} represents the converter switching frequency. In PSC-PWM with an interleave angle approach, the dominant switching harmonics appear as side bands of $2Nm_f \pm 1$ order harmonic component, where $m_f = \frac{f_c}{f_o}$ is the frequency modulation index. ■

CASE STUDY 2.2 Analysis of PSC-PWM without an Interleave Angle

Objective: In this case study, the working principle of PSC-PWM without an interleave angle is studied through MATLAB simulation.

Parameters: The rated parameters of the MMC are as follows: rated apparent power $S_o = 2$ MVA, output line-to-line voltage $V_o = 6$ kV, rated output frequency $f_o = 60$ Hz, rated power factor (PF) = 0.85 (lag), rated DC-bus voltage $V_{dc} = 10$ kV, number of SMs per arm $N = 4$, and arm inductance $L = 5$ mH.

Analysis:
In this study, the carrier signals belonging to the same arm are arranged with a phase-shift angle of 90°. The interleave angle between the upper and the lower arm carrier signals is set to zero degrees. Therefore, only four triangular carrier signals are sufficient to modulate both upper and lower arm submodules. The carrier signals are generated with a frequency 540 Hz and a magnitude $4E$.

The upper and lower arm reference modulation signals are generated for an amplitude modulation index $m_a = 0.95$ and an output frequency $f_o = 60$ Hz. The gating signals for submodules in the upper and the lower arm are generated by comparing the corresponding reference arm modulation signal with the carrier signals. The switching frequency of each submodule is equal to the carrier frequency 540 Hz.

Figure 2.12 shows the performance of the PSC-PWM without an interleave angle. The upper and lower arm output voltage has a five voltage levels with a step of $E = 2500$ V. These

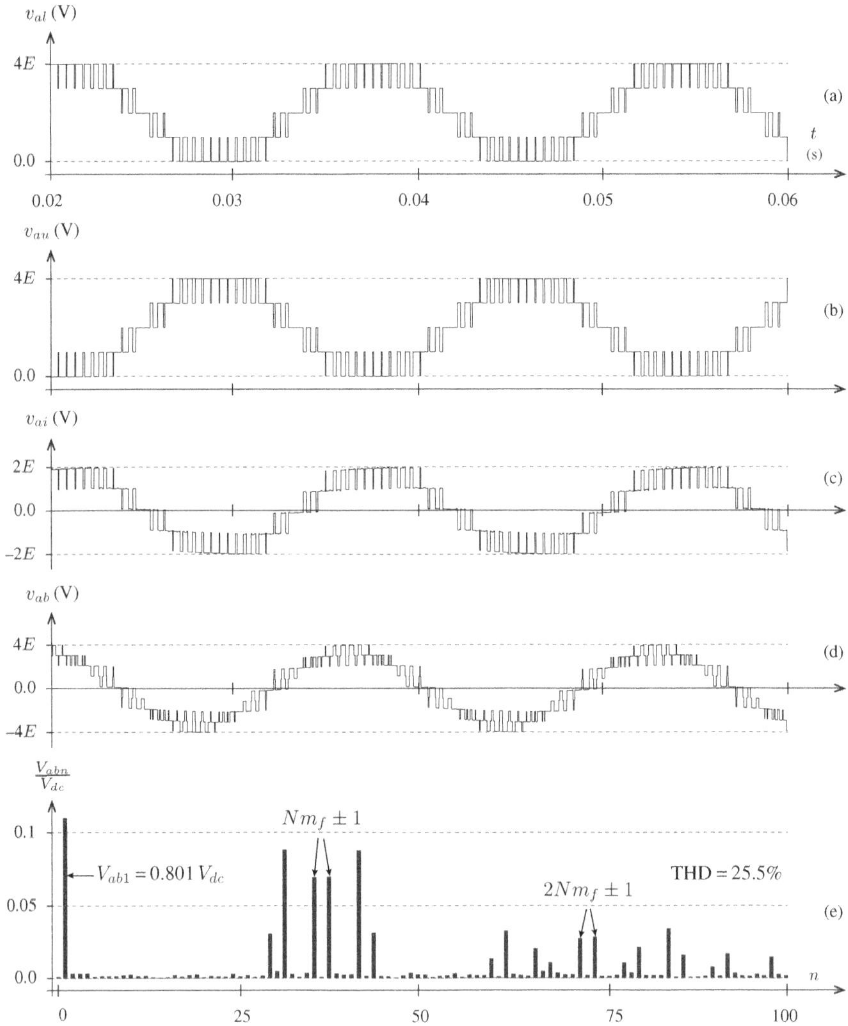

Figure 2.12 Simulation study of PSC-PWM without an interleave angle.

voltages are in opposite phase with each other as shown in Figure 2.12(a)–(b). The difference between the lower and the upper arm voltage gives the output phase voltage with respect to the mid-point of DC-bus. The output phase voltage has only $N + 1 = 5$ voltage levels with a step of $E = 2500$ V as shown in Figure 2.12(c). The line-to-line voltage (v_{ab}) has only $2N + 1 = 9$ voltage levels as shown in Figure 2.12(d). Each voltage step is equal to $\frac{E}{2} = 1250$ V. There is approximately 50% lesser number of voltage levels and twice the voltage step as compared to the PSC-PWM with an interleave angle approach.

The reduction in the voltage levels increases the THD. The line-to-line voltage has a THD of 25.5% as shown in Figure 2.12(e), which is 54% higher than the THD with an interleave angle approach presented in case study 2.1. The switching frequency of modular multilevel converter is equal to

$$f_{sw} = N \times f_c = 4 \times 540 = 2160\,\text{Hz} \tag{2.26}$$

In PSC-PWM without an interleave angle approach, the dominant switching harmonic components appear as a sideband of $Nm_f \pm 1$ order and its multiples.

■

2.3.2 Level-Shifted Carrier Modulation

The LSC-PWM is one of the modulation schemes used to control the multilevel power converters. In LSC-PWM, the triangular carrier signals with an identical peak-to-peak magnitude and frequency are vertically disposed within the linear modulation region. Each triangular carrier signal represents two voltage levels; hence the name “Level-shifted."

The LSC-PWM is implemented based on the output voltage levels and it can be easily applied to any multilevel converter. However, this modulation scheme is less preferred for the CHB, FC, and MMC converters, since it causes an uneven power loss distribution among the submodules and devices. Further, it affects the harmonic distortion of AC output voltage in CHB, capacitor voltage balance in FC, and results in unequal distribution of submodule capacitors voltage ripple and a large magnitude of circulating currents in MMC [26, 27].

Depending on the phase relation between adjacent carriers, the LSC-PWM has three variations described as: (i) phase-disposition (PD), where the triangular carrier signals are vertically arranged in-phase with each other as shown in Figure 2.13(a); (ii) phase-opposition-disposition (POD), where the triangular carrier signals above the sinusoidal reference zero are in-phase and those below the sinusoidal reference zero are out of phase as shown in Figure 2.13(b); and (iii) alternate phase-opposition-disposition (APOD), where all the triangular carriers are disposed alternatively in phase-opposition as shown in Figure 2.13(c). The PD-PWM strategy is widely accepted, as it generates an output voltage with the lowest THD. To control the modular multilevel converter with N submodules in each arm requires an N triangular carrier signals. The upper and lower arm submodules are modulated with identical carrier signals (i.e., $\phi_{ci} = 0$), resulting in an AC output voltage with $N + 1$ voltage levels.

Figure 2.13 shows different type of carrier arrangements in LSC-PWM scheme including PD, POD, and APOD approaches. The LSC-PWM scheme requires four triangular carrier (T_1, T_2, T_3, and T_4) signals to control the five level MMC system with four submodules per arm. These carrier signals are generated for a frequency modulation index of $m_f = 12$. The frequency modulation index is defined as $m_f = \frac{f_c}{f_o}$, where f_c is the carrier frequency and f_o is the reference voltage frequency. The upper and lower arm reference voltages are generated for an operating condition of $m_a = 0.95$ and $f_o = 60$ Hz. The reference arm voltages are compared with the carrier signals to generate the gating signals for submodules of the corresponding arm.

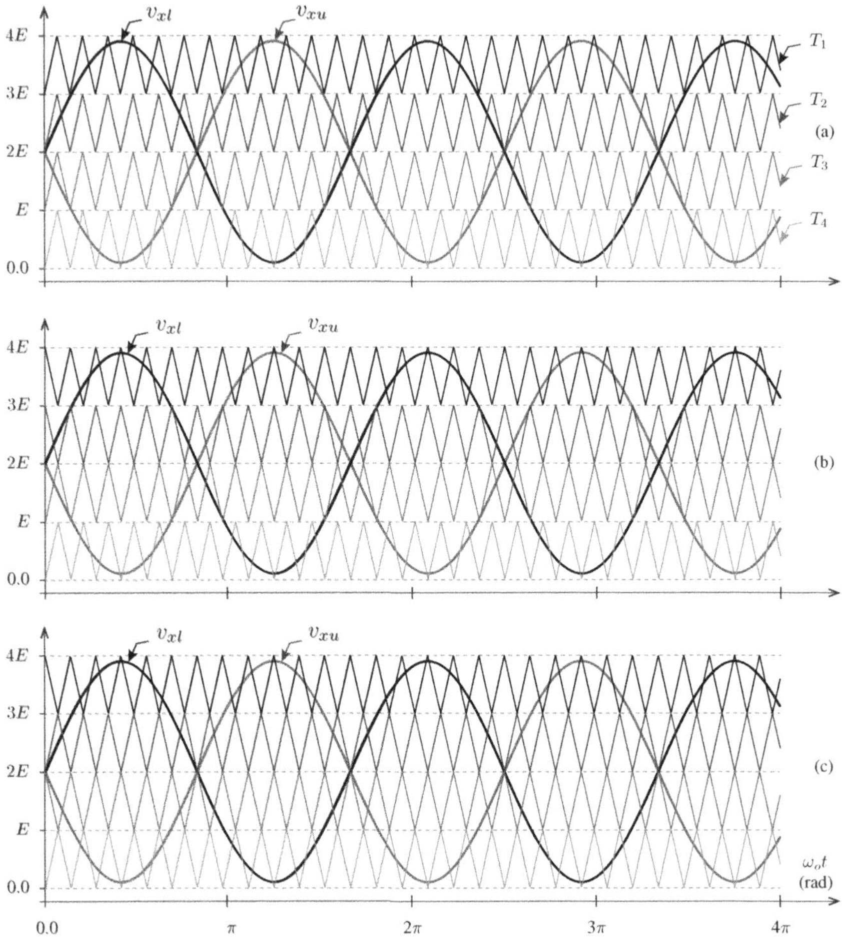

Figure 2.13 Carrier arrangement of LSC-PWM for $m_a = 0.95$, $f_o = 60$ Hz, and $f_c = 540$ Hz: (a) PD-PWM, (b) POD-PWM, and (c) APOD-PWM.

2.3.3 Sampled Average Modulation

The sampled average modulation scheme directly controls the phase voltage of an MMC, and generates the line-to-line voltage implicitly. This approach is computationally less complex and easy to apply for MMC with any number of submodules without any modifications. The three-phase equivalent of sampled average modulation results in the elimination of average zero sequence voltage in each sampling interval, reduces the ripple in the output current, and minimizes the harmonic distortion of the output voltage and current [17]. More importantly, the sampled average approach does not require a complex coordinate transformation from abc-frame to $\alpha\beta$-frame.

In this method, the reference phase voltage is generated by averaging the two nearest voltage levels in each sampling interval. This approach eliminates the volt-sec error in the staircase or nearest voltage level modulation [17]. The volt-sec balance of the reference phase voltage is expressed in terms of two nearest voltage levels and corresponding dwell times as

$$\begin{aligned} v_{rxi}^{n} T_s &= V_{x1} T_{x1} + V_{x2} T_{x2} \\ T_s &= T_{x1} + T_{x2} \end{aligned} \tag{2.27}$$

where v_{rxi}^{n} represents the normalized reference phase voltage with offset, V_{x1}, V_{x2} represent the nearest phase voltage levels, T_{x1}, T_{x2} represent the dwell times of the phase voltage levels, respectively, and T_s represents the sampling time, which is equal to the summation of dwell times T_{x1} and T_{x2}. The implementation of sampled average modulation mainly consists of three steps as follows.

(1) Identification of Two Nearest Voltage Levels: The total DC-bus voltage of each arm is expressed in terms of the number of SMs and rated SM capacitor voltage as

$$V_{dc} = N \times V_C \tag{2.28}$$

The two nearest phase voltage levels are obtained from the normalized reference phase voltage in each sampling interval. The reference phase voltage is given in terms of the number of submodules and its capacitor voltage using equations (2.17) and (2.28) as

$$v_{xi} = \frac{N\,V_C}{2} \times m_a \sin(\omega_o t + \theta_x) \tag{2.29}$$

The normalized reference phase voltage is obtained by dividing equation (2.29) with a submodule rated capacitor voltage as

$$v_{xi}^{n} = \frac{N}{2} \times m_a \sin(\omega_o t + \theta_x) \tag{2.30}$$

The sampled average modulation requires the nearest voltage levels, which are positive integers only. Therefore, an offset value of $\frac{N}{2}$ is added to equation (2.30) and results in

$$v_{rxi}^{n} = \frac{N}{2} \times [1 + m_a \sin(\omega_o t + \theta_x)] \tag{2.31}$$

where v_{xi}^{n} represents the normalized reference phase voltage, v_{rxi}^{n} represents the normalized reference phase voltage with offset, N represents the number of submodules per arm, m_a represents the amplitude modulation index of range 0 to 1, and ω_o is the fundamental angular frequency in rad/sec.

The phase voltage levels are obtained from equation (2.31), where the lower voltage level (V_{x1}) is obtained from the operator `floor` (v_{rxi}^{n}) and the upper voltage level (V_{x2}) is equal to $1 + V_{x1}$. The nearest voltage levels are given by

$$\begin{aligned} V_{x1} &= \text{floor}\,(v_{rxi}^{n}) \\ V_{x2} &= 1 + V_{x1} \end{aligned} \tag{2.32}$$

(2) Calculation of Dwell Times: The voltage levels V_{x1} and V_{x2} are applied for a time duration of T_{x1} and T_{x2}, respectively over a sampling interval T_s. From the volt-sec balance principle given in equation (2.27), the dwell times T_{x1} and T_{x2} are given by

$$\begin{aligned} T_{x2} &= \frac{v_{rxi}^{n} - V_{x1}}{V_{x2} - V_{x1}} \times T_s \\ T_{x1} &= T_s - T_{x2} \end{aligned} \tag{2.33}$$

These dwell times are compared with a symmetrical triangular waveform to generate the pulses g_{t1} and g_{t2} as shown in Figure 2.14. The pulse g_{t1} is applied to the submodules in the upper and lower arms of the corresponding phase to generate the phase voltage level V_{x1}. Similarly, the voltage level V_{x2} is obtained by modulating the submodules in upper and lower arms for a pulse duration of g_{t2}.

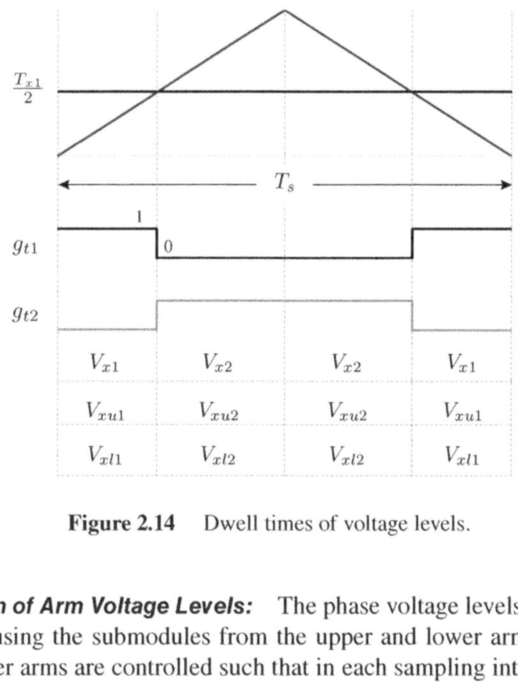

Figure 2.14 Dwell times of voltage levels.

(3) Calculation of Arm Voltage Levels: The phase voltage levels V_{x1} and V_{x2} are generated by using the submodules from the upper and lower arms together. The upper and lower arms are controlled such that in each sampling interval, N-number of submodules out of $2N$ submodules from each phase are used to generate the required phase voltage level (i.e., the number of on-state submodules is constant and equal to N in each sampling interval). According to this criteria, the upper and lower arm voltage levels are given by

$$\begin{aligned} V_{xuk} + V_{xlk} &= N \\ V_{xlk} &= V_{xk} \\ V_{xuk} &= N - V_{xlk} \end{aligned} \tag{2.34}$$

where $k \in \{1, 2\}$ represents the voltage levels.

CASE STUDY 2.3 Analysis of Sampled Average Modulation Scheme

Objective: In this case study, the working principle of sampled average modulation scheme is analytically studied for a modular multilevel converter. The performance of the sampled average modulation is presented through MATLAB simulation.

Parameters: The rated parameters of the MMC are as follows: rated apparent power $S_o = 2$ MVA, output line-to-line voltage $V_o = 6$ kV, rated output frequency $f_o = 60$ Hz, rated power factor (PF) = 0.85 (lag), rated DC-bus voltage $V_{dc} = 10$ kV, number of SMs per arm $N = 4$, arm inductance $L = 5$ mH, and sampling time $T_s = 555.555\ \mu$ s.

Analysis:

The sampled average modulation scheme requires a normalized phase voltage, which is obtained from the equation (2.30) for phase-a with $m_a = 0.95$, $f_o = 60$ Hz, and $\theta_a = 0°$ as

$$v_{ai}^{n} = \frac{4}{2} \times 0.95 \sin(2\pi \times 60\, t) \tag{2.35}$$

The normalized phase voltage has a peak-to-peak value of "4" as shown in Figure 2.15(a). To calculate the nearest voltage levels, an offset value of $\frac{N}{2} = \frac{4}{2} = 2$ is added to the normalized phase voltage as shown in Figure 2.15(b). The resultant phase modulation signal is given by

$$v_{rai}^{n} = \frac{4}{2} \times [1 + 0.95 \sin(2\pi \times 60\, t)] \tag{2.36}$$

The nearest two voltage levels at a sampling instant $t = 0.0267$ s are obtained from the phase modulation signal as

$$\begin{aligned} v_{rai}^{n} &= 0.863 \\ V_{a1} &= \text{floor}\,(0.863) = 0 \\ V_{a2} &= 1 + V_{a1} = 1 \end{aligned} \tag{2.37}$$

The dwell times T_{a1} and T_{a2} corresponding to these two voltage levels are

$$\begin{aligned} T_{a2} &= \frac{0.863 - 0}{1 - 0} \times 555.555 \times 10^{-6} = 479.443\,\mu s \\ T_{a1} &= 555.555 \times 10^{-6} - 479.443 \times 10^{-6} = 76.111\,\mu s \end{aligned} \tag{2.38}$$

In sampled average modulation, the phase voltage is generated by controlling the submodules in upper and lower arm of the corresponding phase. The required upper and lower arm voltage levels are

$$\begin{aligned} V_{au1} + V_{al1} &= 4 \\ V_{al1} &= V_{a1} = 0 \\ V_{au1} &= N - V_{a1} = 4 - 0 = 4 \end{aligned} \tag{2.39}$$

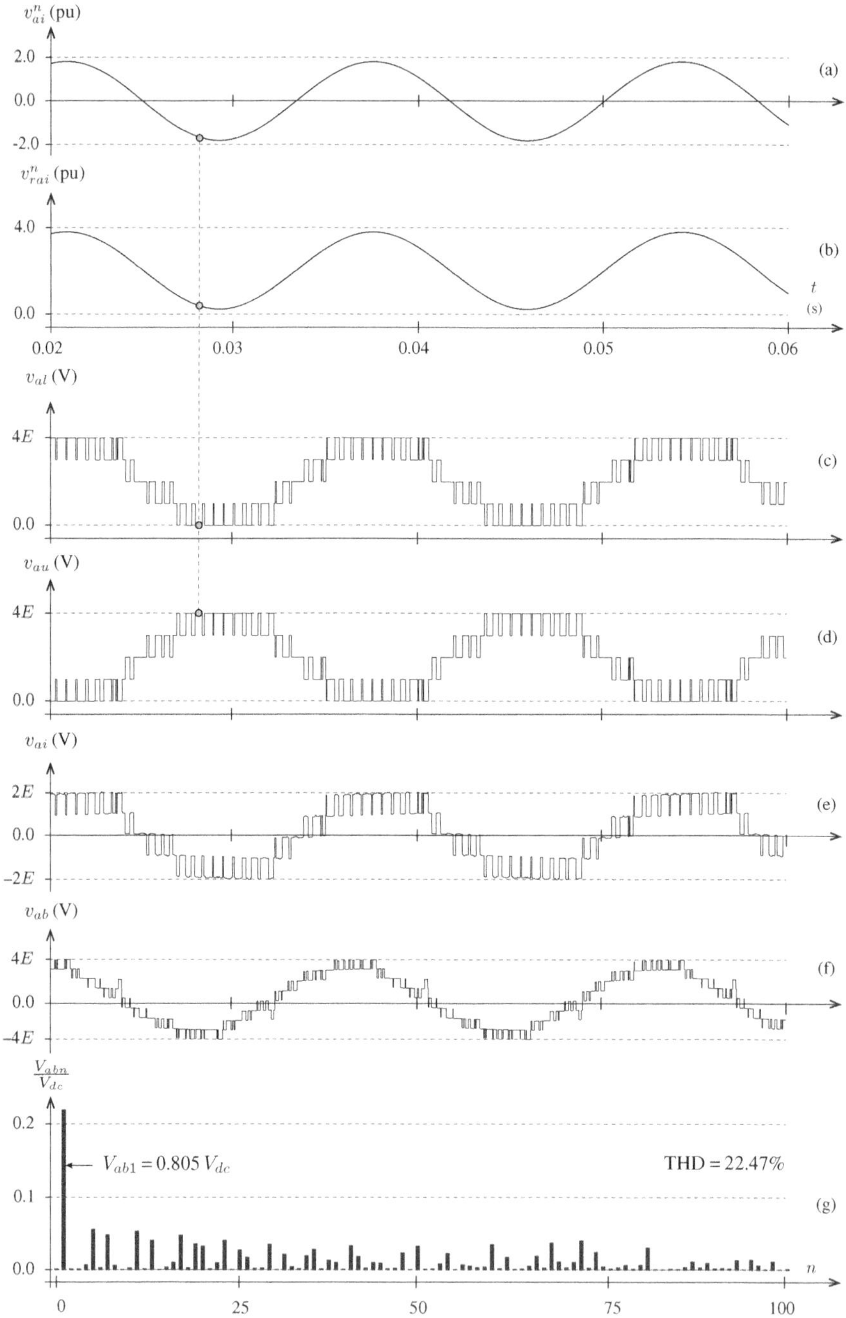

Figure 2.15 Simulation study of sampled average modulation.

Similarly, the second voltage level for upper and lower arms are given by

$$\begin{aligned} V_{au2} + V_{al2} &= 4 \\ V_{al2} &= V_{a2} = 1 \\ V_{au2} &= N - V_{a2} = 4 - 1 = 3 \end{aligned} \tag{2.40}$$

Each voltage level represents the required number of ON-state submodules. Figure 2.15 shows the performance of the sampled average modulation scheme. The voltage levels V_{au1}, V_{al1} are applied for a dwell time T_{a1} and voltage levels V_{au2}, V_{al2} are applied for a dwell time T_{a2} to generate the upper and lower arm voltages. The modular multilevel converter has redundant (multiple) switching combinations to generate these voltage levels. In this approach, the switching state with a minimum number of switchings is applied to the converter. For example, to generate the voltage level $V_{au1} = 4$, the switching state (1,1,1,1), and for voltage level $V_{au2} = 3$, the switching state (1,1,1,0) is applied over a sampling interval. In the next sampling interval, the switching state which involves only one submodule change is used to generate the new voltage level.

The arm voltage has five voltage steps 0, E, $2E$, $3E$, and $4E$. The upper and lower arm voltages are in opposite phase with each other as shown in Figure 2.15(c)–(d). The difference between the lower and upper arm voltage gives the AC output voltage. The AC output voltage has a similar number of voltage levels as that of the arm voltage with a voltage step of $2E$, E, 0, $-E$, $-2E$ as shown in Figure 2.15(e).

The line-to-line voltage waveform has a total of nine voltage levels with a half-wave symmetry as shown in Figure 2.15(f). The PWM waveform has a THD of 22.47% with a fundamental component of 0.805 V_{dc} as shown in Figure 2.15(g). The switching frequency of modular multilevel converter is equal to the sampling frequency of

$$f_{sw} = f_s = 1800\,\text{Hz} \tag{2.41}$$

where f_{sw} represents the switching frequency of MMC, and f_s represents the sampling frequency.

The modular multilevel converter has four submodules per arm, and each submodule is switched at a frequency of

$$f_{SM} = \frac{f_s}{N} = \frac{1800}{4} = 450\,\text{Hz} \tag{2.42}$$

where f_{SM} represents the submodule switching frequency. ■

2.3.4 Space Vector Modulation

The space vector modulation (SVM) is a low switching frequency modulation approach that can be applied to the high-power multilevel converters. The SVM approach has the flexibility of selecting the best switching vector among the redundant switching vectors to enhance the DC-bus utilization, better harmonic performance, and a regulation of common-mode voltage magnitude, etc. However, in spite of these advantages, the SVM is difficult to implement for converters that have higher than three levels, due to a large number of switching vectors. Generally, for an m-level converter, there are m^3 switching vectors and $6(m-1)^2$ triangles in the space vector diagram. The determination of redundant switching vectors and switching sequences is more challenging in SVM scheme.

The implementation of SVM scheme in $\alpha - \beta$ and $\alpha - \beta - 0$ coordinate system involves a coordinate transformation, trigonometric functions, and look-up tables, which increases the computational load of the modulation stage [28]. The implementation of SVM in abc-coordinate system overcomes the above drawbacks, and makes it easier to obtain the switching vectors and duty cycles as if it were an equivalent of two-level SVM [18, 29, 30]. This approach has the following features:

- Switching states, duty cycles, and switching sequences are obtained by simple calculations; thus, no look-up table is required.
- This scheme is computationally less complex and extendable to a modular multilevel converter with any number of voltage levels without any modifications.

The upper and lowers arms of a three-phase MMC system are controlled using an independent SVM scheme, which is referred to as a dual SVM approach. The three-phase upper and lower arm reference voltages are obtained from the reference phase voltages. From the instantaneous value of reference arm voltages, the nearest switching vectors, and corresponding duty cycles are calculated by decomposing the multilevel space vector into several two-level space vectors in the abc-coordinate system [19]. The implementation of dual SVM approach mainly consists of the following three steps:

(1) Synthesis of Reference Arm Voltage Vectors: The normalized reference output vector (v_r) for an m-level modular multilevel converter is defined as

$$\vec{v_r} = \begin{bmatrix} a^* \\ b^* \\ c^* \end{bmatrix} = \frac{m-1}{2} \begin{bmatrix} m_a \sin(\omega_o t) \\ m_a \sin(\omega_o t - \frac{2\pi}{3}) \\ m_a \sin(\omega_o t - \frac{4\pi}{3}) \end{bmatrix} \tag{2.43}$$

where a^*, b^*, and c^* represent the three-phase reference modulation signals.

The upper and lower arm reference vectors are obtained from the normalized reference output vector and expressed as

$$\begin{aligned} \vec{v_{ru}} &= \begin{bmatrix} a_u^* \\ b_u^* \\ c_u^* \end{bmatrix} = \frac{m-1}{2} \begin{bmatrix} 1 - m_a \sin(\omega_o t) \\ 1 - m_a \sin(\omega_o t - \frac{2\pi}{3}) \\ 1 - m_a \sin(\omega_o t - \frac{4\pi}{3}) \end{bmatrix} \\ \vec{v_{rl}} &= \begin{bmatrix} a_l^* \\ b_l^* \\ c_l^* \end{bmatrix} = \frac{m-1}{2} \begin{bmatrix} 1 + m_a \sin(\omega_o t) \\ 1 + m_a \sin(\omega_o t - \frac{2\pi}{3}) \\ 1 + m_a \sin(\omega_o t - \frac{4\pi}{3}) \end{bmatrix} \end{aligned} \tag{2.44}$$

where a_u^*, b_u^*, and c_u^* represent the three-phase upper arm reference modulation signals, a_l^*, b_l^*, and c_l^* represent the three-phase lower arm reference modulation signals, m is the number of voltage levels, m_a is the modulation index in the range $0 \leq m_a \leq 1$, and ω_o is the fundamental angular frequency.

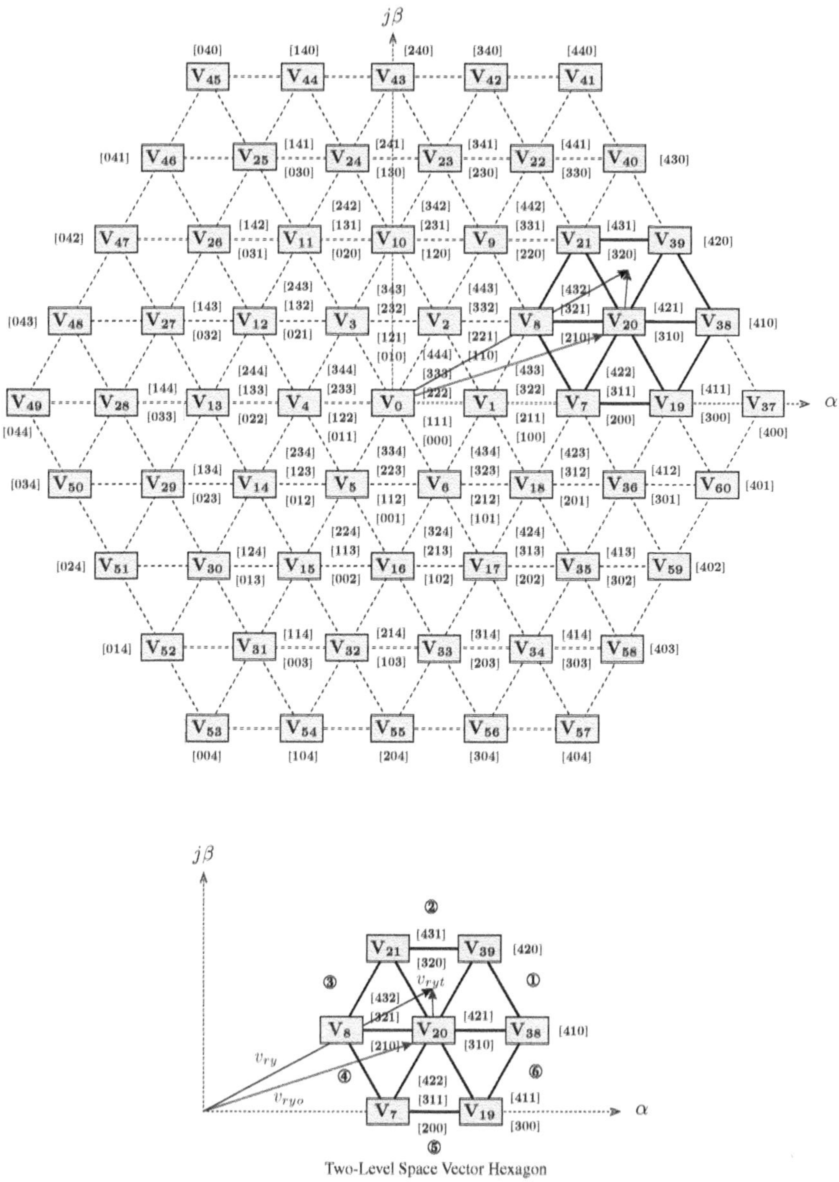

Figure 2.16 Space vector diagram for a 5L-MMC.

The generalized five-level space vector diagram is shown in Figure 2.16. The normalized reference vector (v_{ry}) is decomposed into an offset voltage vector (v_{ryo})

and a two-level voltage vector (v_{ryt}) as

$$\vec{v}_{ry} = \vec{v}_{ryo} + \vec{v}_{ryt} \tag{2.45}$$

The offset voltage vector represents the origin of the two-level space vector diagram, and their coordinates (a_{y0}, b_{y0}, c_{y0}) are given by

$$\begin{aligned} a_{y0} &= \text{floor}\,(a_y^*) \\ b_{y0} &= \text{floor}\,(b_y^*) \\ c_{y0} &= \text{floor}\,(c_y^*) \end{aligned} \tag{2.46}$$

The two-level voltage vector (v_{ryt}) is located in one of the six sectors in the two-level space vector diagram as shown in Figure 2.16. The switching vectors, which forms the two-level space vector diagram are obtained from the coordinates (a_{y0}, b_{y0}, c_{y0}) and (a_{y1}, b_{y1}, c_{y1}). The coordinates (a_{y1}, b_{y1}, c_{y1}) are given by

$$\begin{aligned} a_{y1} &= 1 + a_{y0} \\ b_{y1} &= 1 + b_{y0} \\ c_{y1} &= 1 + c_{y0} \end{aligned} \tag{2.47}$$

The switching vectors, which forms the two-level space vector diagram are (a_{y0}, b_{y0}, c_{y0}), (a_{y1}, b_{y0}, c_{y0}), (a_{y1}, b_{y1}, c_{y0}), (a_{y0}, b_{y1}, c_{y0}), (a_{y0}, b_{y1}, c_{y1}), (a_{y0}, b_{y0}, c_{y1}), (a_{y1}, b_{y0}, c_{y1}), and (a_{y1}, b_{y1}, c_{y1}).

Table 2.8 Switching vectors in each Sector

Sector	$\vec{v}_{y1}$	$\vec{v}_{y2}$	$\vec{v}_{y3}$	$\vec{v}_{y4}$
①	(a_{y0}, b_{y0}, c_{y0})	(a_{y1}, b_{y0}, c_{y0})	(a_{y1}, b_{y1}, c_{y0})	(a_{y1}, b_{y1}, c_{y1})
②	(a_{y0}, b_{y0}, c_{y0})	(a_{y0}, b_{y1}, c_{y0})	(a_{y1}, b_{y1}, c_{y0})	(a_{y1}, b_{y1}, c_{y1})
③	(a_{y0}, b_{y0}, c_{y0})	(a_{y0}, b_{y1}, c_{y0})	(a_{y0}, b_{y1}, c_{y1})	(a_{y1}, b_{y1}, c_{y1})
④	(a_{y0}, b_{y0}, c_{y0})	(a_{y0}, b_{y0}, c_{y1})	(a_{y0}, b_{y1}, c_{y1})	(a_{y1}, b_{y1}, c_{y1})
⑤	(a_{y0}, b_{y0}, c_{y0})	(a_{y0}, b_{y0}, c_{y1})	(a_{y1}, b_{y0}, c_{y1})	(a_{y1}, b_{y1}, c_{y1})
⑥	(a_{y0}, b_{y0}, c_{y0})	(a_{y1}, b_{y0}, c_{y0})	(a_{y1}, b_{y0}, c_{y1})	(a_{y1}, b_{y1}, c_{y1})

(2) Determination of Switching Vectors: In the abc-coordinate system, the normalized reference vector is synthesized by using the nearest four switching vectors. The volt-sec balance of the reference vector is given by

$$\begin{aligned} \vec{v}_{ry} T_s &= \vec{v}_{y1} T_{y1} + \vec{v}_{y2} T_{y2} + \vec{v}_{y3} T_{y3} + \vec{v}_{y4} T_{y4} \\ T_s &= T_{y1} + T_{y2} + T_{y3} + T_{y4} \end{aligned} \tag{2.48}$$

These switching vectors are identified based on the location of the normalized reference vector in two-level space vector diagram. The location of the normalized reference vector is obtained from the coordinates (a_{y0}, b_{y0}, c_{y0}) and (a_{y1}, b_{y1}, c_{y1}). The nearest four switching vectors in different sectors are shown in Table 2.8. The vectors $\vec{v}_{y1}$ and $\vec{v}_{y4}$ are located at the origin of the two-level space vector diagram. The distribution of the duty cycles between the vectors $\vec{v}_{y1}$ and $\vec{v}_{y4}$ will affect the zero sequence voltage component only. In this study, the duty cycles between the vectors $\vec{v}_{y1}$ and $\vec{v}_4$ are equally distributed.

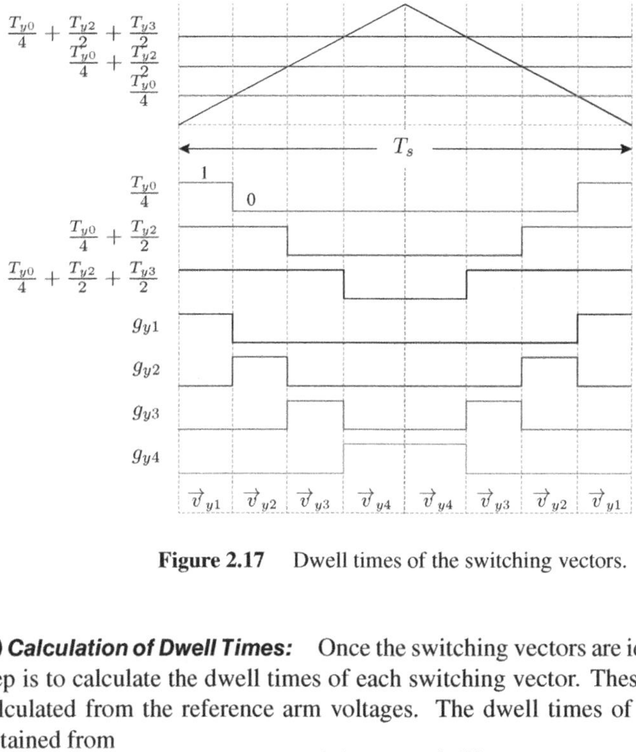

Figure 2.17 Dwell times of the switching vectors.

(3) Calculation of Dwell Times: Once the switching vectors are identified, the next step is to calculate the dwell times of each switching vector. These dwell times are calculated from the reference arm voltages. The dwell times of each arm can be obtained from

$$\begin{aligned} T_{ay} &= (a_y^* - \text{floor}\,(a_y^*)) \times T_s \\ T_{by} &= (b_y^* - \text{floor}\,(b_y^*)) \times T_s \\ T_{cy} &= (c_y^* - \text{floor}\,(c_y^*)) \times T_s \end{aligned} \tag{2.49}$$

The dwell time of each switching vector is given by

$$\begin{aligned} T_{y1} &= 1 - \max\,(T_{ay}, T_{by}, T_{cy}) \\ T_{y2} &= \max\,(T_{ay}, T_{by}, T_{cy}) - \text{med}\,(T_{ay}, T_{by}, T_{cy}) \\ T_{y3} &= \text{med}\,(T_{ay}, T_{by}, T_{cy}) - \min\,(T_{ay}, T_{by}, T_{cy}) \\ T_{y4} &= \min\,(T_{ay}, T_{by}, T_{cy}) \\ T_{y0} &= T_{y1} + T_{y4} \end{aligned} \tag{2.50}$$

These duty cycles are compared with the symmetrical triangular waveform as shown in Figure 2.17, to obtain the pulses g_{y1}, g_{y2}, g_{y3}, and g_{y4} corresponding to the switching vectors $\vec{v}_{y1}$, $\vec{v}_{y2}$, $\vec{v}_{y3}$, and $\vec{v}_{y4}$, respectively. After the switching vectors and duty cycles are calculated, the final output of the modular multilevel converter can be determined by selecting the appropriate sequence of switching vectors. The symmetrical switching sequence is extensively used to achieve the lowest voltage harmonic distortion and current ripple. The symmetrical switching sequence $\vec{v}_{y1} \rightarrow \vec{v}_{y2} \rightarrow \vec{v}_{y3} \rightarrow \vec{v}_{y4} \rightarrow \vec{v}_{y4} \rightarrow \vec{v}_{y3} \rightarrow \vec{v}_{y2} \rightarrow \vec{v}_{y1}$ is applied over a sampling time T_s.

CASE STUDY 2.4 Analysis of Space Vector Modulation Scheme

Objective: In this case study, the implementation of space vector modulation scheme is analytically studied. The performance of space vector modulation is observed through MATLAB simulation.

Parameters: The rated parameters of the MMC are as follows: rated apparent power $S_o = 2$ MVA, output line-to-line voltage $V_o = 6$ kV, rated output frequency $f_o = 60$ Hz, rated power factor (PF) = 0.85 (lag), rated DC-bus voltage $V_{dc} = 10$ kV, number of SMs per arm $N = 4$, arm inductance $L = 5$ mH, and sampling time $T_s = 555.555\ \mu$s.

Analysis:
The MMC generates a five-level output voltage waveform with four submodules per arm. The normalized reference output vector to control the five-level modular multilevel converter is defined for $m_a = 0.95$ and $f_o = 60$ Hz as

$$\vec{v_r} = \begin{bmatrix} a^* \\ b^* \\ c^* \end{bmatrix} = \frac{4}{2} \begin{bmatrix} 0.95\sin(2\pi \times 60\,t) \\ 0.95\sin(2\pi \times 60\,t - \frac{2\pi}{3}) \\ 0.95\sin(2\pi \times 60\,t - \frac{4\pi}{3}) \end{bmatrix} \tag{2.51}$$

The reference output vector is realized using upper and lower arm reference vectors and given by

$$\begin{aligned} \vec{v_{ru}} &= \begin{bmatrix} a_u^* \\ b_u^* \\ c_u^* \end{bmatrix} = \frac{4}{2} \begin{bmatrix} 1 - 0.95\sin(2\pi \times 60\,t) \\ 1 - 0.95\sin(2\pi \times 60\,t - \frac{2\pi}{3}) \\ 1 - 0.95\sin(2\pi \times 60\,t - \frac{4\pi}{3}) \end{bmatrix} \\ \vec{v_{rl}} &= \begin{bmatrix} a_l^* \\ b_l^* \\ c_l^* \end{bmatrix} = \frac{4}{2} \begin{bmatrix} 1 + 0.95\sin(2\pi \times 60\,t) \\ 1 + 0.95\sin(2\pi \times 60\,t - \frac{2\pi}{3}) \\ 1 + 0.95\sin(2\pi \times 60\,t - \frac{4\pi}{3}) \end{bmatrix} \end{aligned} \tag{2.52}$$

The implementation procedure of SVM for upper and lower arms is quite similar. Therefore, the detailed calculations are presented for upper arms only. This procedure is repeated for the lower arms as well. The switching vectors and corresponding duty cycles are presented at a sampling instant $t = 0.0267$ s. The instantaneous value of upper arm reference vector is given by

$$\vec{v_{ru}} = \begin{bmatrix} a_u^* \\ b_u^* \\ c_u^* \end{bmatrix} = \begin{bmatrix} 3.136 \\ 0.113 \\ 2.75 \end{bmatrix} \tag{2.53}$$

The multilevel space vector is divided into several two-level space vector diagrams. The two-level space vector diagram has two zero vectors at the origin. The first zero vector is given by

$$\begin{aligned} a_{u0} &= \text{floor}\,(3.136) = 3 \\ b_{u0} &= \text{floor}\,(0.113) = 0 \\ c_{u0} &= \text{floor}\,(2.750) = 2 \end{aligned} \tag{2.54}$$

The second zero vector is obtained by incrementing the first zero vector by one and it is given by

$$\begin{aligned} a_{u1} &= 1 + 3 = 4 \\ b_{u1} &= 1 + 0 = 1 \\ c_{u1} &= 1 + 2 = 3 \end{aligned} \tag{2.55}$$

The switching vectors which forms the two-level space vector diagram are (3,0,2), (4,0,2), (4,1,2), (3,1,2), (3,1,3), (3,0,3), (4,0,3), and (4,1,3). Among them, only four nearest switching vectors are used to realize the reference vector. The nearest four switching vectors are obtained from the position of two-level space vector.

At $t = 0.0267$ s, the two-level space vector is located in sector-②. From Table 2.8, the four switching vectors corresponding to sector-② are $\vec{v}_{u1} = (3,0,2)$, $\vec{v}_{u2} = (3,1,2)$, $\vec{v}_{u3} = (4,1,2)$, and $\vec{v}_{u4} = (4,1,3)$. The dwell times of three-phase upper arms are

$$\begin{aligned} T_{au} &= (3.136 - 3) \times 555.555 \times 10^{-6} &&= 75.555\,\mu\text{s} \\ T_{bu} &= (0.113 - 0) \times 555.555 \times 10^{-6} &&= 62.777\,\mu\text{s} \\ T_{cu} &= (2.75 - 2) \times 555.555 \times 10^{-6} &&= 416.666\,\mu\text{s} \end{aligned} \tag{2.56}$$

The dwell times of four switching vectors are

$$\begin{aligned} T_{u1} &= 1 - 416.666\,\mu = 0.999\,\text{s} \\ T_{u2} &= 416.666\,\mu - 75.555\,\mu = 341.111\,\mu\text{s} \\ T_{u3} &= 75.555\,\mu - 62.777\,\mu = 12.778\,\mu\text{s} \\ T_{u4} &= 62.777\,\mu\text{s} \\ T_{u0} &= 0.999 + 62.777\,\mu\text{s} \end{aligned} \tag{2.57}$$

Similarly, the switching vectors and dwell times are obtained for the lower arm, using the lower arm reference vector. Figure 2.18 shows the performance of the MMC with SVM scheme. At each sampling interval, the upper and the lower arm reference vectors are positioned exactly in 180° phase. The actual upper and lower arm voltages are also in 180° phase with each other as shown in Figure 2.18(a)–(b). The difference between the upper and lower arm reference vectors gives the reference output voltage vector as shown in Figure 2.18(c). In SVM approach, the line-to-line voltages are defined by the space vector. The phase voltages are derived from the line-to-line voltages. Therefore, the line voltage waveform has symmetrical voltage levels without any unwanted switching actions as shown in Figure 2.18(d). The pulse width modulated voltage waveform has a fundamental component of $0.796\,V_{dc}$ and a THD of 18.22% as shown in Figure 2.18(e). In SVM, each arm is switched at a frequency equal to the sampling frequency (f_s). The switching frequency of modular multilevel converter is

$$f_{sw} = f_s = 1800\,\text{Hz} \tag{2.58}$$

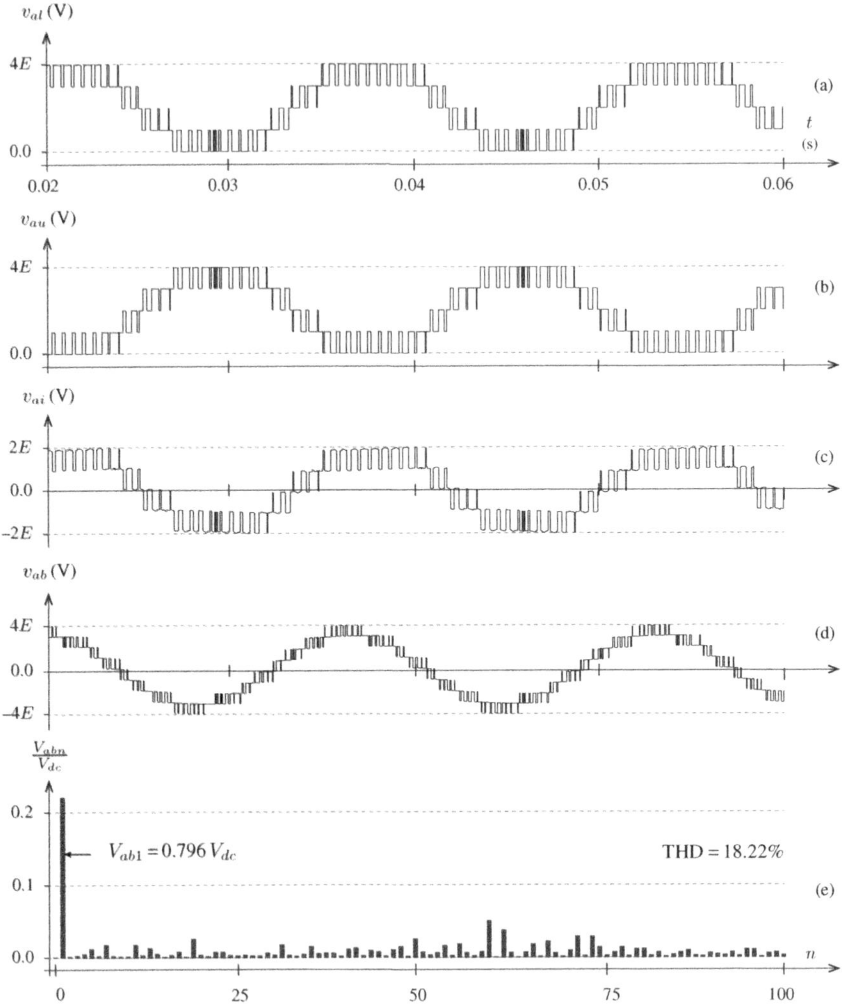

Figure 2.18 Simulation study of space vector modulation.

Each arm consists of four submodules, and each submodule is switching at a frequency of

$$f_{SM} = \frac{f_{sw}}{N} = \frac{1800}{4} = 450\,\text{Hz} \tag{2.59}$$

The upper and lower arm submodules are operating in a complementary manner and switching at the same frequency of 450 Hz.

■

2.3.5 Staircase Modulation

In this method, each leg of the modular multilevel converter is independently controlled using a staircase modulation approach. This method uses the nearest voltage level to generate the desired reference phase voltage; hence it is also termed as nearest level modulation (NLM) [22, 31]. Unlike the sine-triangle PWM method, the NLM is a fundamental switching frequency modulation and avoids the use of triangular carrier signals. The switching states and dwell times are directly computed from the reference phase voltage. This scheme is easy to implement in digital controllers and is most practical PWM methods for high-power MMC-based HVDC systems. The NLM method generates poor quality voltage waveforms if there are smaller number of submodules per arm. Therefore, the staircase modulation approach is mainly applied for the high-voltage applications, where the step number is very large (at least 100–200 steps). From equations (2.16) and (2.17), the upper and lower arm voltages are given by

$$\begin{aligned} v_{xu} &= \frac{V_{dc}}{2} - m_a \times \frac{V_{dc}}{2} \sin(\omega_o t + \theta_x) \\ v_{xl} &= \frac{V_{dc}}{2} + m_a \times \frac{V_{dc}}{2} \sin(\omega_o t + \theta_x) \end{aligned} \tag{2.60}$$

The total DC-bus voltage of each arm is expressed in terms of the number of submodules and the rated submodule capacitor voltage as

$$V_{dc} = N \times V_C \tag{2.61}$$

Equation (2.61) is substituted in equation (2.60) and results in

$$\begin{aligned} v_{xu} &= \frac{N\,V_C}{2} - m_a \times \frac{N\,V_C}{2} \sin(\omega_o t + \theta_x) \\ v_{xl} &= \frac{N\,V_C}{2} + m_a \times \frac{N\,V_C}{2} \sin(\omega_o t + \theta_x) \end{aligned} \tag{2.62}$$

The output voltage has a step value equal to V_C. Normalizing (2.62) with respect to the step value of V_C results in

$$\begin{aligned} v_{xu}^n &= \frac{N}{2} \times [1 - m_a \sin(\omega_o t + \theta_x)] \\ v_{xl}^n &= \frac{N}{2} \times [1 + m_a \sin(\omega_o t + \theta_x)] \end{aligned} \tag{2.63}$$

where v_{xu}^n and v_{xl}^n represents the normalized upper and lower arm voltages. The normalized voltage has steps in the range of 0 to $N+1$, which also represents the number of ON-state submodules in an arm. Each voltage level can be generated by using different switching combinations as shown in Table 2.7. One of the switching states corresponding to the required voltage level is applied for a time duration of T_s. The required voltage level at each sampling interval can be obtained from [22, 31].

$$\begin{aligned} m_{xu}^n &= \text{round}(v_{xu}^n) \\ m_{xl}^n &= \text{round}(v_{xl}^n) \end{aligned} \tag{2.64}$$

where m_{xu}^n, m_{xl}^n represent the nearest voltage level of upper and lower arm instantaneous voltage.

CASE STUDY 2.5 Analysis of Staircase Modulation Scheme

Objective: In this case study, the working principle of staircase modulation is studied for modular multilevel converter. The performance of staircase modulation is observed through MATLAB simulation study.

Parameters: The rated parameters of the MMC are as follows: rated apparent power $S_o = 2$ MVA, output line-to-line voltage $V_o = 6$ kV, rated output frequency $f_o = 60$ Hz, rated power factor (PF) = 0.85 (lag), rated DC-bus voltage $V_{dc} = 10$ kV, number of SMs per arm $N = 4$, arm inductance $L = 5$ mH, and sampling frequency $f_s = 1800$ Hz.

Analysis:
The normalized upper and lower arm reference voltages of phase-a are obtained from equation (2.60) for an amplitude modulation index $m_a = 0.95$, output frequency $f_o = 60$Hz, and an initial angle $\theta_a = 0°$ as shown in Figure 2.19 and is given by

$$\begin{aligned} v_{au}^n &= \frac{4}{2} \times [1 - 0.95\, sin(2\pi \times 60\, t)] \\ v_{al}^n &= \frac{4}{2} \times [1 + 0.95\, sin(2\pi \times 60\, t)] \end{aligned} \tag{2.65}$$

The required voltage level at a sampling instant $t = 0.0267$ s is given by

$$\begin{aligned} m_{au}^n &= \text{round}\,(3.136) = 3.0 \\ m_{al}^n &= \text{round}\,(0.764) = 1.0 \end{aligned} \tag{2.66}$$

Each voltage level is equal to an actual voltage step of E. For example, $m_{au}^n = 3.0$ is equal to an actual voltage of $3E$. Similarly, $m_{al}^n = 1.0$ is equal to E. These voltage levels also represents the required number of ON-state submodules in an arm. Therefore, the required voltage level is obtained by inserting the corresponding number of submodules in the arm for a sampling time $T_s = \frac{1}{f_s} = \frac{1}{1800} = 0.555$ ms. Figure 2.19 shows the operation of phase-a, lower arm with a staircase modulation scheme. At $t = 0.0267$ s, the lower arm requires a voltage level of $m_{al}^n = 1.0$. This voltage level is generated by turning-ON the SM_1 and corresponding output voltage is equal to $V_{H1} = E$ as shown in Figure 2.19(b). The SMs 2, 3 and 4 are turned-OFF, and their output voltages are equal to zero as shown in Figure 2.19(c)–(e). The lower arm voltage (v_{al}) is equal to the summation of each submodule output voltage as shown in Figure 2.19(f). In this approach, each submodule is turned-"ON" and "OFF" only once in one fundamental cycle. Therefore, the submodule switching frequency is equal to $f_{SM} = f_o = 60$ Hz. Similarly, the upper arm requires a voltage level $m_{au}^n = 3.0$. This voltage level is obtained by inserting three submodules out of the four.

Figure 2.20 shows the performance of the MMC with staircase modulation. The normalized upper and lower arm modulation signals are shown in Figure 2.20(a), which consists of positive integers only. The modulation signals are used to modulate the submodules and generate a voltage waveform across the arms. The upper and lower arm voltage has a five-level stepped waveform with a step value of $E = 2500$ V. These voltages are exactly in 180° phase with each other as shown in Figure 2.20(b)–(c). The output phase voltage is obtained by subtracting the upper arm voltage from the lower arm voltage. This voltage waveform has only $N + 1 = 5$ voltage levels of $2E$, E, 0, $-E$, $-2E$ as shown in Figure 2.20(d). The line-to-line

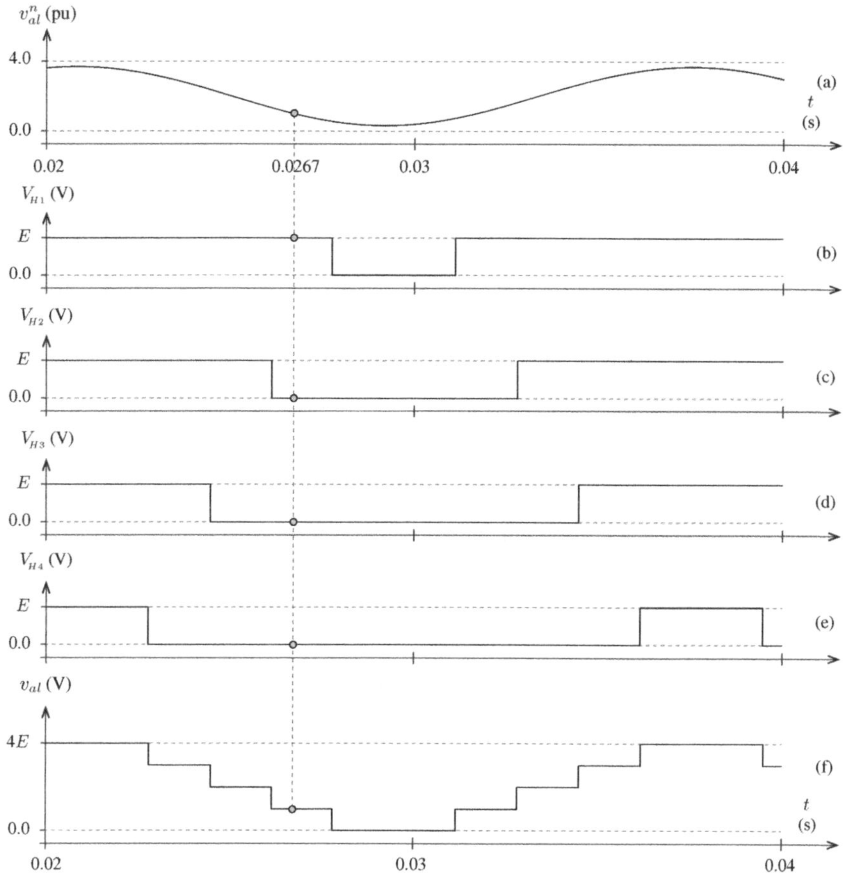

Figure 2.19 Operation of staircase modulation scheme.

voltage waveform has a $2N + 1 = 9$ voltage levels of $4E$, $3E$, $2E$, E, 0, $-E$, $-2E$, $-3E$, and $-4E$ as shown in Figure 2.20(e). The line voltage has a fundamental (60 Hz) voltage component of 0.824 V_{dc} and THD 12.87% as shown in Figure 2.20(f).

The modular multilevel converter is switching at a frequency of

$$f_{sw} = f_o = 60\,\text{Hz} \tag{2.67}$$

Therefore, the lower-order harmonics such as 7th, 11th, and 13th harmonic components become dominant. The individual harmonic components as well the THD can be minimized by increasing the number of voltage levels (i.e., the number of SMs per arm).

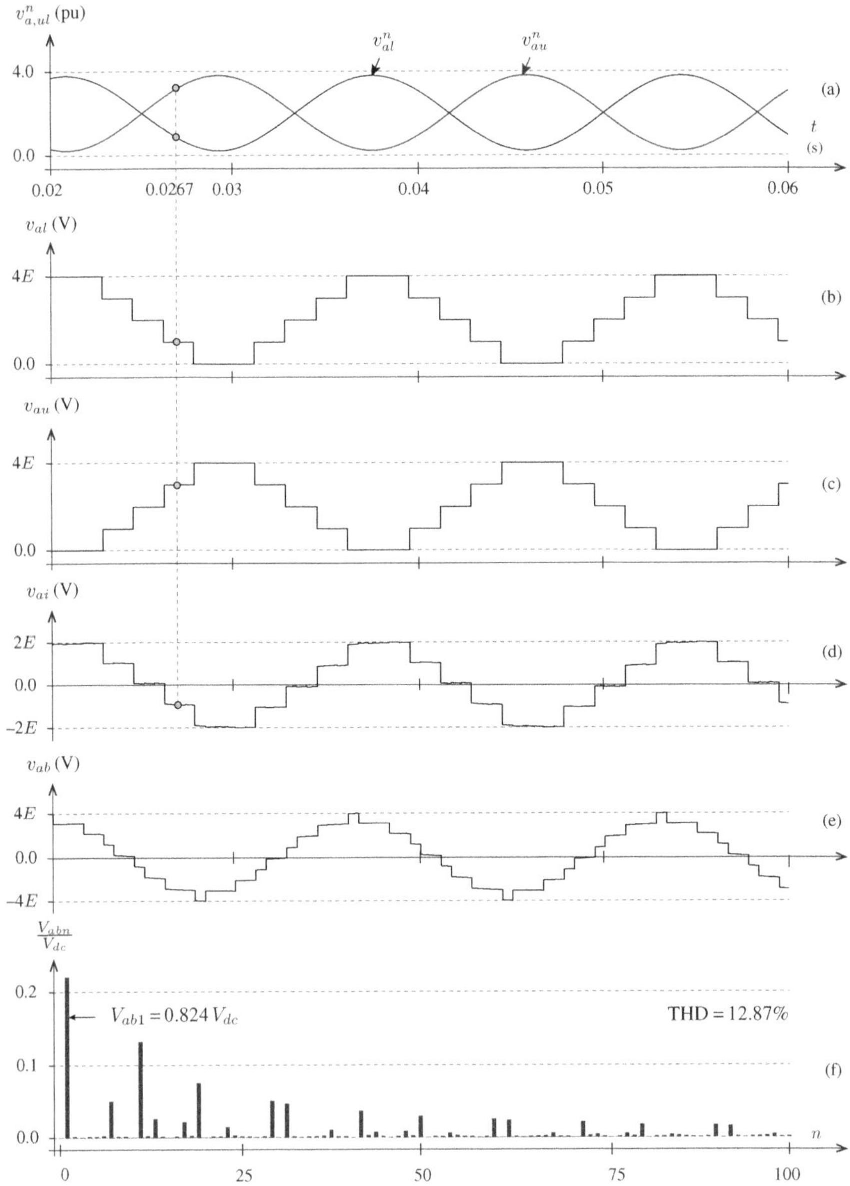

Figure 2.20 Performance of the staircase modulation scheme.

■

2.4 SUMMARY

In this chapter, a comprehensive overview of modular multilevel converter including features, topologies, and working principle are presented. The modular multilevel converter can be realized by using several submodule configurations. The most popular and widely used submodule configurations such as HB, FB, FC, CH, and CD submodules are discussed. The operation, switching states, and the relationship between DC and AC quantities of each submodule circuit are systematically presented. These submodules are compared in terms of component count, rating, power loss, control, and construction complexity. Among them, the half-bridge submodule has a simple construction and control. This submodule is commercially employed in modular multilevel converter based high-voltage direct current (HVDC) and medium-voltage motor drive systems.

PWM schemes are commonly employed to control the modular multilevel converter. The implementation of PWM scheme for MMC is one of the major challenges, due to the large number of submodules in each arm. Considering this fact, the PWM schemes which have less computational complexity and are extendable to any number of submodules per arm are presented. The principle of operation; step-by-step design procedure; and performance of PSC-PWM, LSC-PWM, sampled average, SVM, and staircase modulation schemes are systematically presented with the help of case studies. Among them, the carrier and space vector modulation schemes are used to control the MMC in motor drive applications and staircase modulation is used in HVDC application. A new sampled average modulation approach is also introduced. This approach can be applied to the MMC with any number of submodules without any modifications and it also eliminates the volt-sec balance error present in the staircase modulation. These modulation schemes are further analyzed with the closed loop control methods in Chapter 3.

REFERENCES

1. J. Rodriguez, L. G. Franquelo, S. Kouro, J. I. Leon, R. C. Portillo, M. Á. M. Prats, and M. A. Perez, "Multilevel converters: An enabling technology for high-power applications," *Proc. IEEE*, vol. 97, no. 11, pp. 1786–1817, Nov 2009.

2. S. Kouro, M. Malinowski, K. Gopakumar, J. Pou, L. G. Franquelo, B. Wu, J. Rodriguez, M. A. Perez, and J. I. Leon, "Recent advances and industrial applications of multilevel converters," *IEEE Trans. Ind. Electron.*, vol. 57, no. 8, pp. 2553–2580, Aug 2010.

3. H. Akagi, "Classification, terminology, and application of the modular multilevel cascade converter (mmcc)," *IEEE Trans. Power Electron.*, vol. 26, no. 11, pp. 3119–3130, Nov 2011.

4. Benshaw, "M2l 3000 series medium voltage motor drive." [Online]. Available: http://www.benshaw.com/uploadedFiles/Literature/Benshaw_M2L_MVFD_2.3-6.6kV.pdf

5. Siemens, "Sinamics sm120." [Online]. Available: http://www.industry.siemens.com/drives/global/en/converter/mv-drives/sinamics-sm120-cm/

6. Siemens, "The smart way hvdc plus-one step ahead." [Online]. Available: https://www.energy.siemens.com/us/en/power-transmission/hvdc/hvdcplus/

7. Alstom, "Hvdc maxsine." [Online]. Available: https://www.alstom.com/grid/products-and-services/engineered-energy-solutions/hvdc-transmission-systems/HVDC_Maxsine/

8. Siemens, "The efficient way." [Online]. Available: http://www.energy.siemens.com/ru/pool/hq/power-transmission/FACTS/SVC_PLUS_The%20efficient%20Way.pdf

9. M. Perez, S. Bernet, J. Rodriguez, S. Kouro, and R. Lizana, "Circuit topologies, modeling, control schemes, and applications of modular multilevel converters," *IEEE Trans. Power Electron.*, vol. 30, no. 1, pp. 4–17, Jan 2015.

10. S. Debnath, J. Qin, B. Bahrani, M. Saeedifard, and P. Barbosa, "Operation, control, and applications of the modular multilevel converter: A review," *IEEE Trans. Power Electron.*, vol. 30, no. 1, pp. 37–53, Jan 2015.

11. A. Nami, J. Liang, F. Dijkhuizen, and G. Demetriades, "Modular multilevel converters for hvdc applications: Review on converter cells and functionalities," *IEEE Trans. Power Electron.*, vol. 30, no. 1, pp. 18–36, Jan 2015.

12. A. Dekka, B. Wu, N. Zargari, and R. Fuentes, "Dynamic voltage balancing algorithm for modular multilevel converter: A unique solution," *IEEE Trans. Power Electron.*, vol. 31, no. 2, pp. 952–963, Feb 2016.

13. Y. Xue, Z. Xu, and G. Tang, "Self-start control with grouping sequentially precharge for the c-mmc-based hvdc system," *IEEE Trans. Power Del.*, vol. 29, no. 1, pp. 187–198, Feb 2014.

14. M. Malinowski, K. Gopakumar, J. Rodriguez, and M. Pèrez, "A survey on cascaded multilevel inverters," *IEEE Trans. Ind. Electron.*, vol. 57, no. 7, pp. 2197–2206, Jul 2010.

15. J. Rodriguez, J.-S. Lai, and F. Z. Peng, "Multilevel inverters: a survey of topologies, controls, and applications," *IEEE Trans. Ind. Electron.*, vol. 49, no. 4, pp. 724–738, Aug 2002.

16. B. P. McGrath and D. G. Holmes, "Multicarrier pwm strategies for multilevel inverters," *IEEE Trans. Ind. Electron.*, vol. 49, no. 4, pp. 858–867, Aug 2002.

17. A. Dekka, B. Wu, and N. R. Zargari, "A novel modulation scheme and voltage balancing algorithm for modular multilevel converter," *IEEE Trans. Ind. Appl.*, vol. 52, no. 1, pp. 432–443, Jan 2016.

18. Y. Deng, Y. Wang, K. H. Teo, and R. G. Harley, "A simplified space vector modulation scheme for multilevel converters," *IEEE Trans. Power Electron.*, vol. 31, no. 3, pp. 1873–1886, Mar 2016.

19. A. Dekka, B. Wu, N. R. Zargari, and R. L. Fuentes, "A space-vector pwm-based voltage-balancing approach with reduced current sensors for modular multilevel converter," *IEEE Trans. Ind. Electron.*, vol. 63, no. 5, pp. 2734–2745, May 2016.

20. G. Konstantinou, M. Ciobotaru, and V. Agelidis, "Selective harmonic elimination pulse-width modulation of modular multilevel converters," *IET Power Electron.*, vol. 6, no. 1, pp. 96–107, Jan 2013.

21. Z. Li, P. Wang, H. Zhu, Z. Chu, and Y. Li, "An improved pulse width modulation method for chopper-cell-based modular multilevel converters," *IEEE Trans. Power Electron.*, vol. 27, no. 8, pp. 3472–3481, Aug 2012.

22. P. Meshram and V. Borghate, "A simplified nearest level control (nlc) voltage balancing method for modular multilevel converter (mmc)," *IEEE Trans. Power Electron.*, vol. 30, no. 1, pp. 450–462, Jan 2015.

23. R. Naderi and A. Rahmati, "Phase-shifted carrier pwm technique for general cascaded inverters," *IEEE Trans. Power Electron.*, vol. 23, no. 3, pp. 1257–1269, May 2008.

24. S. Kouro, P. Lezana, M. Angulo, and J. Rodriguez, "Multicarrier pwm with dc-link ripple feedforward compensation for multilevel inverters," *IEEE Trans. Power Electron.*, vol. 23, no. 1, pp. 52–59, Jan 2008.

25. F. Deng and Z. Chen, "Elimination of dc-link current ripple for modular multilevel converters with capacitor voltage-balancing pulse-shifted carrier pwm," *IEEE Trans. Power Electron.*, vol. 30, no. 1, pp. 284–296, Jan 2015.

26. E. Solas, G. Abad, J. Barrena, S. Aurtenetxea, A. Carcar, and L. Zajac, "Modular multilevel converter with different submodule concepts – part i: Capacitor voltage balancing method," *IEEE Trans. Ind. Electron.*, vol. 60, no. 10, pp. 4525–4535, Oct 2013.

27. J. Mei, K. Shen, B. Xiao, L. Tolbert, and J. Zheng, "A new selective loop bias mapping phase disposition pwm with dynamic voltage balance capability for modular multilevel converter," *IEEE Trans. Ind. Electron.*, vol. 61, no. 2, pp. 798–807, Feb 2014.

28. N.-Y. Dai, M.-C. Wong, Y.-H. Chen, and Y.-D. Han, "A 3-d generalized direct pwm algorithm for multilevel converters," *IEEE Power Electron. Lett.*, vol. 3, no. 3, pp. 85–88, Sep 2005.

29. M. Perales, M. Á. M. Prats, R. Portillo, J. Mora, J. Leon, and L. Franquelo, "Three-dimensional space vector modulation in abc coordinates for four-leg voltage source converters," *IEEE Power Electron. Lett.*, vol. 1, no. 4, pp. 104–109, Dec 2003.

30. B. Reddy, V. Somasekhar, and Y. Kalyan, "Decoupled space-vector pwm strategies for a four-level asymmetrical open-end winding induction motor drive with waveform symmetries," *IEEE Trans. Ind. Electron.*, vol. 58, no. 11, pp. 5130–5141, Nov 2011.

31. P. Hu and D. Jiang, "A level-increased nearest level modulation method for modular multilevel converters," *IEEE Trans. Power Electron.*, vol. 30, no. 4, pp. 1836–1842, Apr 2015.

CHAPTER 3

CLASSICAL CONTROL OF MODULAR MULTILEVEL CONVERTER

3.1 INTRODUCTION

The digital control schemes enable safe, reliable, and efficient operation of a modular multilevel converter (MMC). These control schemes provide a superior dynamic and steady-state performance. The control of MMC is quite challenging and involves multiple control objectives as shown in Figure 3.1. The control objectives are categorized into primary and secondary objectives. The submodule (SM) capacitor voltage and output current control are the primary objectives and associated with the operation of MMC. The circulating current control is a secondary objective and associated with the size, reliability, and the efficiency of an MMC. These objectives can be achieved by using both classical and predictive control methods [1, 2].

This chapter deals with the implementation and performance of classical control methods for a modular multilevel converter. The general overview of classical control methods is given in Section 3.2. Classical control methods require an independent controller to control the submodule capacitor voltage, output current, and circulating currents. The submodule capacitor voltage control involves the leg voltage control and voltage balance among the arm submodules. The implementation and control of submodule capacitor voltage is discussed in Section 3.3. The sub-

Modular Multilevel Converters: Analysis, Control, and Applications, First Edition.
Sixing Du, Apparao Dekka, Bin Wu, and Navid Zargari.

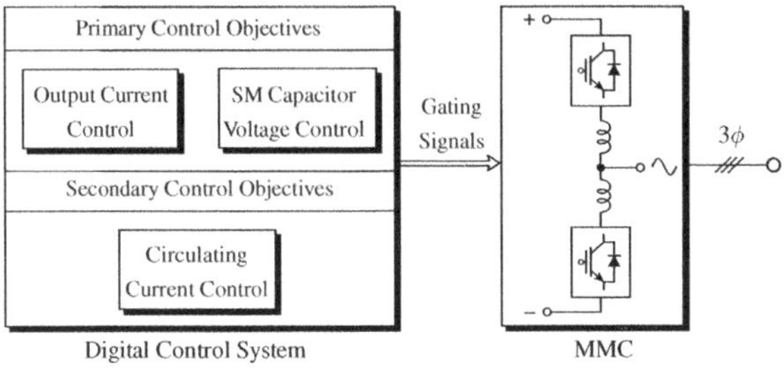

Figure 3.1 Block diagram of MMC control system.

modules in each arm are controlled to generate an output voltage corresponding to the required magnitude of AC output current. The synchronous dq-frame based PI-regulator is employed to achieve close tracking between the actual and reference currents. The fundamentals of synchronous dq-frame approach and output current control are presented in Section 3.4. An output current controller is required to improve the dynamic and steady-state performance of the load. The difference between the submodule capacitors voltage in the upper and lower arm of three converter legs leads to a circulating current. The circulating current increases the RMS (Root Mean Square) value of the arm current, and results in an increase in the power loss and reduction of the system efficiency. Therefore, the suppression of circulating current is necessary. The circulating current suppression methods are given in Section 3.5. Finally, the summary of this chapter is presented in Section 3.6.

3.2 OVERVIEW OF CLASSICAL CONTROL METHOD

The general block diagram of a classical control method is shown in Figure 3.2. In this method, an independent control approach is used to control the submodule capacitors voltage, output currents, and circulating currents [3]. The submodule capacitors voltage control includes a leg voltage control and voltage balance strategy [4]. The voltage balance strategy maintains the submodule capacitors voltage at an identical value within the arm. The voltage balancing can be achieved at either control stage using a closed-loop controller or at modulation stage by using logical functions [5]. The balancing method selects a certain number of submodules out of N submodules in an arm to generate the required voltage level. Therefore, the balancing strategy is often referred to as a submodule selection method. The leg voltage control regulates the average of $2N$ submodule capacitors voltage at a constant value of v_C [6]. This controller generates a compensating signal v^*_{dtx}, which is added to the upper and lower arm modulation signals. Similarly, the output current and circulating currents are controlled using a closed-loop controller, and generate control commands v^*_{xi} and v^*_{xz}, respectively [7]. These control commands are added to the upper and lower arm modulation signals, and results in

$$\begin{aligned} v_{xu}^{*} &= \frac{V_{dc}}{2} - v_{xi}^{*} - v_{xz}^{*} - v_{dtx}^{*} \\ v_{xl}^{*} &= \frac{V_{dc}}{2} + v_{xi}^{*} - v_{xz}^{*} - v_{dtx}^{*} \end{aligned} \tag{3.1}$$

where v_{xu}^{*} and v_{xl}^{*} represent the modulation signals of upper and lower arms. These modulation signals are given to the modulation stage.

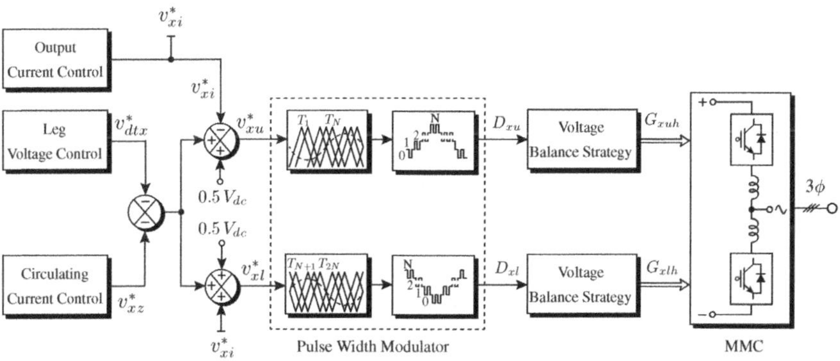

Figure 3.2 Block diagram of the classical control method.

Classical control methods can be used with any modulation scheme (i.e., PSC-PWM, LSC-PWM, NLC, sampled average modulation, and SVM). The implementation of several PWM schemes has been discussed in Chapter 2. In this study, the performance of the classical control method is presented with the PSC-PWM scheme. The arm modulation signals are compared with corresponding arm carrier signals. For example, the upper arm modulation signal is compared with the upper arm carrier signals $(T_1, \ldots, T_N)$. The output of each comparator is added together to obtain a normalized voltage waveform (D_{xu}). The normalized voltage waveform resembles the actual output voltage waveform of the upper arm. Similarly, the lower arm normalized voltage waveform (D_{xl}) is generated by the comparison of lower arm modulation signals with the lower arm carrier signals $(T_{N+1}, \ldots, T_{2N})$. The normalized voltage waveform (D_{xu} and D_{xl}) has voltage levels of $0, 1, 2, \ldots, N$, and represents the required number of submodules to be inserted in the arm. The normalized voltage waveform is applied to the voltage balancing strategy. The balancing strategy selects the submodules based on the instantaneous value of capacitors voltage, and the direction of the arm current. The selected submodules are inserted in the arm to generate the voltage level corresponding to the normalized voltage waveform.

3.3 SUBMODULE CAPACITOR VOLTAGE CONTROL

The operating voltage of modular multilevel converter can be increased by connecting more submodules in series in each arm. These submodules have floating capacitors, and their voltages must be controlled at their nominal value. The control of submodule capacitor voltage is a fundamental requirement to achieve a high-performance current control, to improve output power quality, and to provide a stable and controllable operation. The capacitor voltage control is divided into two parts: the first part is a leg voltage control, and the second part is a submodule capacitor voltage balancing strategy. The leg voltage control regulates the average of $2N$ submodule capacitor voltages in each leg at a value of v_C. The voltage balance between submodules within the arm is achieved by using the capacitor voltage balance strategy [8].

3.3.1 Leg Voltage Control

The block diagram of leg voltage control is shown in Figure 3.3. This control has an outer voltage and an inner current control loop [4, 6]. The outer voltage loop regulates the average of $2N$ submodule capacitor voltages in each leg at a constant value V_C^*, which is equal to the submodule capacitor voltage v_C. The voltage control loop minimizes the error between the reference and the actual average voltage and gives the DC current component command (i_{xd}^*). The DC-bus current command is given by

$$i_{xd}^* = k_{pv}(V_C^* - v_{Cx}) + k_{iv}\int (V_C^* - v_{Cx})dt \tag{3.2}$$

where k_{pv} and k_{iv} are the proportional and integral gain of voltage control loop, respectively.

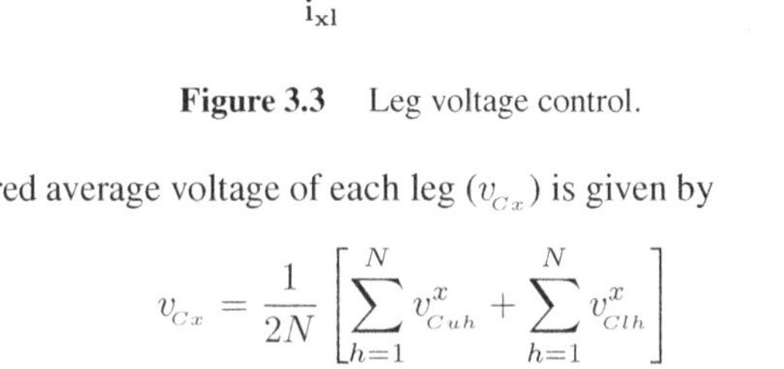

Figure 3.3 Leg voltage control.

The measured average voltage of each leg (v_{Cx}) is given by

$$v_{Cx} = \frac{1}{2N}\left[\sum_{h=1}^{N} v_{Cuh}^x + \sum_{h=1}^{N} v_{Clh}^x\right] \tag{3.3}$$

where v_{Cuh}^x and v_{Clh}^x are the submodule capacitor voltages in the upper and lower arms of phase-x.

The magnitude of DC current component is proportional to the active power demanded by the load plus power losses at steady state. The DC current component is controlled using an inner current loop. This loop minimizes the error between the reference and the actual DC current component and gives a compensating signal of

$$v_{dtx}^{*} = k_{pi}(i_{xd}^{*} - i_{xd}) + k_{ii}\int(i_{xd}^{*} - i_{xd})dt \tag{3.4}$$

where k_{pi} and k_{ii} are the proportional and integral gain of current control loop, respectively.

The actual DC-bus current is estimated from the measured arm currents and is given by

$$i_{xd} = \frac{\mathbf{i_{xu}} + \mathbf{i_{xl}}}{6} \tag{3.5}$$

where $\mathbf{i_{xu}} \in \{i_{au}, i_{bu}, i_{cu}\}$ and $\mathbf{i_{xl}} \in \{i_{al}, i_{bl}, i_{cl}\}$. The compensating signal v_{dtx}^{*} is used to correct the submodule duty cycles, thereby, the average voltage of each leg is maintained constant.

3.3.2 Voltage Balance Strategy

Voltage balancing among the submodules is a vital factor for the reliable operation of modular multilevel converters. The basic principle of voltage balancing approach is to control the charging and discharging of submodule capacitors on the basis of the arm current direction and the instantaneous value of capacitors voltage. For the positive direction of the current, capacitors with the lowest voltage are inserted in the arm to achieve voltage balancing and vice versa. The voltage balancing can be achieved at the control stage by using a closed-loop PI-controller to control each submodule capacitor voltage independently. The output of the PI-controller is added to the modulation signal of the corresponding submodule. This approach is suitable to implement with PSC-PWM only. On the other hand, the voltage balancing can be achieved at a modulation stage by using logical function-based algorithm. In this approach, all the submodules in each arm are controlled together. Hence, this approach is suitable to implement with any PWM scheme. The flow chart of a simple voltage balance strategy based on logical functions is shown in Figure 3.4(a). The implementation of balancing strategy involves the following steps [9]:

- Measure the SM capacitor voltages ($v_{Cyh}^{x} \in \{v_{Cy1}^{x}, v_{Cy2}^{x}, \dots, v_{CyN}^{x}\}$) in each arm, and feed their magnitude value to the comparison logic as shown in Figure 3.4(b).

- In comparison logic, each capacitor voltage is compared with other capacitor voltages. The output of each comparator is added together to obtain an index number (VI_h). The highest index number is assigned to the submodule with the lowest capacitor voltage and vice versa.

- Measure the arm current (i_{xy}) and determine its direction (A_d). For positive direction $A_d = 1$, and for negative direction $A_d = 0$ is assigned.

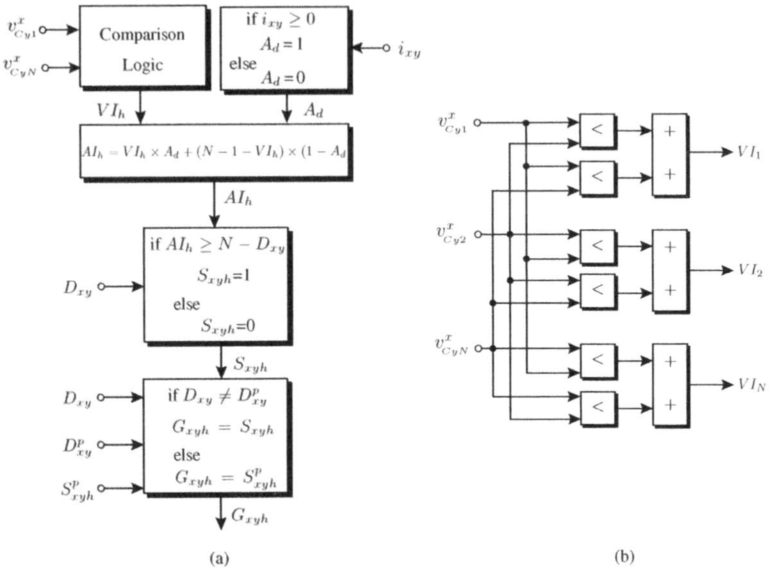

Figure 3.4 Voltage balance strategy: (a) flow chart of balancing strategy, and (b) comparison logic.

- Arrange the submodule index numbers in either ascending or descending order based on the direction of arm current. A simple mathematical expression is presented below to arrange the submodule index numbers.

$$AI_h = VI_h \times A_d + (N - 1 - VI_h) \times (1 - A_d) \tag{3.6}$$

  Equation (3.6) rearranges the submodule index numbers based on the arm current direction. Thereby, the submodules with lowest capacitor voltage are inserted in the arm for the positive current direction and are charged (increasing their voltage). Similarly, the submodules with highest capacitor voltage are inserted in the arm for the negative current direction and are discharged (reducing their voltage).

- Obtain the required number of inserted submodules (D_{xy}) from the modulation stage.

- Compare the actual index number of each submodule (AI_h) with the reference index number ($N - D_{xy}$) to generate the `INSERT` = 1 or `BYPASS` = 0 states for each submodule (S_{xyh}).

- The inserted number of submodules in the present control cycle (D_{xy}) is compared with the previous control cycle D^p_{xy}. The difference in D_{xy} and D^p_{xy} allows to apply the new gating signals (i.e., $G_{xyh} = S_{xyh}$) or the gating signals in previous control cycle are maintained (i.e., $G_{xyh} = S^p_{xyh}$). Therefore, the submodule switching frequency can be maintained constant and equal to the carrier frequency of the PWM scheme.

CASE STUDY 3.1 Implementation of Capacitor Voltage Balancing Strategy

Objective: In this case study, the implementation of voltage balancing strategy is studied. The design steps are discussed through analytical calculations.

Parameters: The rated parameters of the MMC are as follows: rated apparent power $S_o = 2$ MVA, output line-to-line voltage $V_o = 6$ kV, rated output frequency $f_o = 60$ Hz, rated power factor (PF) = 0.85 (lag), rated DC-bus voltage $V_{dc} = 10$ kV, number of SMs per arm $N = 4$, arm inductance $L = 5$ mH, and SM capacitance $C = 2200$ μF.

Analysis:

The nominal value of submodule capacitor voltage is given by

$$v_C = \frac{V_{dc}}{N} = \frac{10,000}{4} = 2500\ \text{V} \tag{3.7}$$

The implementation of balancing strategy involves different steps, which are discussed through the following example. The instantaneous value of submodule capacitor voltages is considered as $v^a_{Cl1} = 2400$ V, $v^a_{Cl2} = 2600$ V, $v^a_{Cl3} = 2300$ V, and $v^a_{Cl4} = 2550$ V. The submodule capacitors voltage is compared with each other, and obtain the index number VI_h. The highest index number is assigned to the SM-3 with the lowest capacitor voltage and the lowest index number is assigned to the SM-2 with the highest capacitor voltage. The assignment of index number for each submodule is given in Table 3.1.

Table 3.1 Generation of index number

SM Capacitors Voltage	VI_h	AI_h for $i_{xy} \geq 0$ & $A_d = 1$	AI_h for $i_{xy} < 0$ & $A_d = 0$
$v^a_{Cl1} = 2400$	$VI_1 = 2$	$AI_1 = 2$	$AI_1 = 1$
$v^a_{Cl2} = 2600$	$VI_2 = 0$	$AI_2 = 0$	$AI_2 = 3$
$v^a_{Cl3} = 2300$	$VI_3 = 3$	$AI_3 = 3$	$AI_3 = 0$
$v^a_{Cl4} = 2550$	$VI_4 = 1$	$AI_4 = 1$	$AI_4 = 2$

For the positive direction of current, the submodules with the lowest voltage are inserted in the arm to charge them. Thereby, the submodule capacitors voltage can reach their nominal value. Similarly, the submodules with the highest voltage are inserted in the arm to discharge them for the negative direction of the current. To meet the above requirement, the submodule index numbers are rearranged using equation (3.6). The rearranged submodule index numbers (AI_h) are given in Table 3.1.

The next step is to obtain the required number of inserted submodules in the arm. This information is obtained from the modulation stage. The modulation stage requires an arm modulation signal and carrier signals. The arm modulation signal consists of an output voltage component (v^*_{xi}), and a leg voltage control command v^*_{dtx}. The phase-a output voltage with an amplitude modulation index $m_a = 0.95$, output frequency $f_o = 60$ Hz, and initial phase angle $\theta_a = 0°$ is given by

$$v_{ai}^{*} = 0.95 \times \frac{10,000}{2} \sin(2\pi \times 60t) \tag{3.8}$$

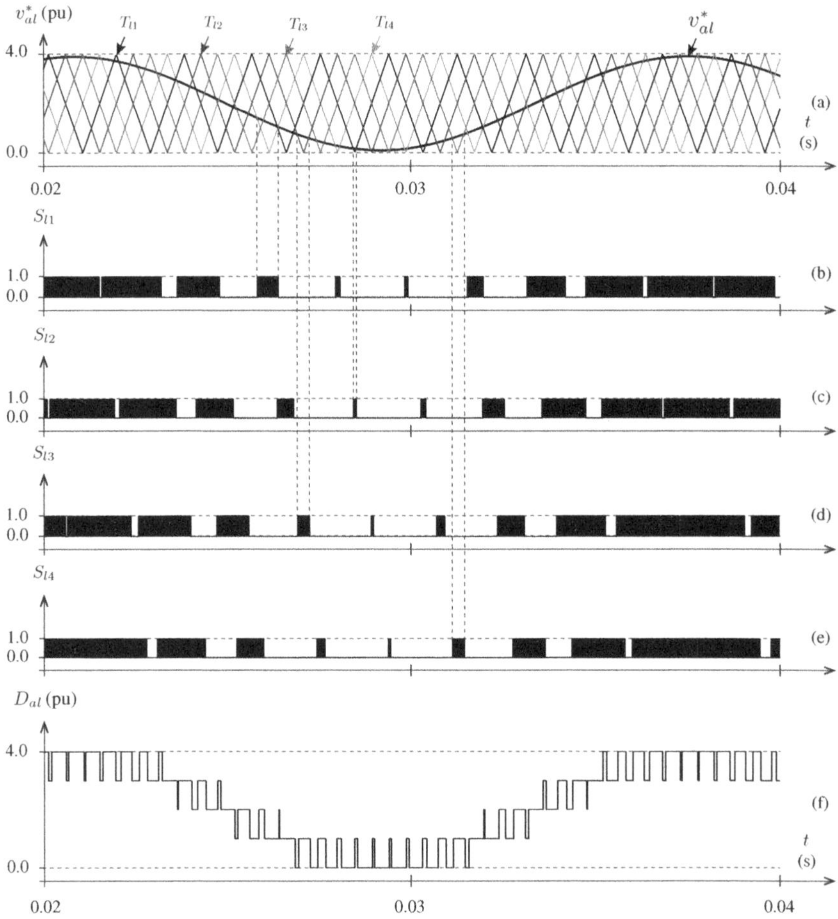

Figure 3.5 Generation of the normalized voltage waveform.

The average voltage control is designed to regulate the average voltage of each leg at 10 kV. The upper and lower arm modulation signals are normalized with respect to the submodule rated capacitor voltage of 2.5 kV. The normalized modulation signal has a maximum value "4" and a minimum value "0." These modulation signals are given to the modulation stage. The carrier signals are arranged as per the PSC-PWM with an interleave angle approach. The arm modulation signal is compared with each carrier signal, and its comparison gives the switching signals S_{l1}, S_{l2}, S_{l3}, and S_{l4} as shown in Figure 3.5. These switching signals are added together to form a normalized voltage waveform (D_{al}). This waveform has voltage levels of "0, 1, 2, 3, 4," which represents the required number of inserted submodules in the arm. The normalized voltage waveform is applied to the voltage balancing strategy.

Let us assume that the number of inserted submodules in the present control cycle is $D_{al} = 3$, and the direction of current is positive $A_d = 1$, which implies that only three submodules out of the four with the lowest voltage are inserted in the arm. The index number of each submodule (AI_h) is compared with the reference index number $N - D_{al} = 4 - 3 = 1$ to identify the inserted submodules. The comparison of submodule index number and reference index number gives the INSERT = 1 or BYPASS = 0 state of the submodule (S_{xyh}). The state of submodules is given in Table 3.2. It is also assumed that the inserted number of submodules in previous control cycle is $D^p_{al} = 3$. Thereby, the submodule gating signals in present control cycle are $G_{xyh} = S_{xyh}$. The Table 3.2 shows that the SM-1, SM-3, and SM-4 have lowest capacitor voltages as compared to the SM-2. Hence, these three submodules are inserted in the arm for the positive current direction to charge them. The SM-2 is bypassed, and there is no effect on its capacitor voltage.

Table 3.2 Generation of gating signals

SM Capacitor Voltage	AI_h for $i_{xy} \geq 0$ & D = 1	S_{xyh} for $i_{xy} \geq 0$ & $A_d = 1$	G_{xyh} for $i_{xy} \geq 0$ & $A_d = 1$
$v^a_{Cl1} = 2400$	$AI_1 = 2$	$S_{al1} = 1$	$G_{al1} = 1$
$v^a_{Cl2} = 2600$	$AI_2 = 0$	$S_{al2} = 0$	$G_{al1} = 0$
$v^a_{Cl3} = 2300$	$AI_3 = 3$	$S_{al3} = 1$	$G_{al1} = 1$
$v^a_{Cl4} = 2550$	$AI_4 = 1$	$S_{al4} = 1$	$G_{al1} = 1$

■

CASE STUDY 3.2 Performance of Submodule Capacitor Voltage Control

Objective: The dynamic performance of the submodule capacitor voltage control approach is studied through MATLAB simulation.

Parameters: The rated parameters of the MMC are as follows: rated apparent power $S_o = 2$ MVA, output line-to-line voltage $V_o = 6$ kV, rated output frequency $f_o = 60$ Hz, rated power factor (PF) = 0.85 (lag), rated DC-bus voltage $V_{dc} = 10$ kV, number of SMs per arm $N = 4$, arm inductance $L = 5$ mH, and SM capacitance $C = 2200\ \mu$F.

Analysis:

In this case study, the importance of submodule capacitor voltage control is presented by disabling the balancing strategy for a short duration of time. Initially, the balancing strategy is enabled from $t = 1.9$ s to $t = 2.0$ s. Figure 3.6 shows that the leg voltage control maintains the average voltage of leg-a at 10 kV. The upper and lower arm of phase-a have four submodules, and the balancing strategy maintains each submodule capacitor voltage at 2.5 kV. From $t = 2.0$ s to $t = 2.05$ s, the balancing strategy disabled. The upper and lower arm submodule capacitors voltage diverges from their nominal value, which will affect the magnitude of output fundamental voltage, device reverse blocking voltage, the reliability of the converter, and device current. The average leg voltage is deviating from their nominal value to due to the difference in unequal submodule capacitors voltage. At $t = 2.05$ s, the balancing strategy is enabled again, and the balancing strategy brings back all the submodule capacitors voltage to their nominal value.

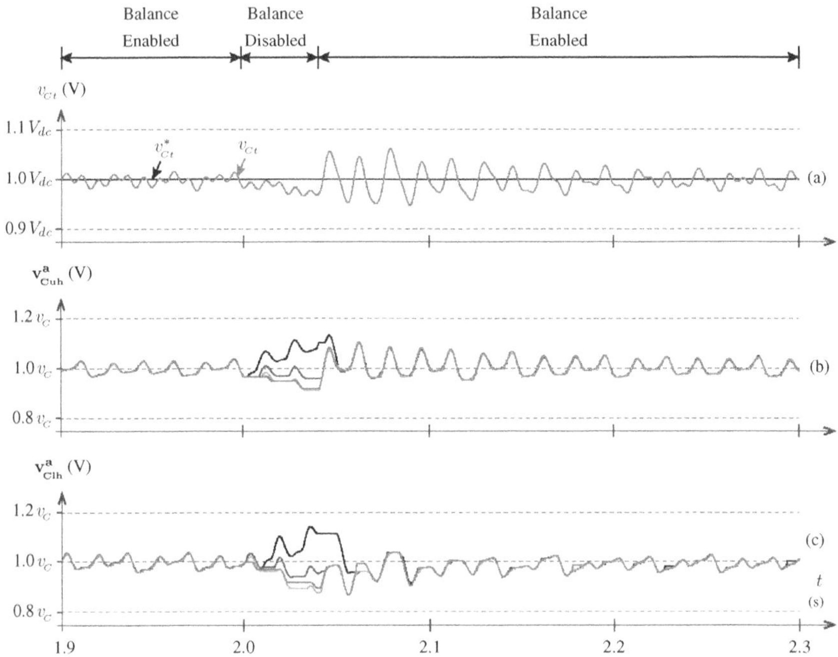

Figure 3.6 Dynamic performance of submodule capacitor voltage control approach.

■

3.4 OUTPUT CURRENT CONTROL

The modular multilevel converter can be used in high-voltage direct current (HVDC) transmission [10], medium-voltage motor drives [11], and static synchronous compensator (STATCOM) [12, 13] applications. Each application requires different techniques to control their output currents. For example, voltage-oriented control (VOC) is used in HVDC application, and field-oriented control (FOC) or direct torque control (DTC) is applied to motor drive systems. These control techniques can be implemented in stationary-abc, stationary $\alpha\beta$, and synchronous-dq reference frames. In this section, the fundamental details of the reference transformation theory are presented.

3.4.1 Reference Frame Theory

The reference frame theory simplifies the modeling and analysis of three-phase circuits, electrical machines, and power converters. The reference frame theory also facilitates the design and simulation of advanced digital control schemes in power engineering. A number of reference frames are presented over the years, in which stationary-abc frame, stationary-$\alpha\beta$ frame, and synchronous-dq reference frames are widely used [14, 15].

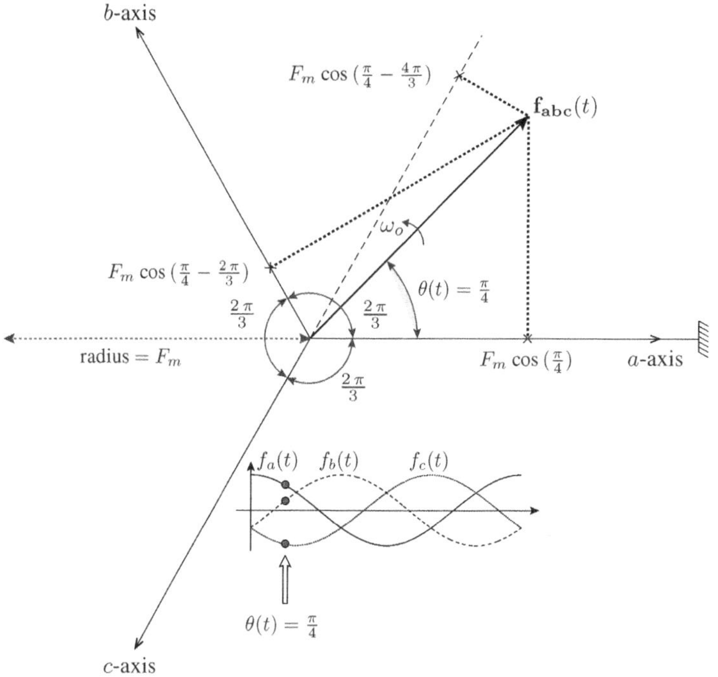

Figure 3.7 Representation of three-phase variables and corresponding space phasor.

(1) Representation of Space Phasor in the Stationary-abc Frame: Let us consider a balanced three-phase electrical system variables as

$$\begin{aligned} f_a(t) &= F_m \cos(\omega_o t + \theta_o) \\ f_b(t) &= F_m \cos(\omega_o t + \theta_o - \frac{2\pi}{3}) \\ f_c(t) &= F_m \cos(\omega_o t + \theta_o - \frac{4\pi}{3}) \end{aligned} \tag{3.9}$$

where F_m, ω_o, θ_o, and t represents the peak amplitude (voltage, current, or flux linkage), angular frequency (rad/sec), initial phase angle (rad), and time value (s) respectively. The three-phase variables can be represented by a space phasor $\mathbf{f_{abc}}(t)$ in a stationary-abc reference frame. The space phasor is given by

$$\mathbf{f_{abc}}(t) = \frac{2}{3}\left[e^{j\,0} f_a(t) + e^{j\,\frac{2\pi}{3}} f_b(t) + e^{j\,\frac{4\pi}{3}} f_c(t)\right] \tag{3.10}$$

which includes the information of amplitude, frequency, and phase angle of three-phase system variables. The relationship between the space phasor and its three-phase variables are illustrated in Figure 3.7. The phase-a axis is considered as a reference point. The space phasor $\mathbf{f_{abc}}(t)$ rotating in a counterclockwise direction at an arbitrary speed ω_o with respect to the reference point in the stationary-abc frame. The three-phase variables $f_a(t)$, $f_b(t)$, and $f_c(t)$ are obtained by projecting the space phasor to the corresponding phase-a, -b, and -c axes, which are $2\pi/3$ apart in space. Since, the abc-axes are stationary in space, each of the three-phase variables complete one fundamental cycle when the space phasor rotates one complete revolution. The equivalent phase angle of the space phasor is given by

$$\theta(t) = \theta_o + \int_0^t \omega_o(\tau)\, d\tau \tag{3.11}$$

Assuming that the length and the rotating speed of space phasor $\mathbf{f_{abc}}(t)$ is constant. The space phasor diagram and corresponding three-phase variables at $\theta(t) = \pi/4$ and initial phase angle $\theta_o = 0°$ is shown in Figure 3.7. The magnitude of $f_a(t)$ is greater than $f_b(t)$, and $f_c(t)$ is negative at this instant. The magnitude of three-phase variables varies with the position of space phasor. But, the tip of space phasor moves along the circumference of the circle which is centered in the complex plane. In other words, the tip of the space phasor is always equal to the radius F_m.

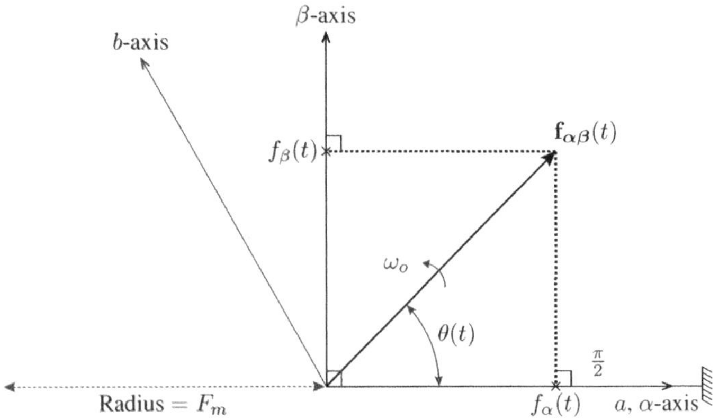

Figure 3.8 Representation of stationary-frame variables and corresponding space phasor.

(2) Transformation between abc/$\alpha\beta$ ***Frame:*** The space phasor in the stationary-abc frame is a complex function and time-varying in nature. This space phasor can be decomposed into real and imaginary components, which is referred to as a stationary-$\alpha\beta$ frame. In this frame, the three-phase time-varying variables are transformed into two-phase time-varying variables. The graphical representation of stationary-abc to stationary-$\alpha\beta$ frame transformation is shown in Figure. 3.8. The space phasor in $\alpha\beta$-frame is given by

$$\mathbf{f}_{\alpha\beta}(t) = f_\alpha(t) + j\, f_\beta(t) \tag{3.12}$$

where $f_\alpha(t)$ and $f_\beta(t)$ are α- and β-axis components of $\mathbf{f}_{\alpha\beta}(t)$. The space phasor rotates at an arbitrary speed of ω_o with respect to the reference point. The magnitude of $f_\alpha(t)$ and $f_\beta(t)$ varies with the position of $\mathbf{f}_{\alpha\beta}(t)$. The position of space phasor can be obtained from the real and imaginary components of $\mathbf{f}_{\alpha\beta}(t)$ as

$$\theta(t) = \tan^{-1}\left[\frac{f_\beta(t)}{f_\alpha(t)}\right] \tag{3.13}$$

The three-phase variables in stationary-abc frame are transformed into stationary-$\alpha\beta$ frame using the following transformation matrix

$$\begin{bmatrix} f_\alpha(t) \\ f_\beta(t) \end{bmatrix} = \underbrace{\frac{2}{3}\begin{bmatrix} 1 & -\frac{1}{2} & -\frac{1}{2} \\ 0 & \frac{\sqrt{3}}{2} & -\frac{\sqrt{3}}{2} \end{bmatrix}}_{\mathbf{T}_{abc\Rightarrow\alpha\beta}} \begin{bmatrix} f_a(t) \\ f_b(t) \\ f_c(t) \end{bmatrix} \tag{3.14}$$

A coefficient of $2/3$ is arbitrarily added to the transformation matrix. The main advantage of using $2/3$ is that the magnitude of two-phase variables is equal to the three-phase variables after the transformation. For a three-phase balanced system $f_a(t) + f_b(t) + f_c(t) = 0$, thus the $f_\alpha(t)$ in stationary-$\alpha\beta$ frame is equal to the $f_a(t)$ in stationary-abc frame. Similarly, the inverse transformation from stationary-$\alpha\beta$ frame to stationary-abc frame is given by

$$\begin{bmatrix} f_a(t) \\ f_b(t) \\ f_c(t) \end{bmatrix} = \underbrace{\begin{bmatrix} 1 & 0 \\ -\frac{1}{2} & \frac{\sqrt{3}}{2} \\ -\frac{1}{2} & -\frac{\sqrt{3}}{2} \end{bmatrix}}_{\mathbf{T}_{\alpha\beta\Rightarrow abc}} \begin{bmatrix} f_\alpha(t) \\ f_\beta(t) \end{bmatrix} \tag{3.15}$$

(3) Transformation between* abc/dq *Frame: Similar to the $abc/\alpha\beta$ transformation, the three-phase system variables in the stationary-abc frame can be transformed into two-phase system variables rotating in the synchronous reference frame. The synchronous reference frame is often referred to as a synchronous-dq frame. In this frame, the d (direct) and q (quadrature) axes are perpendicular to each other as shown in Figure 3.9. The d- and q-axes are not stationary, they rotate at an angular speed of ω_o with respect to the reference point (a-axis). Therefore, the two-phase system variables in the synchronous-dq frame are DC nature in contrast to the AC nature of variables in the stationary-abc and $\alpha\beta$ frames. The dq-axis has an arbitrary position with respect to the abc-frame by a reference frame angle θ and is equal to $\theta = \frac{d\omega_o}{dt}$. The space phasor in synchronous-dq frame is given by

$$\mathbf{f}_{\mathbf{dq}}(t) = f_d(t) + j\,f_q(t) \tag{3.16}$$

The space phasor $\mathbf{f}_{\mathbf{dq}}(t)$ has positioned at an angle ϕ with respect to the d-axis. It is also possible to align the space phasor $\mathbf{f}_{\mathbf{dq}}(t)$ with the d-axis to simplify the analysis. For example, in voltage-oriented control (VOC) of the grid-tied converter, the grid voltage vector is forced to align with the d-axis. The three-phase variables in the stationary-abc frame can be transformed to the synchronous-dq frame by using simple trigonometric functions as

$$\begin{bmatrix} f_d(t) \\ f_q(t) \end{bmatrix} = \underbrace{\frac{2}{3}\begin{bmatrix} \cos\theta & \cos(\theta - \frac{2\pi}{3}) & \cos(\theta - \frac{4\pi}{3}) \\ -\sin\theta & -\sin(\theta - \frac{2\pi}{3}) & -\sin(\theta - \frac{4\pi}{3}) \end{bmatrix}}_{\mathbf{T}_{abc \Rightarrow dq}} \begin{bmatrix} f_a(t) \\ f_b(t) \\ f_c(t) \end{bmatrix} \tag{3.17}$$

Similarly, the transformation from synchronous-dq frame to the stationary-abc frame is given by

$$\begin{bmatrix} f_a(t) \\ f_b(t) \\ f_c(t) \end{bmatrix} = \underbrace{\begin{bmatrix} \cos\theta & -\sin\theta \\ \cos(\theta - \frac{2\pi}{3}) & -\sin(\theta - \frac{2\pi}{3}) \\ \cos(\theta - \frac{4\pi}{3}) & -\sin(\theta - \frac{4\pi}{3}) \end{bmatrix}}_{\mathbf{T}_{dq \Rightarrow abc}} \begin{bmatrix} f_d(t) \\ f_q(t) \end{bmatrix} \tag{3.18}$$

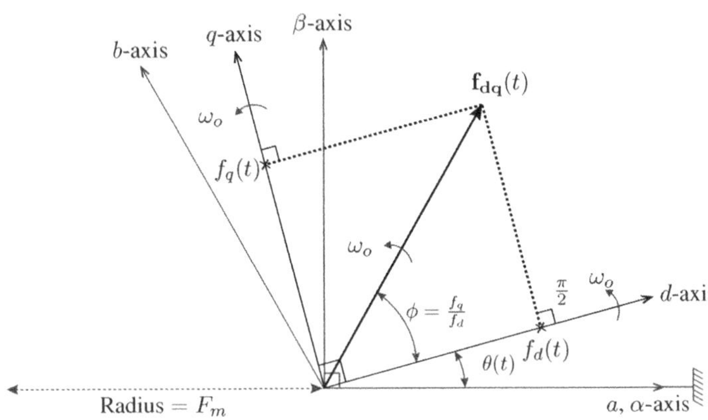

Figure 3.9 Representation of abc and dq frame voltages.

3.4.2 Control of MMC with Passive Load

A modular multilevel converter connected to a passive R-L load is shown in Figure 3.10, where the AC source and rectifier systems are replaced with an ideal DC source of voltage V_{dc}. The DC source (V_{dc}) is connected to each leg of an MMC. The mid-point of each leg is connected to the passive load, which consists of an inductor L_o, and a resistor r_o. The current flowing through the load is equally distributed among the upper and lower arms. The output current in terms of the upper and lower arm current is given by

$$\mathbf{i_{xi}} = \mathbf{i_{xu}} - \mathbf{i_{xl}} \tag{3.19}$$

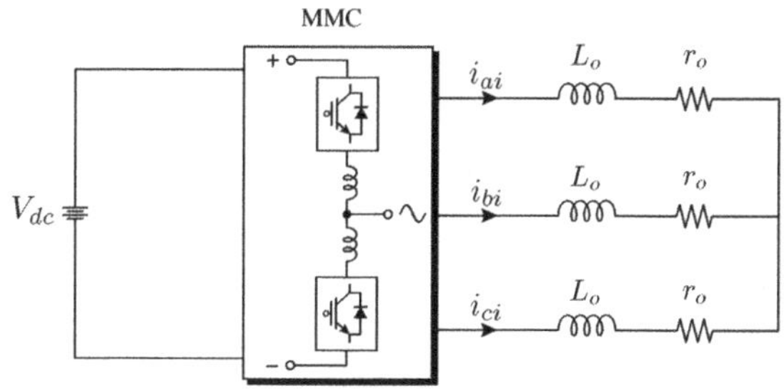

Figure 3.10 MMC with the passive load.

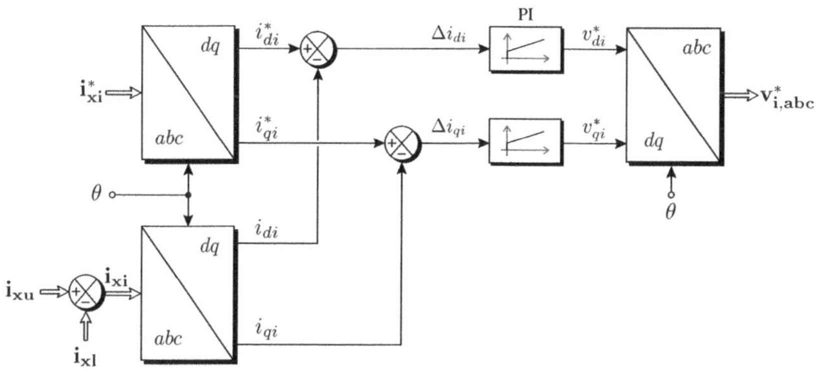

Figure 3.11 Output current control in dq-frame.

The output current ($\mathbf{i_{xi}}$) is controlled using a PI-regulators in the synchronous-dq reference frame. In dq-frame, the time-varying signals become DC variables, and can be controlled using a PI-regulator without any steady-state error. The control block diagram of output current control for MMC with passive load is shown in Figure 3.11. The operator or user defines the three-phase reference output currents ($\mathbf{i}^*_{\mathbf{xi}}$). These currents are transformed into the dq-frame rotating at speed equal to the rated frequency of the load (ω_o). The actual output currents ($\mathbf{i_{xi}}$) flowing through the load are obtained from the measured three-phase upper and lower arm currents using equation (3.19). These currents are transformed into the synchronous dq-frame using the transformation matrix given in equation (3.17). The magnitude of d- and q-axes current components are equal to

$$
\begin{aligned}
i_{di} &= +I\cos\varphi \\
i_{qi} &= -I\sin\varphi
\end{aligned}
\tag{3.20}
$$

where I represents the magnitude of output current (peak value), and φ represents the load power factor angle, and is given by

$$\varphi = \tan^{-1}\left[\frac{\omega_o L_o}{r_o}\right] \tag{3.21}$$

The reference and actual d- and q-axis current components are compared with each other. The current errors Δi_{di} and Δi_{qi} are given to the PI-regulators. The current regulator minimizes the current errors, and generates reference d- and q-axis voltage commands v_{di}^* and v_{qi}^*, respectively. The reference dq voltage commands are transformed to the abc-frame using the transformation matrix given in equation (3.18). The reference voltages are used along with the PWM scheme to control the modular multilevel converter.

CASE STUDY 3.3 Performance of dq-frame Output Current Control

Objective: The dynamic performance of output current controller in the synchronous-dq frame is studied through MATLAB simulation.

Parameters: The rated parameters of the MMC are as follows: rated apparent power $S_o = 2$ MVA, output line-to-line voltage $V_o = 6$ kV, rated output frequency $f_o = 60$ Hz, rated power factor (PF) = 0.85 (lag), rated DC-bus voltage $V_{dc} = 10$ kV, number of SMs per arm $N = 4$, arm inductance $L = 5$ mH, and SM capacitance $C = 2200$ μF.

Analysis:
In this case study, the modular multilevel converter with a passive load is considered. The base values of MMC system are

$$\begin{aligned}
\text{Base power } P_B &= \frac{2\times10^6}{3} = 0.667\,\text{MVA}\\
\text{Base voltage } V_B &= \frac{6\times10^3}{\sqrt{3}} = 3.464\,\text{kV}\\
\text{Base current } I_B &= \frac{0.667\times10^6}{3.464\times10^3} = 192.45\,\text{A (RMS), and } 272.165\,\text{A (peak)}\\
\text{Base current } Z_B &= \frac{3.464\times10^3}{192.456} = 18\,\Omega
\end{aligned} \tag{3.22}$$

The passive load is configured using a resistor and an inductor. The design specifications of passive load are given by

$$\begin{aligned}
r_o &= Z_B\cos\varphi = 18\times0.85 = 15.3\,\Omega\\
L_o &= \frac{Z_B\sin\varphi}{2\pi f_o} = \frac{18\times0.527}{2\times\pi\times60} = 25\,\text{mH}
\end{aligned} \tag{3.23}$$

The dynamic performance of output current control in dq-frame is presented with a step change in the magnitude of reference output current. The simulation results of three-phase output currents in dq and abc-frames are shown in Figure 3.12. Initially, the reference output current magnitude is set to $I^* = 0.4$ pu (108.866 A (peak)). The corresponding d and q-axis current components are

$$\begin{aligned} i_{di}^* &= +108.866 \times 0.85 &&= +92.536\,\mathrm{A}\,(+0.34\,\mathrm{pu}) \\ i_{qi}^* &= -108.866 \times 0.527 &&= -57.372\,\mathrm{A}\,(-0.21\,\mathrm{pu}) \end{aligned} \tag{3.24}$$

The actual d- and q-axis output current components i_{di} and i_{qi} are perfectly following the given reference currents. At $t = 0.05$ s, a step change in the reference current magnitude from $I^* = 0.4$ pu (108.866 A (peak)) to $I^* = 0.8$ pu (217.732 A (peak)) is applied. The reference d- and q-axis current components are also changed from $i_{di}^* = +0.34$ pu to $i_{di}^* = +0.64$ pu, and $i_{qi}^* = -0.21$ pu to $i_{qi}^* = -0.42$ pu. However, the PI-regulators in dq-frame are perfectly controlling the converter to generate the given reference currents at the output. The current components in dq-frame are transformed to abc-frame using $\mathbf{T}_{dq \Rightarrow abc}$ transformation matrix. The peak value of output current in abc-frame is perfectly matching with the reference current magnitude.

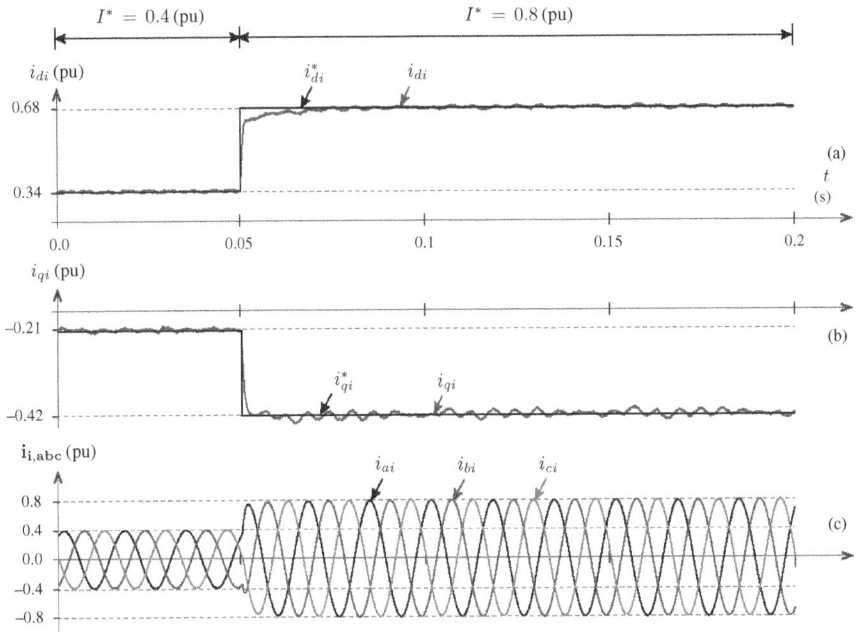

Figure 3.12 Dynamic performance of output current control.

■

3.5 CIRCULATING CURRENT CONTROL

Another major issue in the MMC is the circulating currents among the converter legs. These currents originate from the voltage difference between the upper and lower arms of the converter legs. The circulating current consists mainly of negative

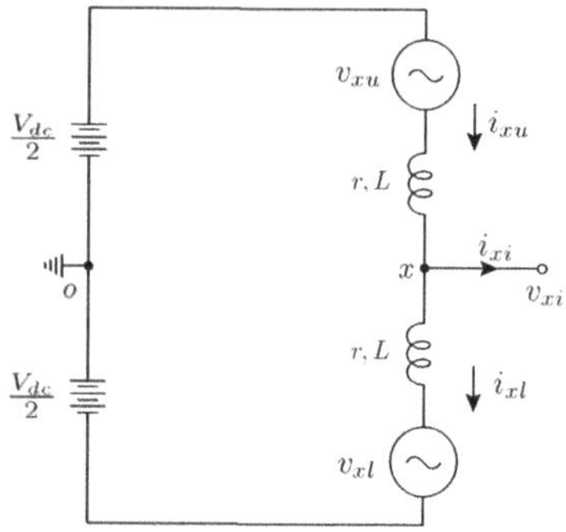

Figure 3.13 Per-phase equivalent circuit of MMC.

sequence components at twice the fundamental frequency [16, 17]. The circulating current does not have any impact on the AC output voltages and currents. However, the improper control of the circulating current increases the peak/RMS value of the arm current, which consequently increases the rating of devices, device power losses, and ripple in submodule capacitor voltages [18, 19]. By properly sizing the arm inductors, the magnitude of the circulating currents can be suppressed to an extent [20]. However, a closed-loop control technique is required to eliminate the circulating currents completely.

Typically, the synchronous reference frame based control approach is used for the complete elimination of circulating currents. In this approach, the circulating currents in the stationary-abc frame are transformed into a synchronous-dq frame rotating at twice the fundamental frequency. In the synchronous-dq frame, the circulating currents become DC signals. The DC signals can be easily controlled using a simple PI-regulators [21, 22]. Alternatively, resonant regulators are employed to control the circulating currents in the stationary-abc frame. In this approach, the resonant regulators are designed to eliminate the specific dominant harmonic frequency component such as second- and fourth-order harmonic components from the circulating currents [23].

3.5.1 Mathematical Model

The per-phase equivalent model of modular multilevel converter is shown in Figure 3.13. The upper and lower arm currents are given by

$$\begin{aligned} i_{xu} &= \frac{1}{3} i_{dc} + i_{xz} + \frac{1}{2} i_{xi} \\ i_{xl} &= \frac{1}{3} i_{dc} + i_{xz} - \frac{1}{2} i_{xi} \end{aligned} \tag{3.25}$$

The arm current consists of DC-bus current (i_{dc}), AC circulating currents (i_{xz}), and AC output current (i_{xi}). The upper and lower arm currents are added together, which gives a common-mode current component flowing through each leg as

$$i_{xzf} = \frac{1}{2}\left(i_{xu} + i_{xl}\right) = \frac{1}{3} i_{dc} + i_{xz} \tag{3.26}$$

The upper and lower arm voltages are given by

$$\begin{aligned} v_{xu} &= \frac{V_{dc}}{2} - v_{xi} - L\,\frac{di_{xu}}{dt} - r\, i_{xu} \\ v_{xl} &= \frac{V_{dc}}{2} + v_{xi} - L\,\frac{di_{xl}}{dt} - r\, i_{xl} \end{aligned} \tag{3.27}$$

From equations (3.26) and (3.27), the common-mode current component is given by

$$2L\,\frac{di_{xzf}}{dt} + 2r\, i_{xzf} = V_{dc} - \left(v_{xu} + v_{xl}\right) \tag{3.28}$$

From equation (3.26) and (3.28), the mathematical model of circulating current is given by

$$L\,\frac{di_{xz}}{dt} + r\, i_{xz} = \frac{V_{dc}}{2} - \frac{1}{2}(v_{xu} + v_{xl}) - r\,\frac{i_{dc}}{3} \tag{3.29}$$

The DC-bus current (i_{dc}) represents the average power supplied by the DC source, which is consumed by the load and submodule capacitors. This component is controlled by a leg voltage control presented in Section 3.3.1. To simplify the analysis, the term $r\,\frac{i_{dc}}{3}$ is neglected. The simplified circulating current model is given by

$$L\,\frac{di_{xz}}{dt} + r\, i_{xz} = \underbrace{\frac{V_{dc}}{2} - \frac{1}{2}(v_{xu} + v_{xl})}_{v_{xz}} \tag{3.30}$$

3.5.2 Control in Synchronous-dq Reference Frame

The circulating currents mainly consist of even-order harmonic components. Among them, the second- and fourth-order harmonic components are dominant. These harmonic components are time-varying in nature, and difficult to control using PI-regulators without any steady-state error. Therefore, the time-varying signals are transformed into DC signals using the $\mathbf{T}_{abc \Rightarrow dq}$ transformation matrix. From equation (3.30), the three-phase circulating current model in the stationary-abc frame is given by

$$\begin{bmatrix} v_{az} \\ v_{bz} \\ v_{cz} \end{bmatrix} = L\,\frac{d}{dt}\begin{bmatrix} i_{az} \\ i_{bz} \\ i_{cz} \end{bmatrix} + r\begin{bmatrix} i_{az} \\ i_{bz} \\ i_{cz} \end{bmatrix} \tag{3.31}$$

In this study, the synchronous-dq frame controller is designed to eliminate the second-order harmonic component from the circulating currents. Therefore, the circulating current model in a synchronous-dq frame rotating at $-2\omega_o$ frequency is given in [21, 22].

$$\begin{bmatrix} v_{dz} \\ v_{qz} \end{bmatrix} = L\,\frac{d}{dt}\begin{bmatrix} i_{dz} \\ i_{qz} \end{bmatrix} + \begin{bmatrix} 0 & -2\omega_o L \\ 2\omega_o L & 0 \end{bmatrix}\begin{bmatrix} i_{dz} \\ i_{qz} \end{bmatrix} + r\begin{bmatrix} i_{dz} \\ i_{qz} \end{bmatrix} \tag{3.32}$$

From the mathematical model given in equation (3.32), the circulating current control in the synchronous-dq frame is designed [21, 22]. The structure of circulating current control is shown in Figure 3.14. The feedback circulating currents are obtained from the measured arm currents. These currents are transformed to the synchronous-dq frame, and give the i_{dz} and i_{qz} current components. The reference current components i_{dz}^* and i_{qz}^* are set to zero. The comparison between the reference and actual currents gives the current errors Δi_{dz} and Δi_{qz}. The current errors are minimized by using PI-regulators. The d- and q-axis current control loops are decoupled by adding the induced speed voltages in the inductor to the current control loops. The PI-regulator generates reference dq voltage commands v_{dz}^* and v_{qz}^*. The reference dq voltages are transformed back to the stationary-abc frame using a $\mathbf{T}_{dq\Rightarrow abc}$ transformation matrix. The resultant reference voltages ($\mathbf{v}_{\mathbf{z,abc}}^*$) are used along with the submodule capacitors voltage and output current controllers to improve the efficiency and reliability of the modular multilevel converter.

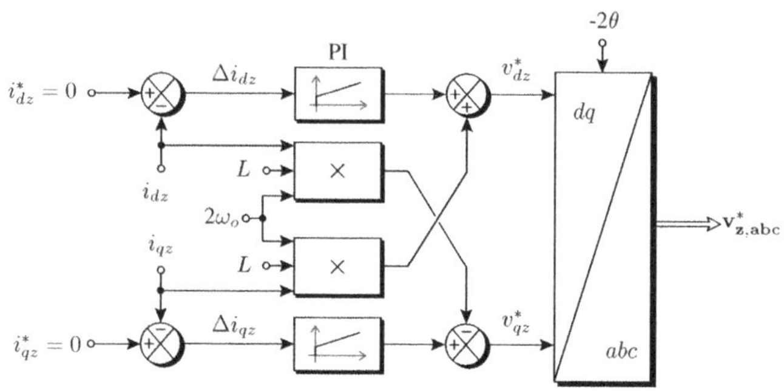

Figure 3.14 Circulating current control in dq-frame.

CASE STUDY 3.4 Performance of dq-frame Circulating Current Control

Objective: The suppression of circulating currents using a synchronous PI-regulator is studied through MATLAB simulation.

Parameters: The rated parameters of the MMC are as follows: rated apparent power $S_o = 2$ MVA, output line-to-line voltage $V_o = 6$ kV, rated output frequency $f_o = 60$ Hz, rated power factor (PF) = 0.85 (lag), rated DC-bus voltage $V_{dc} = 10$ kV, number of SMs per arm $N = 4$, arm inductance $L = 5$ mH, and SM capacitance $C = 2200$ μF.

Analysis:

The circulating currents are one of the major issues in modular multilevel converter. In this case study, the suppression of second-order harmonic component using a synchronous-dq frame based PI-regulator is presented. The feedback d- and q-axis current components are obtained from the measured arm currents in the stationary-abc frame. The signals in the stationary-abc frame are transformed to synchronous-dq frame rotating at $-2\omega_o$ frequency. The reference d- and q-axis current components are set to zero.

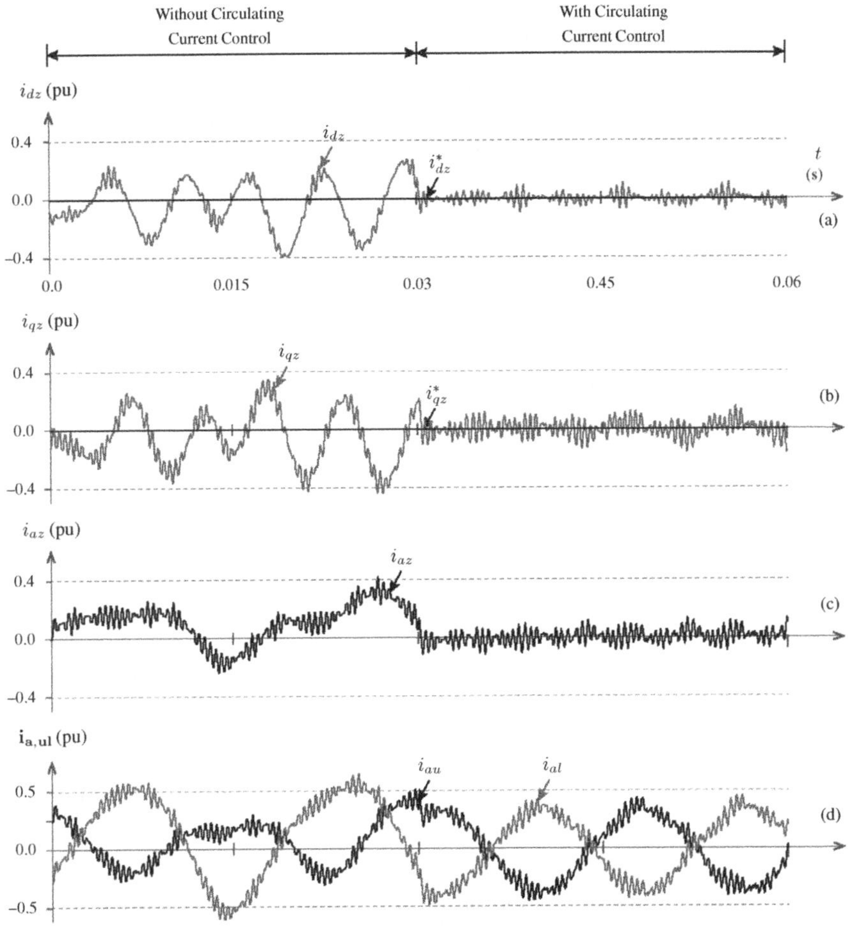

Figure 3.15 Performance of circulating current controller in dq-frame.

The performance of circulating current controller in the synchronous-dq frame is shown in Figure 3.15. Initially, the controller is disabled, and the d- and q-axis current components have a peak value of 0.4 pu. The circulating current (i_{az}) mainly consists of the second-order harmonic component, which increases the peak and RMS value of the arm current flowing through the converter legs. At $t = 0.03$ s, the controller is activated, which significantly reduces the magnitude of d- and q-axis current components (close to zero value). The second-order harmonic component in circulating current is also significantly minimized. Thereby, the converter power losses are reduced and the converter efficiency is improved.

■

3.5.3 Control in Stationary-abc Reference Frame

As mentioned in earlier, circulating current consists of even-order harmonics of frequency $h\omega_o$, where $h \in \{2, 4, 6, 8, \ldots\}$. These harmonic components can be suppressed using multiple resonant current controllers in a stationary-abc frame [23]. Each resonant current controller is specifically designed to eliminate a specific harmonic component. The transfer function of resonant controller is given by

$$\text{Res}h(s) = \frac{k_h\, s}{s^2 + (h\omega_o)^2} \tag{3.33}$$

where k_h is the coefficient of resonant controller and $h\omega_o$ is the resonant frequency corresponding to the harmonic component to be eliminated in circulating current.

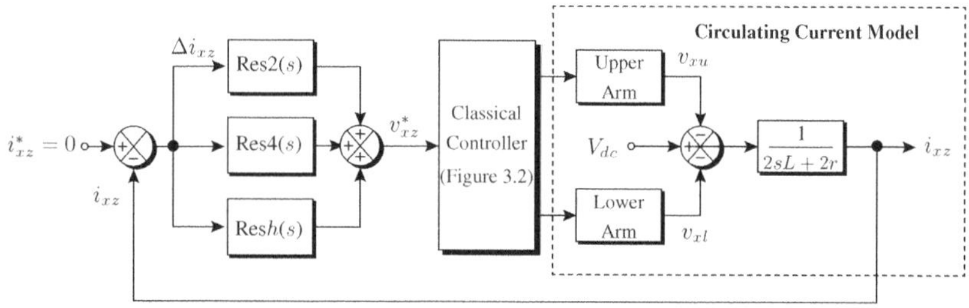

Figure 3.16 Circulating current control in abc-frame.

The magnitude of even-order harmonic component decreases as the order of the harmonic frequency increases. Therefore, the suppression of higher-order even harmonics is not required. Typically, the second- and fourth-order harmonics are the dominant components. Therefore, the resonant controllers are designed to eliminate the second- and fourth-order harmonic components only. The control block diagram of circulating current in the stationary-abc frame is shown in Figure 3.16. The feedback circulating current (i_{xz}) is obtained from the measured arm currents, and the reference circulating current (i^*_{xz}) is set to zero value. The comparison of reference and actual circulating current gives the current error Δi_{xz}, which is given to the parallel resonant controllers Res2(s), Res4(s), ..., Res$h$(s). Each resonant controller eliminates a specific current component from the circulating current corresponding to their resonant frequency and generates a control signal. The output of parallel resonant controllers is added together, which forms a reference voltage command v^*_{xz}. The reference voltage command is used with the pulse width modulator to generate the gating signals for a modular multilevel converter.

3.6 SUMMARY

In this chapter, detailed analysis, implementation, and performance of classical control methods for modular multilevel converter are presented. Classical control methods use closed-loop controllers and algorithms based on logical functions to control the submodule capacitor voltages, AC output current, and circulating currents. The submodule capacitors voltage control is one of the major challenges. In this chapter, a simple control approach which includes the leg voltage control, and balancing of the submodule capacitor voltages within the arm is presented. This approach can be applied with any modulation scheme (PSC-PWM, LSC-PWM, and SVM).

Another important objective is the AC output current control whose control structure depends on the type of the application. For example, the VOC approach for grid-connected application, and FOC for the motor drive system. In this chapter, a simple dq-frame controller for an MMC feeding a passive load is presented. The current control method utilizes a reference frame theory to convert the time-varying signals into DC signals. The DC signals can be easily controlled using PI-regulators without any steady-state error. The circulating current among the converter legs is another major issue in the modular multilevel converter. It mainly consists of even-order harmonic components, where the second- and fourth-order harmonic components are most dominant. To suppress the even-order harmonic components, a simple control approach in both dq- and abc-frames is presented. The suppression of circulating currents improves the efficiency and ensures reliable performance of the modular multilevel converter.

REFERENCES

1. M. Perez, S. Bernet, J. Rodriguez, S. Kouro, and R. Lizana, "Circuit topologies, modeling, control schemes, and applications of modular multilevel converters," *IEEE Trans. Power Electron.*, vol. 30, no. 1, pp. 4–17, Jan 2015.

2. S. Debnath, J. Qin, B. Bahrani, M. Saeedifard, and P. Barbosa, "Operation, control, and applications of the modular multilevel converter: A review," *IEEE Trans. Power Electron.*, vol. 30, no. 1, pp. 37–53, Jan 2015.

3. R. Lizana, M. Perez, S. Bernet, J. Espinoza, and J. Rodriguez, "Control of arm capacitor voltages in modular multilevel converters," *IEEE Trans. Power Electron.*, vol. 31, no. 2, pp. 1774–1784, Feb 2016.

4. M. Hagiwara and H. Akagi, "Control and experiment of pulsewidth-modulated modular multilevel converters," *IEEE Trans. Power Electron.*, vol. 24, no. 7, pp. 1737–1746, Jul 2009.

5. R. Darus, J. Pou, G. Konstantinou, S. Ceballos, R. Picas, and V. Agelidis, "A modified voltage balancing algorithm for the modular multilevel converter: Evaluation for staircase and phase-disposition pwm," *IEEE Trans. Power Electron.*, vol. 30, no. 8, pp. 4119–4127, Aug 2015.

6. S. Du, J. Liu, and T. Liu, "Modulation and closed-loop-based dc capacitor voltage control for mmc with fundamental switching frequency," *IEEE Trans. Power Electron.*, vol. 30, no. 1, pp. 327–338, Jan 2015.

7. R. Lizana, M. Perez, D. Arancibia, J. Espinoza, and J. Rodriguez, "Decoupled current model and control of modular multilevel converters," *IEEE Trans. Ind. Electron.*, vol. 62, no. 9, pp. 5382–5392, Sep 2015.

8. E. Solas, G. Abad, J. Barrena, S. Aurtenetxea, A. Carcar, and L. Zajac, "Modular multilevel converter with different submodule concepts – part i: Capacitor voltage balancing method," *IEEE Trans. Ind. Electron.*, vol. 60, no. 10, pp. 4525–4535, Oct 2013.

9. A. Dekka, B. Wu, N. Zargari, and R. Fuentes, "Dynamic voltage balancing algorithm for modular multilevel converter: A unique solution," *IEEE Trans. Power Electron.*, vol. 31, no. 2, pp. 952–963, Feb 2016.

10. M. Saeedifard and R. Iravani, "Dynamic performance of a modular multilevel back-to-back hvdc system," *IEEE Trans. Power Del.*, vol. 25, no. 4, pp. 2903–2912, Oct 2010.

11. M. Hagiwara, K. Nishimura, and H. Akagi, "A medium-voltage motor drive with a modular multilevel pwm inverter," *IEEE Trans. Power Electron.*, vol. 25, no. 7, pp. 1786–1799, Jul 2010.

12. H. Mohammadi and M. Bina, "A transformerless medium-voltage statcom topology based on extended modular multilevel converters," *IEEE Trans. Power Electron.*, vol. 26, no. 5, pp. 1534–1545, May 2011.

13. M. Hagiwara, R. Maeda, and H. Akagi, "Negative-sequence reactive-power control by a pwm statcom based on a modular multilevel cascade converter (mmcc-sdbc)," *IEEE Trans. Ind. Appl.*, vol. 48, no. 2, pp. 720–729, Mar 2012.

14. B. Wu, Y. Lang, N. Zargari, and S. Kouro, *Power Conversion and Control of Wind Energy Systems*, 1st ed., ser. IEEE Press Series on Power Engineering. Hoboken, NJ: John Wiley & Sons, Inc., Jul 2011.

15. A. Yazdani and R. Iravani, *Voltage-Sourced Converters in Power Systems*. Hoboken, NJ: John Wiley & Sons, Inc., Jan 2010.

16. L. He, K. Zhang, J. Xiong, and S. Fan, "A repetitive control scheme for harmonic suppression of circulating current in modular multilevel converters," *IEEE Trans. Power Electron.*, vol. 30, no. 1, pp. 471–481, Jan 2015.

17. R. Darus, J. Pou, G. Konstantinou, S. Ceballos, and V. G. Agelidis, "Controllers for eliminating the ac components in the circulating current of modular multilevel converters," *IET Power Electron.*, vol. 9, no. 1, pp. 1–8, May 2016.

18. M. Zhang, L. Huang, W. Yao, and Z. Lu, "Circulating harmonic current elimination of a cps-pwm-based modular multilevel converter with a plug-in repetitive controller," *IEEE Trans. Power Electron.*, vol. 29, no. 4, pp. 2083–2097, Apr 2014.

19. J.-W. Moon, C.-S. Kim, J.-W. Park, D.-W. Kang, and J.-M. Kim, "Circulating current control in mmc under the unbalanced voltage," *IEEE Trans. Power Del.*, vol. 28, no. 3, pp. 1952–1959, Jul 2013.

20. Y. Li and F. Wang, "Arm inductance selection principle for modular multilevel converters with circulating current suppressing control," in *Applied Power Electronics Conference and Exposition (APEC), 2013 Twenty-Eighth Annual IEEE*, Mar 2013, pp. 1321–1325.

21. Q. Tu, Z. Xu, and L. Xu, "Reduced switching-frequency modulation and circulating current suppression for modular multilevel converters," *IEEE Trans. Power Del.*, vol. 26, no. 3, pp. 2009–2017, Jul 2011.

22. B. Bahrani, S. Debnath, and M. Saeedifard, "Circulating current suppression of the modular multilevel converter in a double-frequency rotating reference frame," *IEEE Trans. Power Electron.*, vol. 31, no. 1, pp. 783–792, Jan 2016.

23. Z. Li, P. Wang, Z. Chu, H. Zhu, Y. Luo, and Y. Li, "An inner current suppressing method for modular multilevel converters," *IEEE Power Electron. Lett.*, vol. 28, no. 11, pp. 4873–4879, Nov 2013.

CHAPTER 4

MODEL PREDICTIVE CONTROL OF MODULAR MULTILEVEL CONVERTER

4.1 INTRODUCTION

Chapter 3 provides an overview of classical control methods for modular multilevel converters (MMCs). These methods use pulse width modulation (PWM) schemes, closed-loop PI-regulators, and simple algorithms based on logical functions to control the submodule capacitor voltages, output currents, and circulating currents. The classical control methods are easy to implement in digital controllers. However, the dynamic response is greatly limited by the gains of PI-regulator, bandwidth of current controller, switching frequency of the converter, and the type of PWM scheme. Alternatively, the predictive control often referred to as a model predictive control approach is becoming more popular and is widely employed to control the power converters [1–5]. This approach is simple, intuitive, and flexible. The predictive control includes multiple constraints and non-linearities, offers a fast dynamic response, and eliminates the use of PI-regulators and PWM schemes [6–8]. In this chapter, the model predictive control is further extended to control the modular multilevel converter [9].

The block diagram of model predictive control (MPC) approach is shown in Figure 4.1. This approach requires a discrete-time model of the MMC system, which

Modular Multilevel Converters: Analysis, Control, and Applications, First Edition.
Sixing Du, Apparao Dekka, Bin Wu, and Navid Zargari.

is presented in Section 4.2. The discrete-time model is used to predict the future behavior of the control variables for each valid switching state of the converter. The reference control variables ($\mathbf{x}^*(k)$) are extrapolated to $(k+1)^{\text{th}}$- sampling instant. The most widely used extrapolation techniques such as vector angle and Lagrange extrapolation techniques are given in Section 4.3. The predicted ($\mathbf{x}(k+1)$) and the extrapolated ($\widehat{\mathbf{x}}^*(k+1)$) variables are included in a cost function. The cost function consists of multiple control objectives, which are optimized for weighting factors to reach the desired system behavior. The resulting switching state is then applied to the converter. The design of cost function and weighting factors is given in Section 4.4.

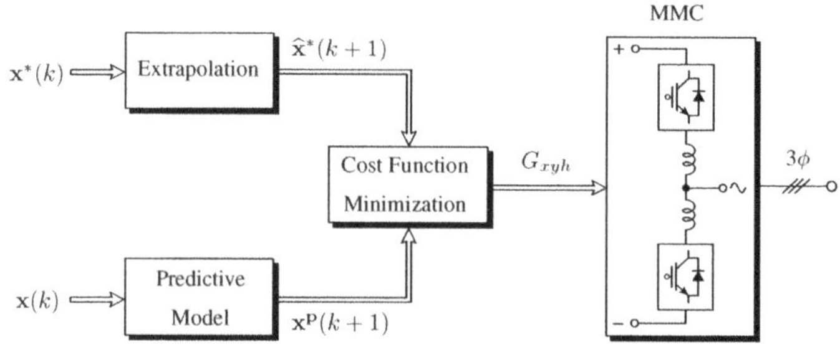

Figure 4.1 Block diagram of model predictive control.

The model predictive control approach can be categorized into direct model predictive control (DMPC) and indirect model predictive control (IMPC) approaches as shown in Figure 4.2. In the DMPC, the control goals are achieved using a single cost function. In the case of the IMPC, the voltage balance strategy is used along with the cost function to achieve the control objectives. The design procedure and control algorithm of direct and indirect model predictive control approaches are presented in Sections 4.5 and 4.6, respectively. Finally, the summary of this chapter is given in Section 4.7.

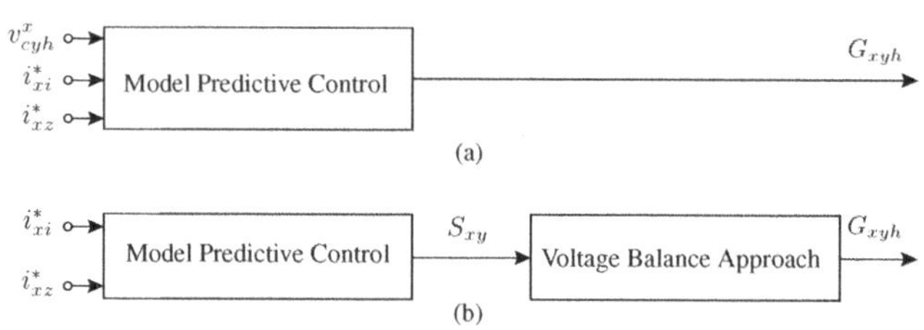

Figure 4.2 MPC schemes for MMC: (a) direct MPC, and (b) indirect MPC.

4.2 MATHEMATICAL MODEL OF MMC

The digital implementation of model predictive control (MPC) approach requires a discrete-time model of the MMC system. The discrete-time (DT) model is obtained from the discretization of continuous-time (CT) model. In this section, the continuous-time and discrete-time model of MMC connected to a simple passive load are presented [10, 11].

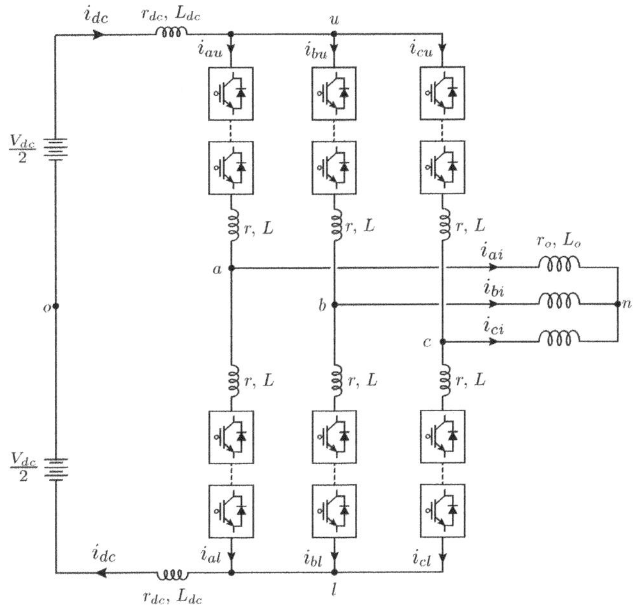

Figure 4.3 Three-phase MMC system.

4.2.1 Continuous-Time Model

The structure of a three-phase MMC connected to a simple passive load is shown in Figure 4.3. Each leg of MMC is composed of two arms, namely upper and lower arms ($y \in \{u, l\}$). Each arm consists of several half-bridge submodules in series with an inductor (L). The power losses in each arm are represented with an equivalent resistance (r). The front-end rectifier of MMC is modeled as a constant DC source (V_{dc}). The rectifier output is connected to the positive and negative bars of each leg of an MMC system through a DC-line. The DC-line is modeled as an inductor (L_{dc}) and a series resistance (r_{dc}). The passive load consists of an inductor (L_o) and a resistor (r_o) connected to the mid-point of each converter leg ($x \in \{a, b, c\}$) and the load neutral point (n). The submodules in each arm are modeled as a voltage source with a value equal to the sum of N submodule capacitors voltage as shown in Figure 4.4.

The arm current (i_{xy}) consists of several current components, which includes the DC-bus current (i_{dc}), the output current (i_{xi}), and circulating current (i_{xz}). From the equivalent model shown in Figure 4.4, the upper and lower arm currents of phase-a are given by

$$\begin{aligned} i_{au} &= \frac{1}{3} i_{dc} + i_{az} + \frac{1}{2} i_{ai} \\ i_{al} &= \frac{1}{3} i_{dc} + i_{az} - \frac{1}{2} i_{ai} \end{aligned} \tag{4.1}$$

Applying Kirchhoff's law to the model shown in Figure 4.4, the upper and lower arm voltages of phase-a are given by

$$\begin{aligned} v_{au} &= \frac{V_{dc}}{2} - L_{dc}\frac{d\,i_{dc}}{dt} - r_{dc}\,i_{dc} - L\frac{d\,i_{au}}{dt} - r\,i_{au} - L_o\frac{d\,i_{ai}}{dt} - r_o\,i_{ai} \\ v_{al} &= \frac{V_{dc}}{2} - L_{dc}\frac{d\,i_{dc}}{dt} - r_{dc}\,i_{dc} - L\frac{d\,i_{al}}{dt} - r\,i_{al} + L_o\frac{d\,i_{ai}}{dt} + r_o\,i_{ai} \end{aligned} \tag{4.2}$$

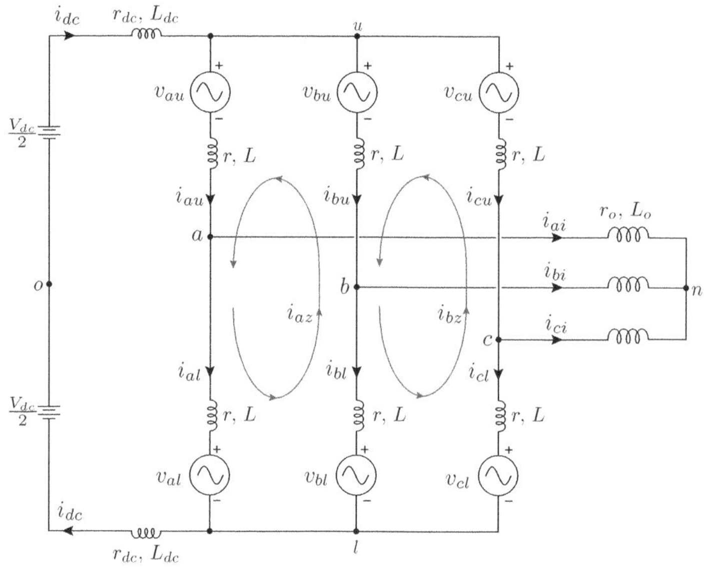

Figure 4.4 Equivalent model of MMC.

(1) Continuous-Time Model of DC-bus Current: The DC-bus current flows among the DC system and the converter legs. The magnitude of DC-bus current varies proportionally with the active power demanded by the load. The DC-bus current is given in terms of three-phase arm currents as

$$i_{dc} = \sum_{x=a,b,c} i_{xu} = \sum_{x=a,b,c} i_{xl} = \frac{1}{2} \sum_{x=a,b,c} (i_{xu} + i_{xl}) \tag{4.3}$$

From equation (4.2), the upper and lower arm voltages of phase-a are added, which results in

$$\begin{aligned} V_{dc} = 2L_{dc}\frac{d\,i_{dc}}{dt} + 2r_{dc}\,i_{dc} + v_{au} + L\frac{d\,(i_{au}+i_{al})}{dt} \\ + r\,(i_{au}+i_{al}) + v_{al} \end{aligned} \tag{4.4}$$

Similarly, the dynamic model of phase-b and -c is given by

$$\begin{aligned} V_{dc} &= 2L_{dc}\frac{d\,i_{dc}}{dt} + 2r_{dc}\,i_{dc} + v_{bu} + L\frac{d\,(i_{bu}+i_{bl})}{dt} + r\,(i_{bu}+i_{bl}) + v_{bl} \\ V_{dc} &= 2L_{dc}\frac{d\,i_{dc}}{dt} + 2r_{dc}\,i_{dc} + v_{cu} + L\frac{d\,(i_{cu}+i_{cl})}{dt} + r\,(i_{cu}+i_{cl}) + v_{cl} \end{aligned} \tag{4.5}$$

The dynamic model of DC-bus current is formulated from (4.3), (4.4), and (4.5) as

$$\frac{d\,i_{dc}}{dt} = \frac{1}{6L_{dc}+2L}\left[3V_{dc} - \sum_{x=a,\,b,\,c} (v_{xu}+v_{xl}) - (6r_{dc}+2r)\,i_{dc}\right] \tag{4.6}$$

(2) Continuous-Time Model of Output Current: The output current flows through the converter legs to the load. The difference between the upper and lower arm current gives the magnitude of the current flowing through the load. From equation (4.1), the phase-a output current is given by

$$i_{ai} = i_{au} - i_{al} \tag{4.7}$$

From equation (4.2), the upper and lower arm voltages are subtracted from each other, which gives the dynamic model of the phase-a current as

$$v_{au} - v_{al} = L\frac{d\,(i_{au}-i_{al})}{dt} + r\,(i_{au}-i_{al}) + 2L_o\frac{d\,i_{ai}}{dt} + 2r_o\,i_{ai} \tag{4.8}$$

Substitute equation (4.7) in (4.8) and the resultant dynamic model of phase-a output current is given by

$$\frac{d\,i_{ai}}{dt} = \frac{1}{L+2L_o}\left[v_{al} - v_{au} - (r+2r_o)\,i_{ai}\right] \tag{4.9}$$

From equation (4.9), the dynamic model of three-phase output currents is given by

$$\frac{d\,\mathbf{i_{xi}}}{dt} = \frac{1}{L+2L_o}\left[\mathbf{v_{xl}} - \mathbf{v_{xu}} - (r+2r_o)\,\mathbf{i_{xi}}\right] \tag{4.10}$$

where,

$$\mathbf{i_{xi}} = \begin{bmatrix} i_{ai} \\ i_{bi} \\ i_{ci} \end{bmatrix}, \ \mathbf{v_{xl}} = \begin{bmatrix} v_{al} \\ v_{bl} \\ v_{cl} \end{bmatrix}, \ \mathbf{v_{xu}} = \begin{bmatrix} v_{au} \\ v_{bu} \\ v_{cu} \end{bmatrix} \tag{4.11}$$

(3) Continuous-Time Model of Circulating Current: The circulating current flows among the converter legs due to the difference between the upper and lower arm voltages. The circulating current mainly consists of even-order harmonic components corresponding to the multiples of twice the fundamental frequency. These circulating currents are controlled to achieve the arm voltage balance, reduction of capacitors voltage ripple, and converter power losses. From equation (4.1), the upper and lower arm currents are added together, which results in

$$i_{au} + i_{al} = \frac{2}{3}\, i_{dc} + 2\, i_{az} \tag{4.12}$$

Equation (4.12) is substituted in (4.4), and results in

$$\begin{aligned} V_{dc} = {} & (2\, L_{dc} + \frac{2}{3}\, L)\frac{d\, i_{dc}}{dt} + (2\, r_{dc} + \frac{2}{3}\, r)\, i_{dc} + v_{au} + 2\, L\, \frac{d\, i_{az}}{dt} \\ & + 2\, r\, i_{az} + v_{al} \end{aligned} \tag{4.13}$$

From equations (4.6) and (4.13), the dynamic model of phase-a circulating current is given by

$$\frac{d\, i_{az}}{dt} = \frac{1}{6L}\left[\sum_{x=a,\, b,\, c} (v_{xu} + v_{xl}) - 3\,(v_{au} + v_{al}) - 6\, r\, i_{az}\right] \tag{4.14}$$

From equation (4.14), the dynamic model of three-phase circulating currents is given by

$$\frac{d\, \mathbf{i_{xz}}}{dt} = \frac{1}{6L}\left[\sum_{x=a,\, b,\, c} (v_{xu} + v_{xl}) - 3\,(\mathbf{v_{xu}} + \mathbf{v_{xl}}) - 6\, r\, \mathbf{i_{xz}}\right] \tag{4.15}$$

where,

$$\mathbf{i_{xz}} = \begin{bmatrix} i_{az} \\ i_{bz} \\ i_{cz} \end{bmatrix} \tag{4.16}$$

4.2.2 Discretization Methods

The model predictive control (MPC) algorithms are implemented in digital control platforms such as DSP, FPGA, and a computer with real-time interface cards. These control platforms deal with the sample-based continuous-time systems, which is also referred to as discrete-time systems. Therefore, in order to implement the MPC, the continuous-time model of MMC must be converted into the discrete-time model. Various discretization methods can be used to convert the continuous-time system into a discrete-time system. These methods can be categorized into three groups: 1) exact methods, 2) quasi-exact methods, and 3) approximate methods. These methods are selected according to the application and the nature of control variables [12, 13].

The approximate discretization methods are widely employed in the real-time implementation of MPC algorithm. These methods are mainly based on the numerical integration or numerical differentiation (finite difference) techniques. The dynamic model of MMC given in Section 4.2.1 consists of ordinary differential equations, which can be solved using different approximation methods, including forward Euler, backward Euler, and trapezoidal methods. Among them, the Euler approximation methods are widely used in the solving of the continuous-time equations, and are described in the following subsections.

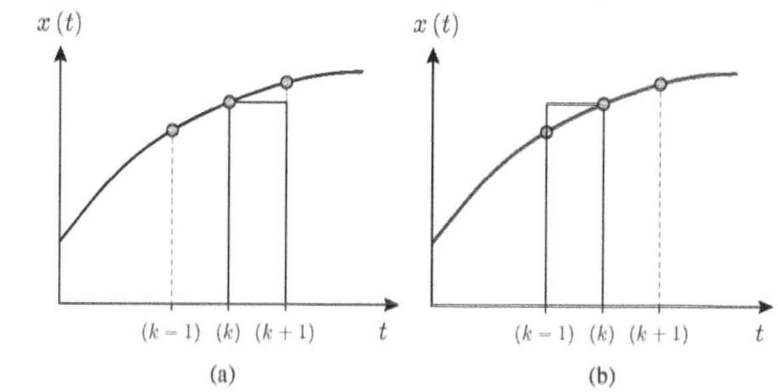

Figure 4.5 Approximate discretization methods: (a) forward Euler, and (b) backward Euler.

(1) Forward Euler Approximation: In this approach, the derivative is approximated using the present sample at kT_s instant and forward sample at kT_s+T_s instant as shown in Figure 4.5(a). Hence, it is referred to as a forward Euler approximation [14]. As a part of analysis, let us consider the first-order differential equation

$$\frac{d\,x(t)}{dt} = u(t) \tag{4.17}$$

The approximation of derivative with forward Euler approach is given by

$$\left\{\frac{d\,x(t)}{dt}\right\}_{t=kT_s} \approx \frac{x(kT_s+T_s)-x(kT_s)}{T_s} \tag{4.18}$$

which is simplified as

$$x(kT_s+T_s) \approx x(kT_s) + T_s\left\{\frac{d\,x(t)}{dt}\right\}_{t=kT_s} \tag{4.19}$$

The value of derivative given in equation (4.17) is substituted in equation (4.19) and results in

$$x(kT_s+T_s) \approx x(kT_s) + T_s\,u(kT_s) \tag{4.20}$$

From equation (4.20), it is observed that the future state of control variable $x(k+1)$ is defined in terms of the present state of control variable $x(k)$ and present input variable $u(k)$. Therefore, the forward Euler approach is referred to as an explicit method. By applying the Z-transform on both sides of equation (4.20), we can obtain the discrete-time model of

$$z\,X(z) = X(z) + T_s\,U(z) \Rightarrow \frac{X(z)}{U(z)} = T_s\,\frac{1}{z-1} \tag{4.21}$$

(2) Backward Euler Approximation: As shown in Figure 4.5(b), with backward Euler approach, the derivative is approximated using the present sample at kT_s instant and past sample at $kT_s - T_s$ instant. The approximation of derivative with backward Euler approach is given by [3, 15]

$$\left\{\frac{d\,x(t)}{dt}\right\}_{t=kT_s} \approx \frac{x(kT_s) - x(kT_s - T_s)}{T_s} \tag{4.22}$$

The samples in equation (4.22) are forward-shifted by one sample to predict the future behavior of the control variable using an MPC. The resultant discrete-time solution is given by

$$x(kT_s + T_s) \approx x(kT_s) + T_s\left\{\frac{d\,x(t)}{dt}\right\}_{t=kT_s+T_s} \tag{4.23}$$

The value of derivative given in equation (4.17) is substituted in equation (4.23) and results in

$$x(kT_s + T_s) \approx x(kT_s) + T_s\,u(kT_s + T_s) \tag{4.24}$$

From equation (4.24), it is observed that the future state of control variable $x(k+1)$ is defined in terms of the present state of control variable $x(k)$ and future state of input variable $u(k+1)$. Therefore, it is an implicit method. The backward Euler approach provides a stable solution even for a large sampling time. From equations (4.21) and (4.24), it is observed that the backward Euler approach requires an extrapolated input $u(k+1)$ compared to the forward Euler approach. By applying the Z-transform on both sides of equation (4.24), we can obtain the discrete-time model of

$$z\,X(z) = X(z) + z\,T_s\,U(z) \Rightarrow \frac{X(z)}{U(z)} = T_s\frac{z}{z-1} \tag{4.25}$$

4.2.3 Discrete-Time Model

The continuous-time model of MMC is presented in the Section 4.2.1. In this section, the conversion from continuous time to discrete time using the forward Euler approximation is presented [9].

(1) Discrete-Time Model of DC-bus Current: Consider the continuous-time model of DC-bus current given in (4.6) as

$$\frac{d\,i_{dc}}{dt} = \frac{1}{6L_{dc} + 2L}\left[3V_{dc} - \sum_{x=a,\,b,\,c}(v_{xu} + v_{xl}) - (6r_{dc} + 2r)\,i_{dc}\right] \tag{4.26}$$

According to the forward Euler approach, the derivative of DC-bus current is approximated as

$$\begin{aligned}\frac{i_{dc}^{p}(k+1)-i_{dc}^{m}(k)}{T_s} &= \frac{3V_{dc}^{m}(k)}{6L_{dc}+2L}-\frac{\sum_{x=a,\,b,\,c}(v_{xu}^{p}(k)+v_{xl}^{p}(k))}{6L_{dc}+2L}\\ &\quad -\frac{6r_{dc}+2r}{6L_{dc}+2L}\,i_{dc}^{m}(k)\end{aligned} \tag{4.27}$$

which is simplified, and its equivalent discrete-time model is given by

$$i_{dc}^{p}(k+1)=\Phi_{dc}\,i_{dc}^{m}(k)+\Gamma_{dc}\left[3\,V_{dc}^{m}(k)-\sum_{x=a,\,b,\,c}(v_{xu}^{p}(k)+v_{xl}^{p}(k))\right] \tag{4.28}$$

where T_s is the sampling time, and

$$\Phi_{dc}=1-\frac{(6r_{dc}+2r)T_s}{6L_{dc}+2L},\quad \Gamma_{dc}=\frac{T_s}{6L_{dc}+2L} \tag{4.29}$$

(2) Discrete-Time Model of Output Current: Similarly, the discrete-time model of a three-phase output current model is obtained from the continuous-time model. The continuous-time model of three-phase output current is given by

$$\frac{d\,\mathbf{i_{xi}}}{dt}=\frac{1}{L+2L_o}\left[\mathbf{v_{xl}}-\mathbf{v_{xu}}-(r+2r_o)\,\mathbf{i_{xi}}\right] \tag{4.30}$$

and its equivalent discrete-time model is given by

$$\mathbf{i_{xi}^{p}}(k+1)=\Phi_o\,\mathbf{i_{xi}^{m}}(k)+\Gamma_o\left[\mathbf{v_{xl}^{p}}(k)-\mathbf{v_{xu}^{p}}(k)\right] \tag{4.31}$$

where,

$$\Phi_o=1-\frac{(r+2r_o)T_s}{L+2L_o},\quad \Gamma_o=\frac{T_s}{L+2L_o} \tag{4.32}$$

(3) Discrete-Time Model of Circulating Current: The continuous-time model of circulating current is given by

$$\frac{d\,\mathbf{i_{xz}}}{dt}=\frac{1}{6L}\left[\sum_{x=a,\,b,\,c}(v_{xu}+v_{xl})-3\,(\mathbf{v_{xu}}+\mathbf{v_{xl}})-6\,r\,\mathbf{i_{xz}}\right] \tag{4.33}$$

and its discrete-time equivalent model is obtained using a forward Euler approach as

$$\begin{aligned}\mathbf{i_{xz}^{p}}(k+1) &= \Phi_z\,\mathbf{i_{xz}^{m}}(k)+\Gamma_z\sum_{x=a,\,b,\,c}(v_{xu}^{p}(k)+v_{xl}^{p}(k))-\\ &\quad 3\,\Gamma_z\,(\mathbf{v_{xu}^{p}}(k)+\mathbf{v_{xl}^{p}}(k))\end{aligned} \tag{4.34}$$

where,

$$\Phi_z=1-\frac{6r\,T_s}{6L},\quad \Gamma_z=\frac{T_s}{6L} \tag{4.35}$$

(4) Discrete-Time Model of Submodule Capacitor Voltage: The half-bridge submodule has a floating capacitor (C) and two semiconductor devices (S_1 and $\overline{S}_1$) as shown in Figure 4.6. The floating capacitor voltage is controlled by proper turning ON/OFF of the semiconductor devices. The devices S_1 and $\overline{S}_1$ are operated in a complementary manner. Therefore, one independent gating signal is sufficient to control the half-bridge submodule. The continuous-time model of half-bridge submodule capacitor voltage is expressed as

$$\frac{d\,v^{x}_{Cyh}}{dt} = \frac{i^{x}_{Cyh}}{C} \tag{4.36}$$

The discrete-time model of submodule capacitor voltage is given by

$$v^{px}_{Cyh}(k+1) = v^{mx}_{Cyh}(k) + \frac{T_s}{C}\,i^{px}_{Cyh}(k) \tag{4.37}$$

The current flowing through the submodule capacitor is given in terms of the arm current and the switching state as

$$i^{px}_{Cyh}(k) = G_{xyh}\,i^{m}_{xy}(k) \tag{4.38}$$

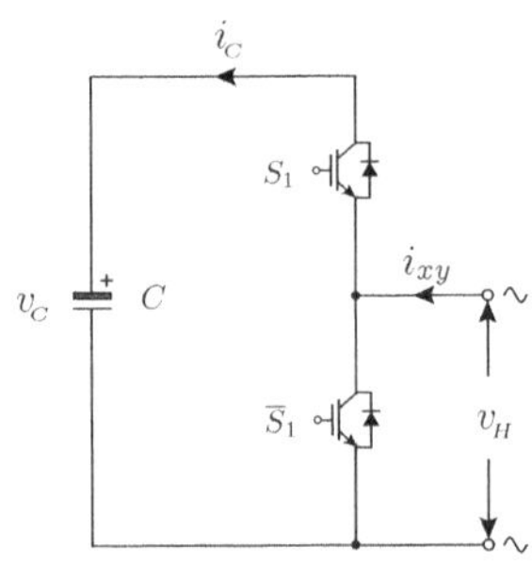

Figure 4.6 Configuration of half-bridge submodule.

(5) Discrete-Time Model of Arm Voltage: Each arm has N submodules and their equivalent output voltage is referred to as an arm voltage. The predicted submodule output voltage is given by

$$v^{px}_{Syh}(k) = G_{xyh}\,v^{mx}_{Cyh}(k) \tag{4.39}$$

The arm voltage is given by

$$v^{p}_{xy}(k) = v^{px}_{Sy1}(k) + \cdots + v^{px}_{SyN}(k) \quad = \sum_{h=1}^{N} v^{px}_{Syh}(k) \tag{4.40}$$

The arm voltage can be expressed in terms of switching state of the arm and the sum of submodule capacitors voltage as

$$v_{xy}^{p}(k) = \frac{S_{xy}(k)}{N} \sum_{h=1}^{N} v_{Cyh}^{mx}(k)$$
$$S_{xy}(k) = G_{xy1} + G_{xy2} + \cdots + G_{xyN} \tag{4.41}$$

where, $S_{xy}(k)$ represents the arm voltage level, which also represents the number of submodules to be inserted in an arm.

4.3 EXTRAPOLATION TECHNIQUES

The minimization of cost function requires extrapolated reference currents at $(k + 1)^{\text{th}}$-sampling instant. The extrapolation is employed in the MPC to minimize the tracking error between the reference and the actual currents [14]. This is needed particularly when the sampling time (T_s) is large (small sampling frequency (f_s)). It should be noted that the extrapolation is required for the implementation of MPC algorithms in the stationary frames (abc and $\alpha\beta$) only. In the synchronous-dq frame, the signals are in DC nature and thus the extrapolation technique is not required for steady-state operation. Some of the most popular extrapolation techniques are presented in the following subsections [1, 16, 17].

4.3.1 Vector Angle Extrapolation

In vector angle extrapolation approach, the reference variables in stationary-abc coordinate system are first converted into a stationary-$\alpha\beta$ frame and then converted into the polar coordinates ($re^{j\,\theta}$) from cartesian coordinate system ($x + jy$). The discrete-time phase shift $\omega^* N_p T_s$ is added to the phase angle of the reference variable to obtain the extrapolated signal. The vector angle extrapolation for a predicted horizon N_p is given by [1]

$$\widehat{x^*}(k + N_p) = x^*(k)e^{j(\theta + \omega^* N_p T_s)} \tag{4.42}$$

The extrapolated reference polar coordinates are converted back to the required reference frame (stationary-abc or stationary-$\alpha\beta$ or synchronous-dq) to use with the MPC cost function. This approach is mainly suitable for a three-phase system with sinusoidal signals only.

4.3.2 Lagrange Extrapolation

The lagrange extrapolation approach is simple and applicable for both sinusoidal and non-sinusoidal signals. The lagrange extrapolation estimates the future sample using the present and past samples of the control variable. The lagrange extrapolation for a predictive horizon N_p is given by [1, 18]

$$\widehat{x^*}(k+N_p) = \sum_{l=0}^{n}(-1)^{n-l}\frac{(n+1)!}{l!(n+1-l)!}\,x^*(k+N_p-1+l-n) \tag{4.43}$$

By solving equation (4.43), the generic expression for the lagrange extrapolation is given by

$$\widehat{x^*}(k+N_p) = a_m\, x^*(k) + a_{m-1}\, x^*(k-1) + \cdots + a_{m-n}\, x^*(k-n) \tag{4.44}$$

where a_m is the coefficient for reference variable $x^*(k)$.

These coefficients vary with the order of extrapolation (n) and predicted horizon (N_p). The third-order lagrange extrapolation for a predictive horizon $N_p = 1$ and $N_p = 2$ is given by

$$\begin{aligned} \widehat{x^*}(k+1) &= 4\,x^*(k) - 6\,x^*(k-1) + 4\,x^*(k-2) - x^*(k-3) \\ \widehat{x^*}(k+2) &= 10\,x^*(k) - 20\,x^*(k-1) + 15\,x^*(k-2) - 4\,x^*(k-3) \end{aligned} \tag{4.45}$$

4.4 COST FUNCTION AND WEIGHT FACTORS

The predicted and extrapolated control variables at $(k+1)^{\text{th}}$-sampling instant are included in a cost function. The main cost function consists of several control variables corresponding to the control objectives. Each control variable is associated with a weighting factor based on the priority of the control objective [1, 19]. The design of the cost function and weighting factors is given in this section.

4.4.1 Formulation of Cost Function

The definition of the cost function is an important design step in the model predictive control. The cost function mainly consists of the control variables, thats need to follow their reference. The cost function is simple and flexible to handle multiple control variables. A generic definition of the cost function is given by

$$g(k) = \begin{cases} \left|\widehat{x^*}(k+1) - x^p(k+1)\right| & \text{(Absolute cost function)} \\ \left(\widehat{x^*}(k+1) - x^p(k+1)\right)^2 & \text{(Quadratic cost function)} \end{cases} \tag{4.46}$$

where $\widehat{x^*}$ and x^p are the extrapolated reference and predicted variables at $(k+1)^{\text{th}}$-sampling instant. The absolute cost function measures the error between the $\widehat{x^*}$ and x^p variables, and represents a positive value while the quadratic cost function gives the square of the error value.

The modular multilevel converter has several control variables that can be represented as control objectives in a cost function. The control objectives are

- Output current, $g_o(k)$

- Submodule capacitors voltage, $g_c(k)$
- Circulating current, $g_z(k)$
- DC-bus current, $g_{dc}(k)$

The definition of cost function for the above control objectives is given by

$$g(k) = \lambda_o \, g_o(k) + \lambda_c \, g_c(k) + \lambda_z \, g_z(k) + \lambda_{dc} \, g_{dc}(k) \tag{4.47}$$

(1) Output Current Control: Most power conversion applications employ an inner current loop to control the output currents. The control of output currents in stationary-abc reference frame requires a complex control structure. With model predictive control, the output currents can be easily controlled in stationary-abc, stationary-$\alpha\beta$, and as well in synchronous-dq frames. The definition of output current cost function is given by

$$\begin{aligned} g_o(k) &= |\widehat{\mathbf{i}^*_{\mathbf{xi}}} - \mathbf{i}^p_{\mathbf{xi}}| \\ &\in \begin{cases} |\widehat{i^*_{ai}} - i^p_{ai}| + |\widehat{i^*_{bi}} - i^p_{bi}| + |\widehat{i^*_{ci}} - i^p_{ci}| & \text{(Stationary-}abc\text{ frame)} \\ |\widehat{i^*_{\alpha i}} - i^p_{\alpha i}| + |\widehat{i^*_{\beta i}} - i^p_{\beta i}| & \text{(Stationary-}\alpha\beta\text{ frame)} \\ |\widehat{i^*_{di}} - i^p_{di}| + |\widehat{i^*_{qi}} - i^p_{qi}| & \text{(Synchronous-}dq\text{ frame)} \end{cases} \end{aligned} \tag{4.48}$$

(2) Submodule Capacitor Voltage Control: The control of submodule capacitor voltages is one of the objectives related to the operation of an MMC. This objective is achieved with the closed loop PI-regulators and incorporated in the outer control loop in the classical control methods. The cascaded control structure can be eliminated by simultaneously incorporating the output currents and submodule capacitor voltages in the cost function. The definition of submodule capacitors voltage cost function is given by

$$\begin{aligned} g_c(k) &= |V^*_C - v^{px}_{Cyh}| \\ &\in \left\{ |V^*_C - v^{px}_{Cy1}| + \cdots + |V^*_C - v^{px}_{CyN}| \right. \end{aligned} \tag{4.49}$$

(3) Circulating Current Control: The circulating current increases the device power losses and ripple in submodule capacitors voltage, resulting in over sizing of the devices and submodule capacitors. Therefore, the control of circulating current is a technical requirement of MMC. These currents are effectively controlled in the synchronous frame or the stationary reference frames (abc and $\alpha\beta$-frames) using predictive control. The cost function of circulating current control is given by

$$\begin{aligned} g_z(k) &= |\widehat{\mathbf{i}^*_{\mathbf{xz}}} - \mathbf{i}^p_{\mathbf{xz}}| \\ &\in \begin{cases} |\widehat{i^*_{az}} - i^p_{az}| + |\widehat{i^*_{bz}} - i^p_{bz}| + |\widehat{i^*_{cz}} - i^p_{cz}| & \text{(Stationary-}abc\text{ frame)} \\ |\widehat{i^*_{\alpha z}} - i^p_{\alpha z}| + |\widehat{i^*_{\beta z}} - i^p_{\beta z}| & \text{(Stationary-}\alpha\beta\text{ frame)} \\ |\widehat{i^*_{dz}} - i^p_{dz}| + |\widehat{i^*_{qz}} - i^p_{qz}| & \text{(Synchronous-}dq\text{ frame)} \end{cases} \end{aligned} \tag{4.50}$$

(4) DC-bus Current Control: The DC-bus current is controlled to achieve the power balance between the input and output of the MMC. This current is regulated as per the requirement of leg voltage. In predictive control, the DC-bus current is regulated by using the following cost function

$$g_{dc}(k) = |\widehat{i^*}_{dc} - i^p_{dc}| \tag{4.51}$$

4.4.2 Selection of Weight Factors

In the previous subsection, we noted that the output current, submodule capacitors voltage, circulating current, and DC-bus current objectives are included in a single cost function through the weight factors. For simple analysis, the cost function in stationary-abc reference frame is considered and is given by

$$\begin{aligned} g(k) &= \lambda_o\, g_o(k) + \lambda_c\, g_c(k) + \lambda_z\, g_z(k) + \lambda_{dc}\, g_{dc}(k) \\ &= \lambda_o\, |\widehat{\mathbf{i}^*_{\mathbf{xi}}}(k+1) - \mathbf{i}^{\mathbf{p}}_{\mathbf{xi}}(k+1)| + \lambda_c\, |V^*_C(k+1) - v^{px}_{Cyh}(k+1)| \\ &+ \lambda_z |\, \widehat{\mathbf{i}^*_{\mathbf{xz}}}(k+1) - \mathbf{i}^{\mathbf{p}}_{\mathbf{xz}}(k+1)| + \lambda_{dc}\, |\widehat{i^*}_{dc}(k+1) - i^p_{dc}(k+1)| \end{aligned} \tag{4.52}$$

where λ_o, λ_c, λ_z, and λ_{dc} are the weight factors of output current, submodule capacitors voltage, circulating current, and DC-bus current objectives, respectively.

The cost function is easy to design with multiple control objectives. However, in the multi-variable cost function, the design of weight factors is one of the major challenges. The weight factors provide the relative importance of one control objective over another and need to be tuned to achieve optimum performance (much like PI-regulators are tuned in the classical control methods). The control objective with a larger weight factor implies more penalty to minimize the error of corresponding control objective. Since the error value varies with the operating conditions, the complete operation range needs to be considered when selecting the optimal weighting factors.

In classical control methods, the gains of PI regulator are systematically tuned with a design procedure. These gains are further modified using an empirical procedure to meet the control requirements. Similarly, the weight factors in MPC can be tuned using an empirical procedure. The use of empirical procedure requires prior knowledge of the system behavior and is a time-consuming process. Thereby, the empirical procedure is difficult to apply for a multi-objective cost function. On the other hand, the per-unit conversion method is commonly employed in the selection of weight factors [20]. In this approach, the per unit value of error term is considered. For example, consider the cost function given by equation (4.52). The error terms are divided with the corresponding rated or base value. Again, the overall cost function is multiplied with the rated or base current value such that the weight factor of the main cost function (output current control) becomes unity.

$$
\begin{aligned}
g(k) = I_B^* &\times \left[\frac{|\mathbf{i}_{\mathbf{xi}}^*(k+1) - \mathbf{i}_{\mathbf{xi}}(k+1)|}{I_B^*}\right] \\
+ I_B^* &\times \left[\frac{|V_C^*(k+1) - v_{Cyh}^{px}(k+1)|}{V_C^*}\right] \\
+ I_B^* &\times \left[\frac{|\widehat{\mathbf{i}_{\mathbf{xz}}^*}(k+1) - \mathbf{i}_{\mathbf{xz}}^{\mathbf{p}}(k+1)|}{I_{xz}^*}\right] \\
+ I_B^* &\times \left[\frac{\widehat{i_{dc}^*}(k+1) - i_{dc}^p(k+1)|}{I_{dc}^*}\right]
\end{aligned}
\tag{4.53}
$$

By using this approach, the weight factors of main and sub-cost functions are given by

$$
\lambda_o = \frac{I_B^*}{I_B^*},\ \lambda_c = \frac{I_B^*}{V_C^*},\ \lambda_z = \frac{I_B^*}{I_{xz}^*},\ \lambda_{dc} = \frac{I_B^*}{I_{dc}^*} \tag{4.54}
$$

4.5 DIRECT MODEL PREDICTIVE CONTROL

The direct model predictive control (DMPC) allows the simultaneous control of output current, circulating current, and submodule capacitor voltage in a modular multilevel converter (MMC). These control objectives are included in a single cost function and evaluated for all possible switching states [21, 22]. Each leg of the MMC is controlled using an independent predictive algorithm. Therefore, the direct model predictive control approach requires evaluating 2^{2N} switching states per phase to control the MMC with N submodules per arm. The number of switching states drastically increases with the number of submodules, which in turn increases the computational burden on the digital controller.

As mentioned earlier, the model predictive control approach can be implemented in any reference frame. In this section, the design and performance of the direct model predictive control are presented in the stationary-abc frame to control an MMC connected to a passive load. The block diagram of a direct model predictive control for a simultaneous control of output current, circulating current, and submodule capacitors voltage in an MMC is shown in Figure 4.7. The implementation of direct model predictive control involves various steps, such as synthesis of feedback and reference signals, prediction of future behavior, optimization of the cost function, and selection of switching states. The step-by-step design procedure of direct model predictive control is presented in the following subsections.

4.5.1 Design Procedure

The design procedure of direct model predictive control involves the following steps:

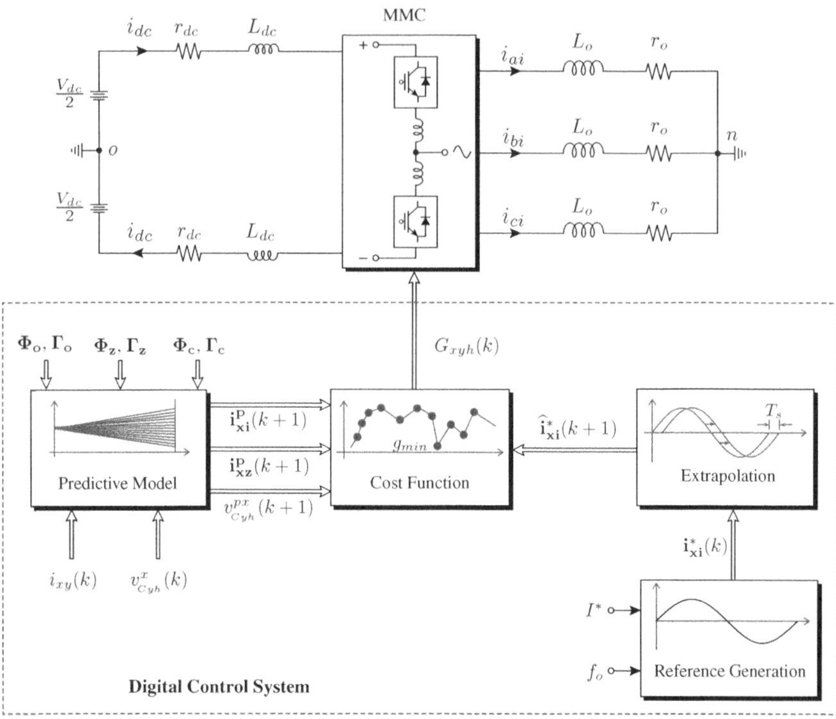

Figure 4.7 Block diagram of direct model predictive control.

(1) Measurement and Synthesis of Feedback Signals: The three-phase output currents ($\mathbf{i_{xi}}$), DC-bus current (i_{dc}), and circulating currents ($\mathbf{i_{xz}}$) are synthesized from the measured upper ($\mathbf{i_{xu}}$) and lower arm currents ($\mathbf{i_{xl}}$) as

$$
\begin{aligned}
\mathbf{i_{xi}}(k) &= \mathbf{i_{xu}}(k) - \mathbf{i_{xl}}(k) \\
i_{dc}(k) &= \frac{1}{2} \sum_{x=a,b,c} (\mathbf{i_{xu}}(k) + \mathbf{i_{xl}}(k)) \\
\mathbf{i_{xz}}(k) &= \frac{\mathbf{i_{xu}}(k) + \mathbf{i_{xl}}(k)}{2} - \frac{i_{dc}(k)}{3}
\end{aligned}
\tag{4.55}
$$

and, the measured submodule capacitors voltage (v^x_{Cyh}) is used in the predictive block.

(2) Calculation of Reference Currents: The reference output currents are generated with a required magnitude (I^*) and frequency (f_o). In the case of motor drive applications, the reference currents are obtained from the torque/speed and flux controllers [23, 24]. In grid-connected systems, the reference currents are obtained from the active and reactive power demands [25]. The reference output currents are defined as

$$\mathbf{i}^*_{\mathbf{xi}}(k) = I^* \sin(2\pi f_o t + \theta_x) \tag{4.56}$$

where $\theta_x \in \left\{0, -\frac{2\pi}{3}, -\frac{4\pi}{3}\right\}$. The circulating current contributes additional power losses, and effects the rating of passive and active elements in an MMC. The magnitude of circulating currents is minimized by setting their reference value to zero ($\mathbf{i}^*_{\mathbf{xz}}(k) = 0$) and the reference value of submodule capacitors voltage is set to $V_C^*(k) = V_{dc}/N$.

(3) Extrapolation of Reference Currents: To minimize the steady-state tracking error, the sinusoidal reference output currents are extrapolated to $(k+1)^{\text{th}}$-sampling instant using a third-order lagrange extrapolation technique. The extrapolated reference currents are given by

$$\widehat{\mathbf{i}^*_{\mathbf{xi}}}(k+1) = 4\,\mathbf{i}^*_{\mathbf{xi}}(k) - 6\,\mathbf{i}^*_{\mathbf{xi}}(k-1) + 4\,\mathbf{i}^*_{\mathbf{xi}}(k-2) - \mathbf{i}^*_{\mathbf{xi}}(k-3) \tag{4.57}$$

The circulating currents and submodule capacitors voltage are DC signals and the extrapolation of DC signals is not required. The reference circulating currents and submodule capacitor voltage at $(k+1)^{\text{th}}$-sampling instant are equal to their values at $(k)^{\text{th}}$-sampling instant (i.e., $\mathbf{i}^*_{\mathbf{xz}}(k+1) = \mathbf{i}^*_{\mathbf{xz}}(k)$ and $V_C^*(k+1) = \mathbf{V}^*_{\mathbf{C}}(k)$).

(4) Prediction of Future Behaviour of MMC: By using the measured quantities and off-line computed system parameters, the future behavior of the output and the circulating currents are predicted with the following discrete-time model of

$$\begin{aligned} \mathbf{i}^{\mathbf{p}}_{\mathbf{xi}}(k+1) &= \Phi_o\, \mathbf{i}^{\mathbf{m}}_{\mathbf{xi}}(k) + \Gamma_o\left[\mathbf{v}^{\mathbf{p}}_{\mathbf{xl}}(k) - \mathbf{v}^{\mathbf{p}}_{\mathbf{xu}}(k)\right] \\ \mathbf{i}^{\mathbf{p}}_{\mathbf{xz}}(k+1) &= \Phi_z\, \mathbf{i}^{\mathbf{m}}_{\mathbf{xz}}(k) + \Gamma_z \sum_{x=a,\,b,\,c} \left(v^p_{xu}(k) + v^p_{xl}(k)\right) \\ &\quad - 3\,\Gamma_z\left(\mathbf{v}^{\mathbf{p}}_{\mathbf{xu}}(k) + \mathbf{v}^{\mathbf{p}}_{\mathbf{xl}}(k)\right) \end{aligned} \tag{4.58}$$

where "p" denotes the predicted quantity and "m" denotes the measured quantity. These currents are expressed in terms of measured currents and predicted arm voltages.

The predictive model of arm voltages is formulated by combining equations (4.39) and (4.40) as

$$v^p_{xy}(k) = G_{xy1}\, v^{mx}_{Cy1}(k) + \cdots + G_{xyN}\, v^{mx}_{CyN}(k) \tag{4.59}$$

The predictive model of submodule capacitors voltage is formulated by using equations (4.37) and (4.38) as

$$v^{px}_{Cyh}(k+1) = v^{mx}_{Cyh}(k) + \frac{T_s}{C}\left[G_{xyh}\, i^m_{xy}(k)\right] \tag{4.60}$$

(5) Minimization of Cost Function: To meet the required goals, the converter cost function is defined as

$$\begin{aligned} g(k) &= \lambda_o\left|\widehat{\mathbf{i}^*_{\mathbf{xi}}}(k+1) - \mathbf{i}^{\mathbf{p}}_{\mathbf{xi}}(k+1)\right| + \lambda_c\left|V_C^*(k+1) - v^{px}_{Cyh}(k+1)\right| \\ &\quad + \lambda_z\left|\widehat{\mathbf{i}^*_{\mathbf{xz}}}(k+1) - \mathbf{i}^{\mathbf{p}}_{\mathbf{xz}}(k+1)\right| \end{aligned} \tag{4.61}$$

The primary control objective is to force the actual output currents to follow their reference, and thus $\lambda_o = 1$. The weight factor λ_c is adjusted to achieve the desired response in the secondary control objectives. The value of λ_c is calculated using the per unit method and with the $\lambda_z > 0$, the reduction in the circulating currents is achieved. During each iteration, the cost function is evaluated over switching states of 2^{2N}. The optimal switching state which minimizes the cost function is selected and applied to the converter.

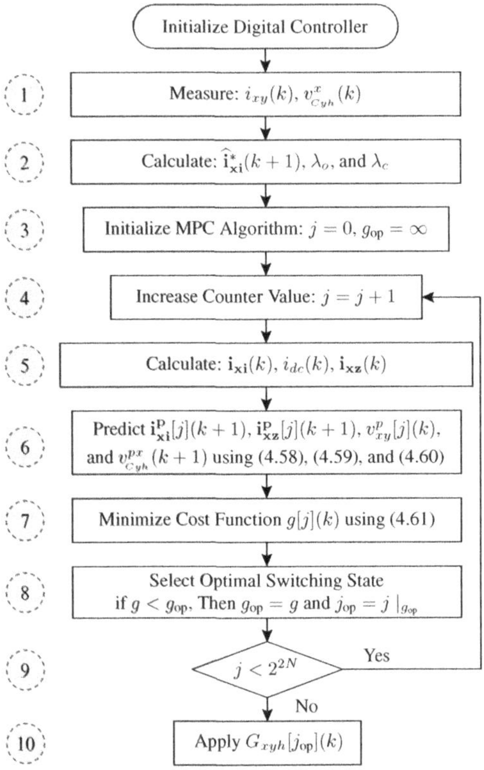

Figure 4.8 Flowchart of direct model predictive control.

4.5.2 Control Algorithm

The flowchart of a direct model predictive control algorithm for MMC is shown in Figure 4.8. The arm currents and submodule capacitors voltage are measured in the block ①. The $(k+1)^{\text{th}}$-sampling instant reference output currents and weight factors λ_o, λ_c are calculated in ②. The iterative loop is performed between the blocks ④ and ⑨. In blocks ⑤–⑧, the MMC arm voltages $v^p_{xy}[j](k)$, output currents $\mathbf{i}^p_{\mathbf{xi}}[j](k+1)$, circulating currents $\mathbf{i}^p_{\mathbf{xz}}[j](k+1)$, and submodule capacitors voltage

$v^{px}_{Cyh}(k+1)$ are predicted, and minimization of cost function $g[j](k)$ is performed for all 2^{2N} possible switching states. An optimal switching state j_{op} and corresponding switching signals are selected and applied directly to the MMC in blocks ⑧–⑩. With direct model predictive control approach, the simultaneous control of output current, circulating currents, and submodule capacitors voltage are achieved.

CASE STUDY 4.1 Dynamic Performance of Direct MPC

Objective: The dynamic performance of the direct model predictive control (DMPC) is studied for MMC through MATLAB simulations.

Parameters: The rated parameters of the MMC are as follows: rated apparent power $S_o = 5$ MVA, output line-to-line voltage $V_o = 4.16$ kV, rated output frequency $f_o = 60$ Hz, rated DC-bus voltage $V_{dc} = 7.354$ kV, number of SMs per arm $N = 4$, arm inductance $L = 0.5$ mH, load inductance $L_o = 1.83$ mH, load resistance $r_o = 3.46\ \Omega$, and SM capacitance $C = 9000\ \mu$F.

Analysis:
The dynamic performance of the direct model predictive control is presented with a step change in the magnitude of reference output current. In this case study, the output currents and submodule capacitor voltages are included in a cost function. The cost function is defined as

$$g(k) = \lambda_o \left|\widehat{\mathbf{i}^*_{\mathbf{xl}}}(k+1) - \mathbf{i}^{\mathbf{p}}_{\mathbf{xl}}(k+1)\right| + \lambda_c \left|V^*_C(k+1) - v^{px}_{Cyh}(k+1)\right| \tag{4.62}$$

The rated (base) value of output current is

$$I_B = \frac{5 \times 10^6}{\sqrt{3} \times 4.16 \times 10^3} = 693.93\ \text{A (RMS) and } 981.366\ \text{A (peak)} \tag{4.63}$$

The reference value of submodule capacitor voltage is

$$V^*_C(k+1) = \frac{7.354 \times 10^3}{4} = 1.8385\ \text{kV} \tag{4.64}$$

The output currents and submodule capacitors voltage are predicted using the discrete-time model of MMC. The predicted and reference variables are included in a cost function. Among them, the output current control is the primary objective and its weight factor is equal to $\lambda_o = 1$. At $t = 0.05$ s, the magnitude of reference output current is changed from 0.4 pu to 0.8 pu. The weight factor of the submodule capacitor voltage control varies with the magnitude of reference output current and is given by

$$\lambda_c = \frac{0.4 \times 981.366}{1.8385 \times 10^3} = 0.2135 \quad (\text{for } I^* = 0.4\ \text{pu}) \tag{4.65}$$

Similarly, the weight factor of the submodule capacitor voltage control for $I^* = 0.8$ pu is given by

$$\lambda_c = \frac{0.8 \times 981.366}{1.8385 \times 10^3} = 0.427 \tag{4.66}$$

The simulation results are presented in Figure 4.9, which proves the effectiveness of direct model predictive control approach. The actual output currents are perfectly following their reference output currents. During the step change, the direct model predictive control takes around 0.5 ms to reach the steady-state reference value as shown in Figure 4.9(a). The number of voltage steps in the output line-to-line voltage is proportionally increased with the magnitude of reference output current as shown in Figure 4.9(b). Along with the output current, the submodule capacitors voltage in each arm is perfectly regulated at its reference value of 1.8385 kV as shown in Figure 4.9(c).

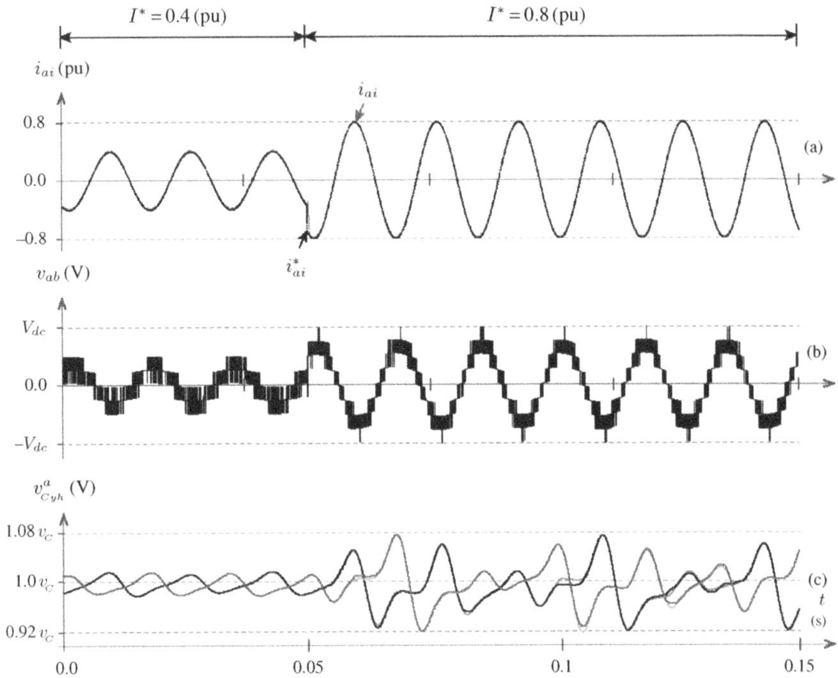

Figure 4.9 Dynamic performance of the direct model predictive control.

■

CASE STUDY 4.2 Minimization of Circulating Currents with DMPC

Objective: The minimization of circulating currents with the direct model predictive control is studied through MATLAB simulations.

Parameters: The rated parameters of the MMC are as follows: rated apparent power $S_o = 5$ MVA, output line-to-line voltage $V_o = 4.16$ kV, rated output frequency $f_o = 60$ Hz, rated DC-bus voltage $V_{dc} = 7.354$ kV, number of SMs per arm $N = 4$, arm inductance $L = 0.5$ mH, load inductance $L_o = 1.83$ mH, load resistance $r_o = 3.46\ \Omega$, and SM capacitance $C = 9000\ \mu$F.

Analysis:
The circulating current flows among the converter legs, which increases the device power losses, and affects the rating of submodule capacitors and devices. Thereby, the minimization

of circulating currents is mandatory to improve the reliability and efficiency of MMC. The cost function of circulating current is given by

$$g_z(k) = \lambda_z |\, \widehat{\mathbf{i}^*_{\mathbf{xz}}}(k+1) - \mathbf{i}^{\mathbf{P}}_{\mathbf{xz}}(k+1)| \tag{4.67}$$

which is included in the main cost function $g(k)$ given by equation (4.62).

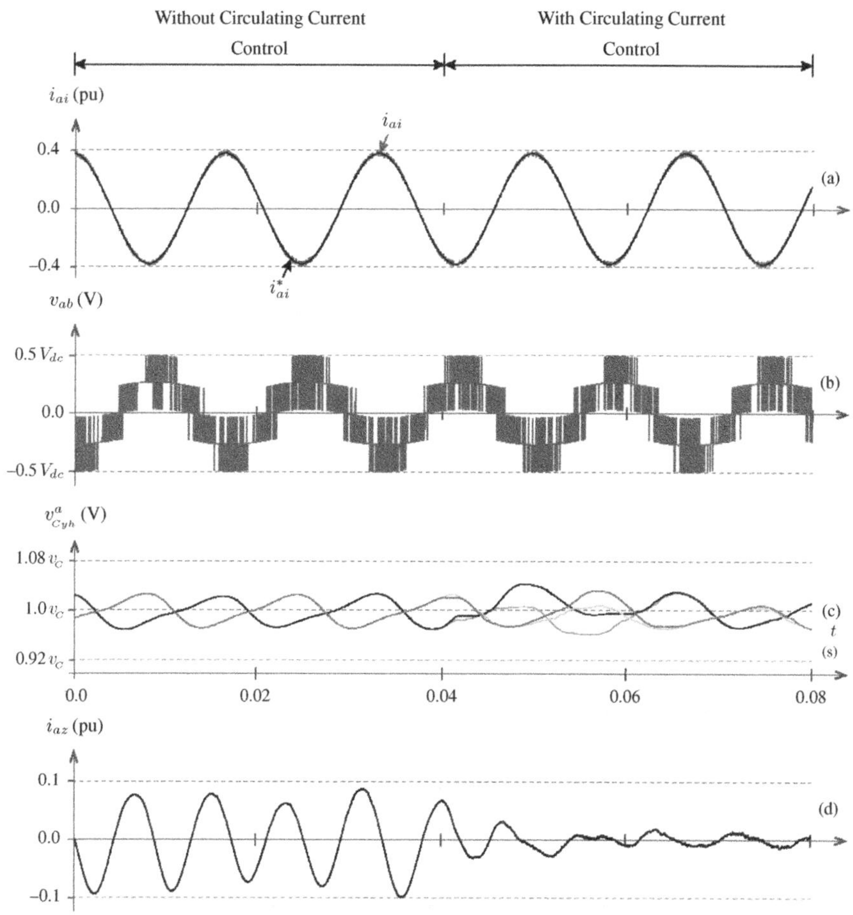

Figure 4.10 Circulating current control with DMPC.

The weight factor for the circulating current minimization is set to $\lambda_z > 0$. The steady-state performance of the direct model predictive control approach with a circulating current control is shown in Figure 4.10. Initially, the circulating current control is deactivated by assigning the weight factor $\lambda_z = 0$. At $t = 0.05$ s, the circulating current control is activated by assigning the weight factor $\lambda_z = 0.2$. During this study, the magnitude of reference output current is maintained constant at 0.4 pu. The actual output current is perfectly tracking their reference output current as shown in Figure 4.10(a). The output line-to-line voltage has a peak value of $\pm\frac{V_{dc}}{2}$, which is proportional to their reference current as shown in Figure 4.10(b). The upper and lower arm submodule capacitor voltage are perfectly balanced at 1.8385 kV as shown in Figure 4.10(c). The magnitude of circulating current is significantly minimized from 0.1 pu to 0.02 pu, which improves the efficiency and the reliability of MMC as shown in Figure 4.10(d). Overall, the direct model predictive control effectively controls the multiple objectives with a single cost function. ■

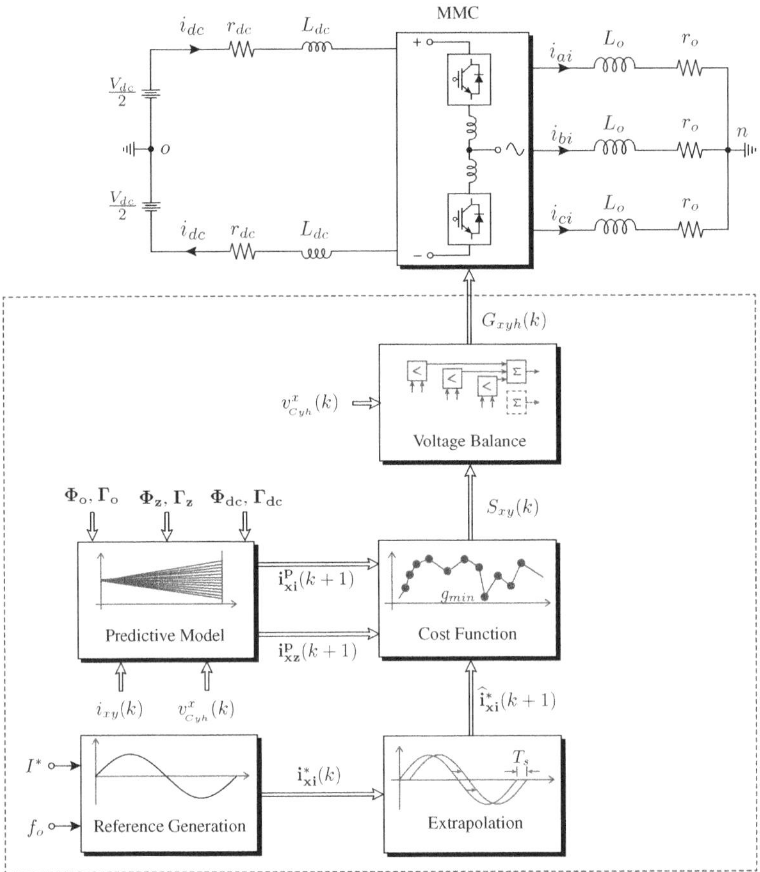

Figure 4.11 Block diagram of an indirect model predictive control.

4.6 INDIRECT MODEL PREDICTIVE CONTROL

In direct model predictive control approach, all the control objectives of MMC are simultaneously achieved using a single cost function. The cost function is optimized for 2^{2N} switching states per phase. The switching states are quite high and increase with the number of submodules per arm, which will further increase the computational complexity of the real-time implementation. The computational complexity of direct model predictive control can be reduced by using either multiple cost functions or using the voltage balance strategy with the predictive algorithm. The resultant approach is referred to as an indirect model predictive control (IMPC) approach [26–28]. In this section, the indirect model predictive control approach, which combines the voltage balance strategy with a predictive algorithm is presented. The voltage balance strategy is used to achieve the voltage balancing between submodules within the arm while the predictive control algorithm is used to control the output and circulating currents. In the indirect model predictive control approach, the predictive algorithm requires only $N+1$ switching states to optimize the cost function [29, 30]. Therefore, the computational complexity of an indirect model predictive control approach is significantly minimized in comparison to the direct model predictive control approach. Also, the real-time implementation of the indirect model predictive control approach is quite easy and simple.

The block diagram of the indirect model predictive control approach for MMC with a passive load is shown in Figure 4.11. The implementation of the indirect model predictive control approach has two stages: the first stage of implementation involves predictive algorithm while the second stage involves the voltage balance strategy. The implementation of voltage balance strategy is already discussed in Section 3.3.2, while the implementation of the predictive algorithm involves reference current generation, extrapolation, prediction of future behavior of the output and circulating currents, formulation of cost function, and weight factors selection. These steps are discussed below in detail.

4.6.1 Design Procedure

The design of indirect model predictive control approach involves the following steps:

(1) Measurement and Synthesis of Feedback Signals: The three-phase output currents ($\mathbf{i_{xi}}$), DC-bus current (i_{dc}), and circulating currents ($\mathbf{i_{xz}}$) are synthesized from the measured upper ($\mathbf{i_{xu}}$) and lower arm currents ($\mathbf{i_{xl}}$). The arm current components are given by

$$\begin{aligned} \mathbf{i_{xi}}(k) &= \mathbf{i_{xu}}(k) - \mathbf{i_{xl}}(k) \\ i_{dc}(k) &= \frac{1}{2} \sum_{x=a,b,c} (\mathbf{i_{xu}}(k) + \mathbf{i_{xl}}(k)) \\ \mathbf{i_{xz}}(k) &= \frac{\mathbf{i_{xu}}(k) + \mathbf{i_{xl}}(k)}{2} - \frac{i_{dc}(k)}{3} \end{aligned} \tag{4.68}$$

and, the measured submodule capacitor voltage (v^x_{Cyh}) is used in the predictive block.

(2) Calculation of Reference Currents: The reference output currents are defined as

$$\mathbf{i}^*_{\mathbf{xi}}(k) = I^* \sin(2\pi f_o t + \theta_x) \tag{4.69}$$

where $\theta_x \in \left\{0, -\frac{2\pi}{3}, -\frac{4\pi}{3}\right\}$. The value of reference circulating currents is set to $\mathbf{i}^*_{\mathbf{xz}}(k) = 0$.

(3) Extrapolation of Reference Currents: To minimize the steady-state tracking error, the sinusoidal reference output currents are extrapolated to $(k+1)^{\text{th}}$-sampling instant using the third-order Lagrange extrapolation technique. The extrapolated reference currents are given by

$$\widehat{\mathbf{i}^*_{\mathbf{xi}}}(k+1) = 4\,\mathbf{i}^*_{\mathbf{xi}}(k) - 6\,\mathbf{i}^*_{\mathbf{xi}}(k-1) + 4\,\mathbf{i}^*_{\mathbf{xi}}(k-2) - \mathbf{i}^*_{\mathbf{xi}}(k-3) \tag{4.70}$$

where as the circulating current is a DC signal. Therefore, the extrapolation of DC signal is not required. The reference circulating current at $(k+1)^{\text{th}}$-sampling instant is equal to its value at $(k)^{\text{th}}$-sampling instant (i.e, $\mathbf{i}^*_{\mathbf{xz}}(k+1) = \mathbf{i}^*_{\mathbf{xz}}(k)$).

(4) Prediction of Future Behaviour of MMC: By using the measured quantities and off-line computed system parameters, the future behavior of the output currents and circulating currents are predicted with the following discrete-time model of

$$\begin{aligned}
\mathbf{i}^{\mathbf{P}}_{\mathbf{xi}}(k+1) &= \Phi_o\, \mathbf{i}^{\mathbf{m}}_{\mathbf{xi}}(k) + \Gamma_o\, \left[\mathbf{v}^{\mathbf{P}}_{\mathbf{xl}}(k) - \mathbf{v}^{\mathbf{P}}_{\mathbf{xu}}(k)\right] \\
\mathbf{i}^{\mathbf{P}}_{\mathbf{xz}}(k+1) &= \Phi_z\, \mathbf{i}^{\mathbf{m}}_{\mathbf{xz}}(k) + \Gamma_z \sum_{x=a,\,b,\,c} \left(v^p_{xu}(k) + v^p_{xl}(k)\right) \\
&\quad - 3\,\Gamma_z\left(\mathbf{v}^{\mathbf{P}}_{\mathbf{xu}}(k) + \mathbf{v}^{\mathbf{P}}_{\mathbf{xl}}(k)\right)
\end{aligned} \tag{4.71}$$

where "p" denotes the predicted quantity and "m" denotes the measured quantity. These currents are expressed in terms of measured currents and predicted arm voltages.

The predictive model of arm voltage is expressed in terms of the arm switching vector and the sum of submodule capacitors voltage as

$$\begin{aligned}
v^p_{xy}(k) &= \frac{S_{xy}(k)}{N} \sum_{h=1}^{N} v^{mx}_{Cyh}(k) \\
S_{xy}(k) &= G_{xy1} + G_{xy2} + \cdots + G_{xyN}
\end{aligned} \tag{4.72}$$

where, $S_{xy}(k)$ represents the arm voltage level, which also represents the number of submodules to be inserted in an arm.

(5) Minimization of Cost Function: To meet the required goals, the converter cost function is defined as

$$g(k) = \lambda_o \left|\widehat{\mathbf{i}^*_{\mathbf{xi}}}(k+1) - \mathbf{i}^{\mathbf{P}}_{\mathbf{xi}}(k+1)\right| + \lambda_z \left|\widehat{\mathbf{i}^*_{\mathbf{xz}}}(k+1) - \mathbf{i}^{\mathbf{P}}_{\mathbf{xz}}(k+1)\right| \tag{4.73}$$

The primary control objective is to force the actual output current to follow their reference current, and thus $\lambda_o = 1$, whereas the $\lambda_z > 0$ to achieve the reduction in the circulating currents and is selected by the empirical method. During each iteration, the cost function is evaluated over a switching vectors of $N + 1$. The optimal switching vector which minimizes the cost function is selected and given to the voltage balance strategy.

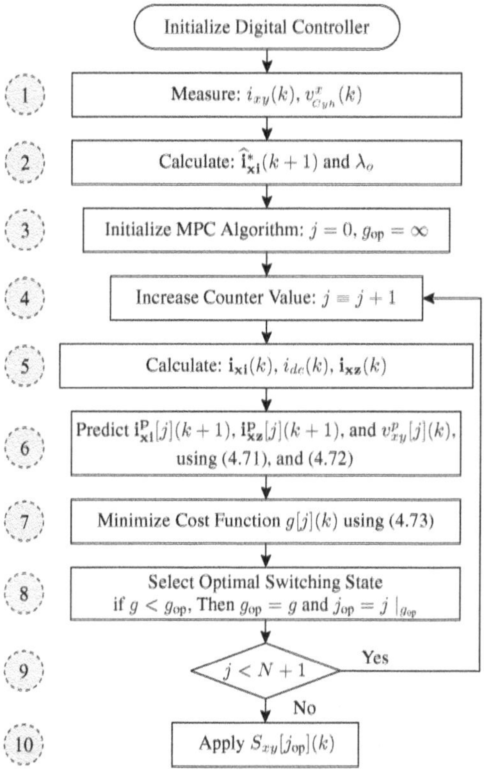

Figure 4.12 Flowchart of indirect model predictive control.

4.6.2 Control Algorithm

The flowchart of the predictive algorithm in an indirect model predictive control is shown in Figure 4.12. The arm currents and submodule capacitor voltages are measured in block ①. The $(k+1)^{\mathrm{th}}$-state reference output currents, and weight factor λ_o are calculated in ②. The iterative loop is performed between blocks ④ and ⑨. In blocks ⑤–⑧, the MMC arm voltages $v^{p}_{xy}[j](k)$, output currents $\mathbf{i}^{\mathrm{P}}_{\mathbf{xi}}[j](k+1)$, and circulating currents $\mathbf{i}^{\mathrm{P}}_{\mathbf{xz}}[j](k+1)$ are predicted, and minimization of cost function $g[j](k)$ is performed for all $N + 1$ possible voltage levels. An optimal voltage level j_{op} is selected and applied to the voltage balance strategy in blocks ⑧–⑩. The voltage balance strategy selects the submodules based on the instantaneous value of submodule capacitors voltage and the direction of arm current. The selected submodules are inserted in the arm to meet the control objectives.

4.7 SUMMARY

The model predictive control approach is a simple concept and easy to implement in a wide range of power converters and applications. It achieves a fast dynamic response, eliminates the use of PI-regulators and PWM modulators, and can control multiple objectives in comparison to the classical control methods. In this chapter, the basic concept, step-by-step design procedure, and implementation of model predictive control approach is systematically described for an MMC. The MPC approach requires a discrete-time model of the MMC for the real-time implementation. The continuous-time and discrete-time models of MMC with a passive load are presented. The derivation of the discrete-time model from the continuous-time model using a forward Euler discretization method is presented. The real-time implementation of MPC involves technical challenges such as the extrapolation of reference signals, formulation of the cost function, and the selection of weight factors are discussed. The flexibility in the formulation of cost function allows controlling multiple objectives such as output currents, circulating currents, and submodule capacitors voltage in an MMC.

Based on the design procedure, the MPC approach is further categorized into direct and indirect methods. The step-by-step design procedure and implementation of direct and indirect model predictive control approaches are discussed. The DMPC uses a single cost function to achieve all the control objectives of an MMC. This approach requires evaluating 2^{2N} switching states per phase, which increases the computational complexity of the real-time implementation. On the other hand, the indirect model combines the voltage balance strategy with a predictive algorithm. In this approach, the output currents and circulating currents are controlled using a predictive algorithm, while the voltage balance between submodules within the arm is achieved using a voltage balance strategy. The indirect model predictive control (IMPC) requires evaluating only $N+1$ voltage levels, which are quite low in comparison to 2^{2N} switching states in a direct model predictive control approach. Therefore, the indirect model predictive control approach is easy to implement in real time.

REFERENCES

1. J. Rodriguez and P. Cortes, *Predictive Control of Power Converters and Electrical Drives*, 1st ed. Chichester, UK: IEEE Wiley press, Mar 2012.
2. J. Rodriguez, M. P. Kazmierkowski, J. R. Espinoza, P. Zanchetta, H. Abu-Rub, H. A. Young, and C. A. Rojas, "State of the art of finite control set model predictive control in power electronics," *IEEE Trans. Ind. Informat.*, vol. 9, no. 2, pp. 1003–1016, May 2013.
3. J. Rodriguez, J. Pontt, C. Silva, P. Correa, P. Lezana, P. Cortes, and U. Ammann, "Predictive current control of a voltage source inverter," *IEEE Trans. Ind. Electron.*, vol. 54, no. 1, pp. 495–503, Feb 2007.
4. C. Xia, T. Liu, T. Shi, and Z. Song, "A simplified finite-control-set model-predictive control for power converters," *IEEE Trans. Ind. Electron.*, vol. 10, no. 2, pp. 991–1002, May 2014.

5. R. Vargas, J. Rodriguez, C. Rojas, and M. Rivera, "Predictive control of an induction machine fed by a matrix converter with increased efficiency and reduced common-mode voltage," *IEEE Trans. Energy Convers.*, vol. 29, no. 2, pp. 473–485, Jun 2014.

6. P. Lezana, R. Aguilera, and D. Quevedo, "Model predictive control of an asymmetric flying capacitor converter," *IEEE Trans. Ind. Electron.*, vol. 56, no. 6, pp. 1839–1846, Jun 2009.

7. M. Perez, P. Cortes, and J. Rodriguez, "Predictive control algorithm technique for multilevel asymmetric cascaded h-bridge inverters," *IEEE Trans. Ind. Electron.*, vol. 55, no. 12, pp. 4354–4361, Dec 2008.

8. P. Cortes, A. Wilson, S. Kouro, J. Rodriguez, and H. Abu-Rub, "Model predictive control of multilevel cascaded h-bridge inverters," *IEEE Trans. Ind. Electron.*, vol. 57, no. 8, pp. 2691–2699, Aug 2010.

9. J. Bocker, B. Freudenberg, A. The, and S. Dieckerhoff, "Experimental comparison of model predictive control and cascaded control of the modular multilevel converter," *IEEE Trans. Power Electron.*, vol. 30, no. 1, pp. 422–430, Jan 2015.

10. B. Riar, T. Geyer, and U. Madawala, "Model predictive direct current control of modular multilevel converters: Modeling, analysis, and experimental evaluation," *IEEE Trans. Power Electron.*, vol. 30, no. 1, pp. 431–439, Jan 2015.

11. J. Wang, R. Burgos, and D. Boroyevich, "Switching-cycle state-space modeling and control of the modular multilevel converter," *IEEE Trans. Emerg. Sel. Topics Power Electron.*, vol. 2, no. 4, pp. 1159–1170, Dec 2014.

12. G. F. Franklin, J. D. Powell, and M. L. Workman, *Digital control of dynamic systems*, 3rd ed. Addison-wesley Menlo Park, 1998.

13. H. Miranda, P. Cortes, J. I. Yuz, and J. Rodriguez, "Predictive torque control of induction machines based on state-space models," *IEEE Trans. Ind. Electron.*, vol. 56, no. 6, pp. 1916–1924, Jun 2009.

14. V. Yaramasu, B. Wu, M. Rivera, M. Narimani, S. Kouro, and J. Rodriguez, "Generalised approach for predictive control with common-mode voltage mitigation in multilevel diode-clamped converters," *IET Power Electron.*, vol. 8, no. 8, pp. 1440–1450, Jul 2015.

15. V. Yaramasu, M. Rivera, B. Wu, and J. Rodriguez, "Model predictive current control of two-level four-leg inverters-part i: Concept, algorithm, and simulation analysis," *IEEE Trans. Power Electron.*, vol. 28, no. 7, pp. 3459–3468, Jul 2013.

16. O. Kukrer, "Discrete-time current control of voltage-fed three-phase pwm inverters," *IEEE Trans. Power Electron.*, vol. 11, no. 2, pp. 260–269, Mar 1996.

17. V. Yaramasu, B. Wu, M. Rivera, J. Rodriguez, and A. Wilson, "Cost-function based predictive voltage control of two-level four-leg inverters using two step prediction horizon for standalone power systems," in *Applied Power Electronics Conference and Exposition (APEC), 2012 Twenty-Seventh Annual IEEE*, Feb 2012, pp. 128–135.

18. V. Yaramasu and B. Wu, "Predictive control of a three-level boost converter and an npc inverter for high-power pmsg-based medium voltage wind energy conversion systems," *IEEE Trans. Power Electron.*, vol. 29, no. 10, pp. 5308–5322, Oct 2014.

19. S. A. Davari, D. A. Khaburi, and R. Kennel, "An improved fcs-mpc algorithm for an induction motor with an imposed optimized weighting factor," *IEEE Trans. Power Electron.*, vol. 27, no. 3, pp. 1540–1551, Mar 2012.

20. M. Narimani, B. Wu, V. Yaramasu, and N. Reza Zargari, "Finite control-set model predictive control (fcs-mpc) of nested neutral point-clamped (nnpc) converter," *IEEE Trans. Power Electron.*, vol. 30, no. 12, pp. 7262–7269, Dec 2015.

21. J. Qin and M. Saeedifard, "Predictive control of a modular multilevel converter for a back-to-back hvdc system," *IEEE Trans. Power Del.*, vol. 27, no. 3, pp. 1538–1547, Jul 2012.

22. L. Ben-Brahim, A. Gastli, M. Trabelsi, K. A. Ghazi, M. Houchati, and H. Abu-Rub, "Modular multilevel converter circulating current reduction using model predictive control," *IEEE Trans. Ind. Electron.*, vol. 63, no. 6, pp. 3857–3866, Jun 2016.

23. T. Geyer and S. Mastellone, "Model predictive direct torque control of a five-level anpc converter drive system," *IEEE Trans. Ind. Appl.*, vol. 48, no. 5, pp. 1565–1575, Sep 2012.

24. M. Preindl and S. Bolognani, "Model predictive direct speed control with finite control set of pmsm drive systems," *IEEE Trans. Power Electron.*, vol. 28, no. 2, pp. 1007–1015, Feb 2013.

25. V. Yaramasu and B. Wu, "Model predictive decoupled active and reactive power control for high-power grid-connected four-level diode-clamped inverters," *IEEE Trans. Ind. Electron.*, vol. 61, no. 7, pp. 3407–3416, Jul 2014.

26. J.-W. Moon, J.-S. Gwon, J.-W. Park, D.-W. Kang, and J.-M. Kim, "Model predictive control with a reduced number of considered states in a modular multilevel converter for hvdc system," *IEEE Trans. Power Del.*, vol. 30, no. 2, pp. 608–617, Apr 2015.

27. M. Vatani, B. Bahrani, M. Saeedifard, and M. Hovd, "Indirect finite control set model predictive control of modular multilevel converters," *IEEE Trans. Smart Grid*, vol. 6, no. 3, pp. 1520–1529, May 2015.

28. K. Ilves, L. Harnefors, S. Norrga, and H.-P. Nee, "Predictive sorting algorithm for modular multilevel converters minimizing the spread in the submodule capacitor voltages," *IEEE Trans. Power Electron.*, vol. 30, no. 1, pp. 440–449, Jan 2015.

29. P. Liu, Y. Wang, W. Cong, and W. Lei, "Grouping-sorting-optimized model predictive control for modular multilevel converter with reduced computational load," *IEEE Trans. Power Electron.*, vol. 31, no. 3, pp. 1896–1907, Mar 2016.

30. Z. Gong, P. Dai, X. Yuan, X. Wu, and G. Guo, "Design and experimental evaluation of fast model predictive control for modular multilevel converters," *IEEE Trans. Ind. Electron.*, vol. 63, no. 6, pp. 3845–3856, Jun 2016.

PART II

ADVANCED MODULAR MULTILEVEL CONVERTERS

CHAPTER 5

PASSIVE CROSS-CONNECTED MODULAR MULTILEVEL CONVERTERS

5.1 INTRODUCTION

The conventional modular multilevel converters (MMCs) have been commercialized for high-voltage direct current (HVDC) transmission applications by `Siemens`, `ABB`, and `Alstom` [1–4]. These converters are extended to medium-voltage motor drives and high-voltage DC–DC power conversion applications and require a variable frequency operation (zero to line frequency) [5–7]. The low/zero output frequency operation results in long charging and discharging periods of submodule capacitors and leads to a larger voltage ripple (voltage fluctuations) [8]. These voltage ripples cause excessive power losses and voltage stress on the submodule semiconductor devices and capacitors.

The ripple in submodule capacitors voltage can be minimized by using either larger capacitors or by injecting a high-frequency common-mode voltage (CMV) and circulating currents into the system [9–14]. The former approach is implemented at the design stage, requires large capacitors to minimize the voltage ripple, and increases the cost and physical size of an MMC. However, it can increase the motor neutral voltage, resulting in increased motor insulation requirement and bearing current in MMC-based motor drive systems. These approaches are highly effective but

Modular Multilevel Converters: Analysis, Control, and Applications, First Edition.
Sixing Du, Apparao Dekka, Bin Wu, and Navid Zargari.

have limited practical applications due to the inherent side effects on the system performance.

In this chapter, a new configuration of MMC to deal with the submodule capacitors voltage ripple at low/zero frequency operation is presented. This new topology is named "passive cross-connected modular multilevel converter" (PC-MMC), since it uses a capacitor connection across the mid-point of the upper and lower arms. The circuit configuration and switching states of PC-MMC are presented in Section 5.2. The modeling and operation of PC-MMC with a phase-shifted carrier modulation is presented in Section 5.3. The performance of the PC-MMC at low-frequency is presented in Section 5.4. The generalized control scheme is presented in Section 5.5 and a summary is presented in Section 5.6.

5.2 PASSIVE CROSS-CONNECTED MMC

The new circuit configuration of a modular multilevel converter is referred to as a passive cross-connected modular multilevel converter (PC-MMC), in which a passive element is connected across the mid-point of the upper and lower arm of each leg. The passive element is a simple capacitor, which provides a path for the power transfer between the upper and lower arms. This path is used to minimize the ripple in submodule capacitors voltage during low/zero frequency operation. The circuit configuration and operation of PC-MMC are presented in the following subsections.

5.2.1 Configuration of Power Circuit

Figure 5.1 shows the circuit configuration of a three-phase PC-MMC system with N submodules per arm. Each leg consists of two arms named as upper and lower. The upper and lower arms form an AC output terminal "x," where $x \in \{a, b, c\}$ represents the output phase. Each arm is further divided into two halves and referred to as a sub-arm. The mid-point of upper and lower arms of each leg is connected through a capacitor "C_x." The cross-connected capacitor in each leg provides a path for power transfer between the upper and lower arms. The power flowing through the cross-connected capacitor mainly consists of high-frequency oscillating components. In each switching cycle, one arm releases the power while another arm absorbs it.

The arm inductor is divided into two parts and connected at both ends of each arm as shown in Figure 5.1. The arm inductor "L" is designed to mitigate the switching ripples in the arm current, minimize the circulating and inrush currents, and prevent the direct connection of cross-connected capacitor with the submodules. The PC-MMC system requires twice the number of arm inductors compared to the conventional MMC. However, their total equivalent inductance remains constant, and is equal to the inductance used in the conventional MMC. The PC-MMC can be realized using different submodule configurations similar to the conventional MMC as discussed in Chapter 2 [15]. The submodule selection process depends on the application requirements. For example, the high-voltage DC–DC conversion system requires a DC-side fault-blocking capability, which can be achieved with full-bridge

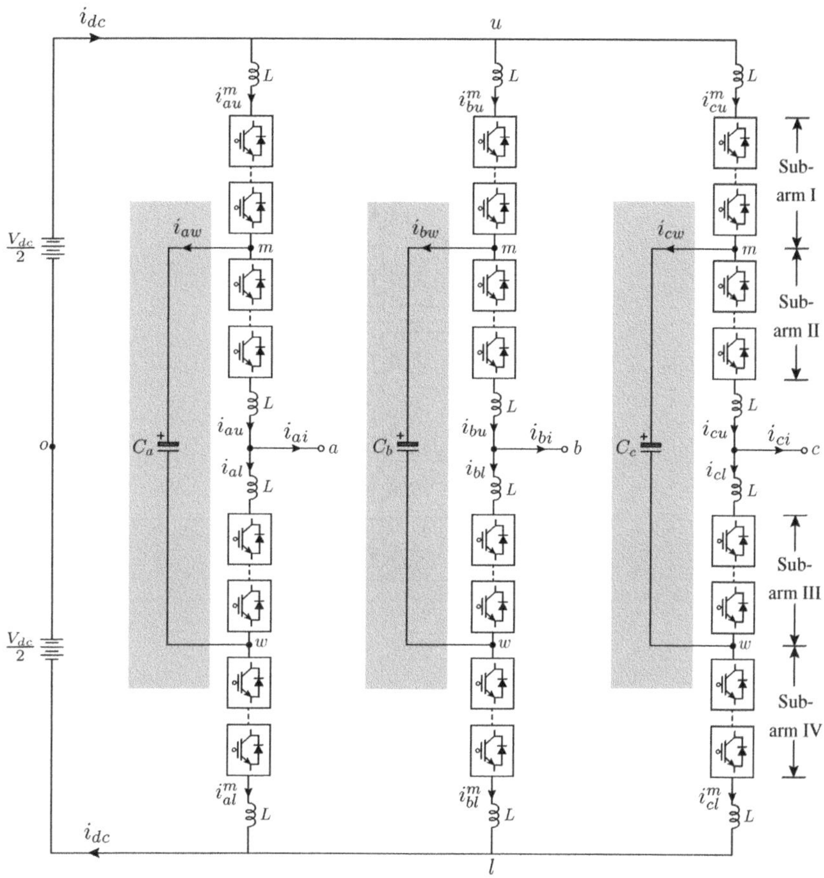

Figure 5.1 Power circuit configuration of PC-MMC.

(FB) and double clamp (CD) submodules. Similarly, the medium-voltage motor drive requires high efficiency and simple configuration, which can be obtained by using half-bridge (HB) submodules. In this section, the operation of the PC-MMC system with the half-bridge (HB) submodule is presented.

5.2.2 Switching States and Output Voltage

The configuration of converter leg in a PC-MMC system is shown in Figure 5.2. Each converter leg is divided into two arms, and each arm consists of four half-bridge (HB) submodules and two arm inductors. The half-bridge submodule has a simple construction and requires one independent gating signal to control the devices.

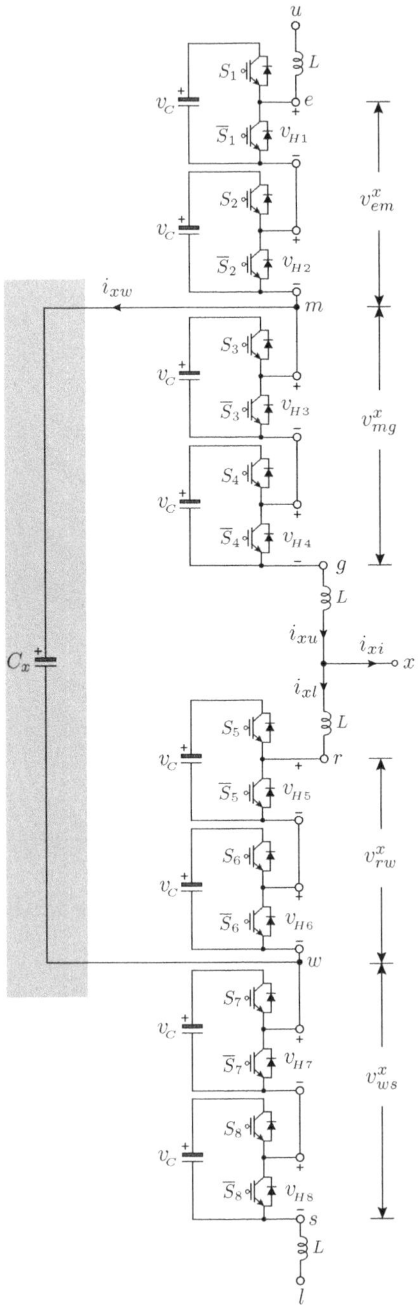

Figure 5.2 Configuration of converter leg in a PC-MMC.

The PC-MMC consists of four half-bridge submodules in each arm and generates five voltage levels across the arm terminals of "e" and "g" in the upper arm and "r" and "s" in the lower arm. The switching states and output voltage levels of the upper arm are shown in Table 5.1. When switches "S_1," "S_2," "S_3," and "S_4" are turned ON, each half-bridge submodule generates a voltage of "v_C," and the resultant arm voltage is $v^x_{eg} = 4\,v_C$. Similarly, when switches "S_1," "S_2," "S_3," and "S_4" are turned OFF, the voltage across the arm is $v^x_{eg} = 0$. There are several redundant switching states to generate the remaining voltage levels of "v_C," "$2\,v_C$," and "$3\,v_C$" as shown in Table 5.1. The redundant switching states are used to achieve the voltage balancing in a modular multilevel converter.

Table 5.1 Switching states and output voltage levels of an arm

S_1	S_2	S_3	S_4	v_{H1}	v_{H2}	v_{H3}	v_{H4}	v^x_{em}	v^x_{mg}	v^x_{eg}	**Voltage Level**
0	0	0	0	0	0	0	0	0	0	0	0
1	0	0	0	v_C	0	0	0	v_C	0	v_C	
0	1	0	0	0	v_C	0	0	v_C	0	v_C	1
0	0	1	0	0	0	v_C	0	0	v_C	v_C	
0	0	0	1	0	0	0	v_C	0	v_C	v_C	
1	1	0	0	v_C	v_C	0	0	$2v_C$	0	$2v_C$	
1	0	1	0	v_C	0	v_C	0	v_C	v_C	$2v_C$	
1	0	0	1	v_C	0	0	v_C	v_C	v_C	$2v_C$	2
0	1	1	0	0	v_C	v_C	0	v_C	v_C	$2v_C$	
0	1	0	1	0	v_C	0	v_C	v_C	v_C	$2v_C$	
0	0	1	1	0	0	v_C	v_C	0	$2v_C$	$2v_C$	
1	1	1	0	v_C	v_C	v_C	0	$2v_C$	v_C	$3v_C$	
1	1	0	1	v_C	v_C	0	v_C	$2v_C$	v_C	$3v_C$	3
1	0	1	1	v_C	0	v_C	v_C	v_C	$2v_C$	$3v_C$	
0	1	1	1	0	v_C	v_C	v_C	v_C	$2v_C$	$3v_C$	
1	1	1	1	v_C	v_C	v_C	v_C	$2v_C$	$2v_C$	$4v_C$	4

The entire arm voltage v^x_{eg} is uniformly distributed between the sub-arms (v^x_{em} and v^x_{mg}). For example, the arm voltage $v^x_{eg} = 2\,v_C$ can be obtained by using the sub-arm voltage combinations of ($2\,v_C$, 0), (v_C, v_C), and (0, $2\,v_C$). The sub-arm voltage $v^x_{em} = v_C$ or $v^x_{mg} = v_C$ can be generated using the redundant switching states. The selection of redundant switching states between the sub-arms gives a degree of freedom to ensure the overall energy stored in each arm is constant. The above objective can be achieved by controlling the current flowing through the cross-connected capacitor (i_{xw}). Typically, the PC-MMC is designed with an even number of submodules in each arm. Therefore, it is easy to form a mid-point to cross-connect the upper and lower arms through a capacitor. The number of voltage levels generated by the entire arm is given by

$$v_{Lev} = N + 1 \tag{5.1}$$

where v_{Lev} represents the arm voltage level, and N represents the number of submodules per arm.

5.3 PRINCIPLE OF OPERATION

The switching of submodules in the upper and lower arm of each leg must follow certain guidelines to generate the required voltage waveform while maintaining the power balance between the converter input and output. According to this criterion, the modulation signals of the upper and lower arms are defined. These modulation signals are used with the phase-shifted carrier modulation scheme (PSC-PWM) to generate the switching signals for submodules. The performance of the PC-MMC with a PSC-PWM is presented in the following subsections.

5.3.1 Modeling of PC-MMC

Each leg of PC-MMC is divided into four sub-arms, and each sub-arm has an inductor and two half-bridge submodules. The total DC-bus voltage V_{dc} is equally distributed between the four sub-arms. The submodules in each sub-arm are controlled to produce a phase voltage v_{xi}. From Figure 5.1, the voltage across the upper and lower arms is given by

$$\begin{aligned} v_{em}^{x} + v_{mg}^{x} &= \frac{V_{dc}}{2} - v_{xi} - L\frac{d\left(i_{xu} + i_{xw}\right)}{dt} - L\frac{d\,i_{xu}}{dt} \\ v_{rw}^{x} + v_{ws}^{x} &= \frac{V_{dc}}{2} + v_{xi} - L\frac{d\left(i_{xl} + i_{xw}\right)}{dt} - L\frac{d\,i_{xl}}{dt} \end{aligned} \tag{5.2}$$

where v_{em}^{x}, v_{mg}^{x} represents the voltage across upper sub-arms of phase-x and v_{rw}^{x}, v_{ws}^{x} represents the voltage across lower sub-arms of phase-x.

The converter leg current consists of DC and AC current components. These current components are estimated from the measured currents i_{xw}, i_{xu}, and i_{xl}, and are given by

$$\begin{aligned} i_{xd} &= \frac{1}{2}\left(i_{xu} + i_{xl} + i_{xw}\right) \\ i_{xh} &= \frac{1}{2} i_{xw} \end{aligned} \tag{5.3}$$

where $i_{xd} \in \{i_{ad}, i_{bd}, i_{cd}\}$ and $i_{xh} \in \{i_{ah}, i_{bh}, i_{ch}\}$ are the DC and high-frequency components in the arm current, $i_{xu} \in \{i_{au}, i_{bu}, i_{cu}\}$ is the upper arm current, $i_{xl} \in \{i_{al}, i_{bl}, i_{cl}\}$ is the lower arm current, and $i_{xw} \in \{i_{aw}, i_{bw}, i_{cw}\}$ is the cross-connected capacitor current.

From Figure 5.2, the output current is given by

$$i_{xi} = i_{xu} - i_{xl} \tag{5.4}$$

where $i_{xi} \in \{i_a, i_b, i_c\}$ represents the output current flowing through the load.

The current components are given in equations (5.3) and (5.4) produce a voltage drop across the arm inductor "L." The voltage drop due to DC current component is approximately zero while the voltage drop due to the fundamental current component is added to the phase voltage. The resultant voltage components are given by

$$\begin{aligned} L\,\frac{d\,i_{xd}}{dt} &\approx 0 \\ 2\,L\,\frac{d\,i_{xi}}{dt} + v_{xi} &= V_f \sin\left(\omega_o\, t + \theta_x\right) \end{aligned} \tag{5.5}$$

where V_f, ω_o, and θ_x are the peak value, fundamental angular frequency in (rad), and the initial angle of fundamental voltage component, respectively.

Substitute equations (5.3), (5.4), and (5.5) in equation (5.2), which gives the upper and lower arm voltage as

$$\begin{aligned} v^x_{em} + v^x_{mg} &= \frac{V_{dc}}{2} - V_f \sin\left(\omega_o\, t + \theta_x\right) \\ v^x_{rw} + v^x_{ws} &= \frac{V_{dc}}{2} + V_f \sin\left(\omega_o\, t + \theta_x\right) \end{aligned} \tag{5.6}$$

The arm voltage is generated by controlling the submodules in the sub-arms. The arm voltage given in equation (5.6) is equally distributed between the four sub-arms. The sub-arms voltage of phase-x are given by

$$\begin{aligned} v^x_{em} &= \frac{V_{dc}}{4} - \frac{V_f}{2} \sin\left(\omega_o\, t + \theta_x\right) - V_h \sin\left(\omega_h\, t\right) \\ v^x_{mg} &= \frac{V_{dc}}{4} - \frac{V_f}{2} \sin\left(\omega_o\, t + \theta_x\right) + V_h \sin\left(\omega_h\, t\right) \\ v^x_{rw} &= \frac{V_{dc}}{4} + \frac{V_f}{2} \sin\left(\omega_o\, t + \theta_x\right) - V_h \sin\left(\omega_h\, t\right) \\ v^x_{ws} &= \frac{V_{dc}}{4} + \frac{V_f}{2} \sin\left(\omega_o\, t + \theta_x\right) + V_h \sin\left(\omega_h\, t\right) \end{aligned} \tag{5.7}$$

where V_h and ω_h are the peak value and angular frequency of high-frequency voltage corresponding to the high-frequency current component.

The sub-arm voltage consists of DC, fundamental, and high-frequency voltage components. The high-frequency voltage component (V_h) is used to control the ripple in submodule capacitors voltage at low/zero frequency operation. This voltage component is set to zero ($V_h = 0$) at high-frequency (high-speed) operation. The magnitude of fundamental and high-frequency voltage components are given by

$$\begin{aligned} V_f &= \frac{m_a}{2}\, V_{dc} \\ V_h &= \frac{1 - m_a}{4}\, V_{dc} \end{aligned} \tag{5.8}$$

where m_a represents the amplitude modulation index of range 0 to 1.

5.3.2 Phase-Shifted Carrier Modulation for PC-MMC

The phase-shifted carrier modulation (PSC-PWM) requires N-triangular carrier signals for a PC-MMC with N submodules per arm. These triangular carrier signals are identical in magnitude and frequency, and horizontally arranged with a phase angle shift of $360°/N$. The pulse width modulator requires a modulation signal along with triangular carrier signals. The modulation signals given in equation (5.7) are normalized with respect to half of the DC-bus voltage $V_{dc}/2$. The resultant normalized modulation signals are given by

$$
\begin{aligned}
m_{u1}^{x} &= \frac{1}{2} - \frac{m_a}{2}\sin(\omega_o t + \theta_x) - \frac{1-m_a}{2}\sin(\omega_h t) \\
m_{u2}^{x} &= \frac{1}{2} - \frac{m_a}{2}\sin(\omega_o t + \theta_x) + \frac{1-m_a}{2}\sin(\omega_h t) \\
m_{l1}^{x} &= \frac{1}{2} + \frac{m_a}{2}\sin(\omega_o t + \theta_x) - \frac{1-m_a}{2}\sin(\omega_h t) \\
m_{l2}^{x} &= \frac{1}{2} + \frac{m_a}{2}\sin(\omega_o t + \theta_x) + \frac{1-m_a}{2}\sin(\omega_h t)
\end{aligned}
\tag{5.9}
$$

The operation of five-level PC-MMC with PSC-PWM scheme is shown in Figure 5.3. The modulation signals m_{u1}^{a} and m_{u2}^{a} for phase-a, upper two sub-arms are generated for an amplitude modulation index $m_a = 0.7$, fundamental frequency $f_o = 30$ Hz, and high frequency $f_h = 180$ Hz. The five-level PC-MMC require four triangular signals. These carrier signals are generated with a carrier frequency $f_c = 120$ Hz as shown in Figure 5.3(a). The carrier signals T_{u1} and T_{u2} are arranged with initial angles $0°$ and $180°$, respectively, and T_{u3} and T_{u4} are arranged with initial angles $90°$ and $270°$, respectively.

The modulation signal m_{u1}^{a} is compared with the carrier signals T_{u1} and T_{u2} to generate the gating signals for the devices S_{u1} and S_{u2}, respectively, as shown in Figure 5.3(b)–(c). These gating signals are applied to the SM-1 and SM-2, and their output voltages are added together resulting in a stepped waveform v_{em}^{a} with a voltage step of E as shown in Figure 5.3(d). Similarly, the modulation signal m_{u2}^{a} is compared with the carriers T_{u3} and T_{u4} to generate the gating signals for the devices S_{u3} and S_{u4} as shown in Figure 5.3(e)–(f). The output voltages of SM-3 and SM-4 are added together and results in a sub-arm voltage v_{mg}^{a} as shown in Figure 5.3(g). For example, at $\omega_o t_1$, the SM-1 is turned ON while the SM-2, SM-3, and SM-4 are turned OFF. The sub-arm voltages are $v_{em}^{a} = E$ and $v_{mg}^{a} = 0$. The first and second sub-arm voltages are added together, which gives the arm voltage v_{eg}^{a}. The arm voltage has a total of five voltage steps with a maximum value of "$4E$" and a minimum value of "0" as shown in Figure 5.3(h). Each voltage step is equal to the rated value of submodule capacitor voltage "E."

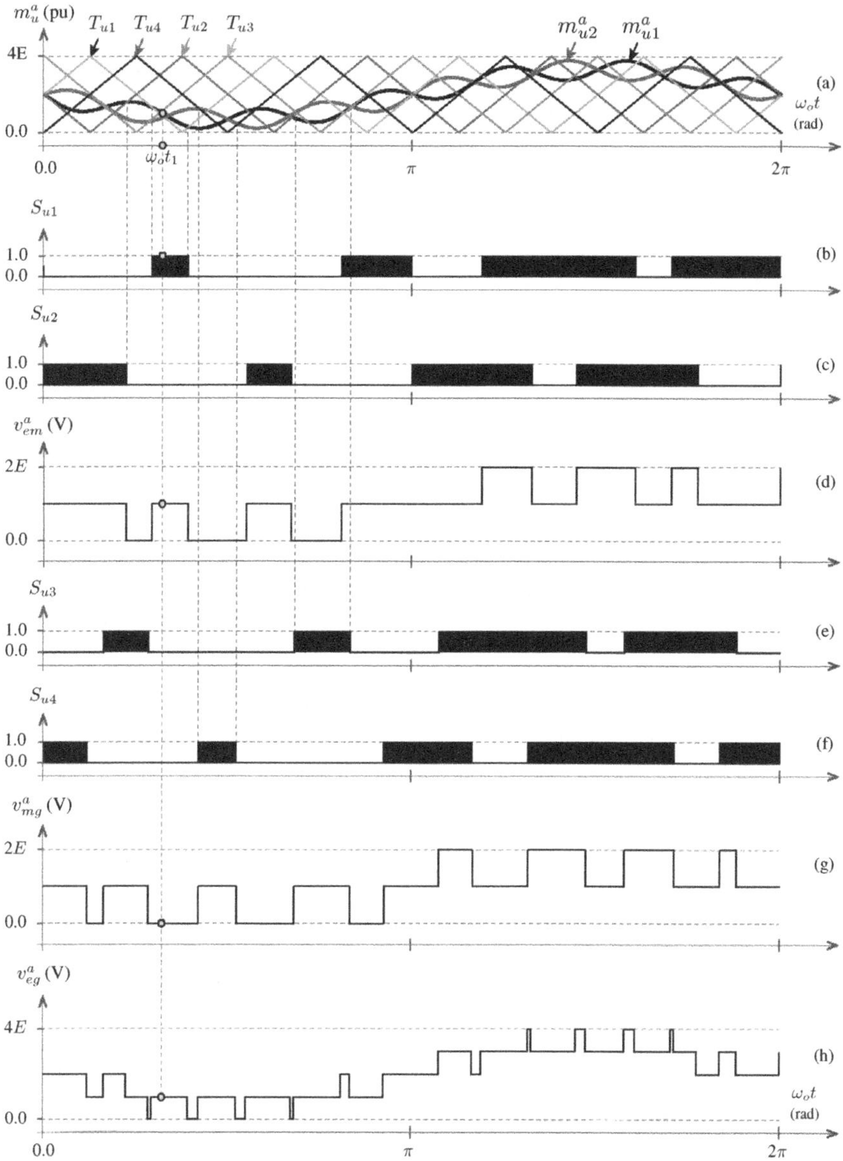

Figure 5.3 Implementation of PSC-PWM for a PC-MMC.

CASE STUDY 5.1 Analysis of PC-MMC with a PSC-PWM Scheme

Objective: In this case study, the performance of the PC-MMC with PSC-PWM scheme is studied through MATLAB simulation.

Parameters: The system parameters of the PC-MMC are as follows: rated DC-bus voltage $V_{dc} = 7$ kV, fundamental frequency $f_o = 30$ Hz, high-frequency $f_h = 180$ Hz, number of SMs per arm $N = 16$, arm inductor $L = 2$ mH, and carrier frequency $f_c = 1000$ Hz.

Analysis:

The PC-MMC modulation signal consists of high-frequency voltage component along with the fundamental voltage component. The high-frequency voltage component is used to minimize the voltage ripple at low/zero frequency. In this study, the high-frequency voltage component of $f_h = 180$ Hz is added to the modulation signals.

The performance of the PC-MMC with a PSC-PWM scheme is shown in Figure 5.4. The output voltage of top-half upper arm consists of a fundamental voltage component along with the high-frequency voltage component as shown in Figure 5.4(a). The sub-arm voltage has a maximum value of "$8E$" and a minimum value of "0." Figure 5.4(b) shows the bottom-half upper arm voltage, which has the same fundamental and high-frequency voltage components as that of the top-half upper arm. The fundamental voltage component in the top and bottom-half upper arms are in-phase, while the high-frequency voltage components are in opposite polarity. The entire upper arm voltage is obtained by adding the sub-arm voltages ($v_{eg}^a = v_{em}^a + v_{mg}^a$) as shown in Figure 5.4(c). The PC-MMC generates a 13-voltage level arm voltage at a modulation index of $m_a = 0.7$. The PC-MMC with 16 submodules per arm can generate a 17-level voltage waveform at unity modulation index.

The harmonic spectrum of the top-half upper arm voltage is shown in Figure 5.4(d). The normalized root mean square (RMS) values of DC component (v_d), fundamental voltage component (v_{xf}), and high-frequency voltage component (v_{xh}) exactly follow the modulation signals. Beside these voltage components, the top-half upper arm also introduces small harmonic components around the equivalent switching frequency of 4800 Hz. Figure 5.4(e) shows the harmonic spectrum of bottom-half upper arm voltage. The RMS value of voltage components is similar to the top-half upper arm voltage. The upper arm voltage consists of twice the magnitude of DC voltage component (v_d) and fundamental voltage component (v_{xf}) while the high-frequency voltage component (v_{xh}) is completely eliminated as shown in Figure 5.4(f). Also, a small amount of higher-order harmonics appears around the equivalent switching frequency.

The performance of the PC-MMC is analyzed at different modulation indices (m_a) as shown in Figure 5.5. The sub-arm voltage has a constant magnitude of 0.25 pu of DC component as shown in Figure 5.5(a) while the arm voltage has 0.5 pu magnitude of DC component as shown in Figure 5.5(b). In sub-arm voltage, the fundamental voltage component v_{xf} linearly increased while the high-frequency voltage component v_{xh} linearly decreased with the modulation index (m_a). The variation of fundamental and high-frequency voltage components are complementary in nature and follow the definition of modulation signals given in equation (5.9). The arm voltage consists of fundamental component only, and its magnitude linearly varies with m_a. The higher-order harmonics are very small in magnitude, and can be eliminated using the arm inductors.

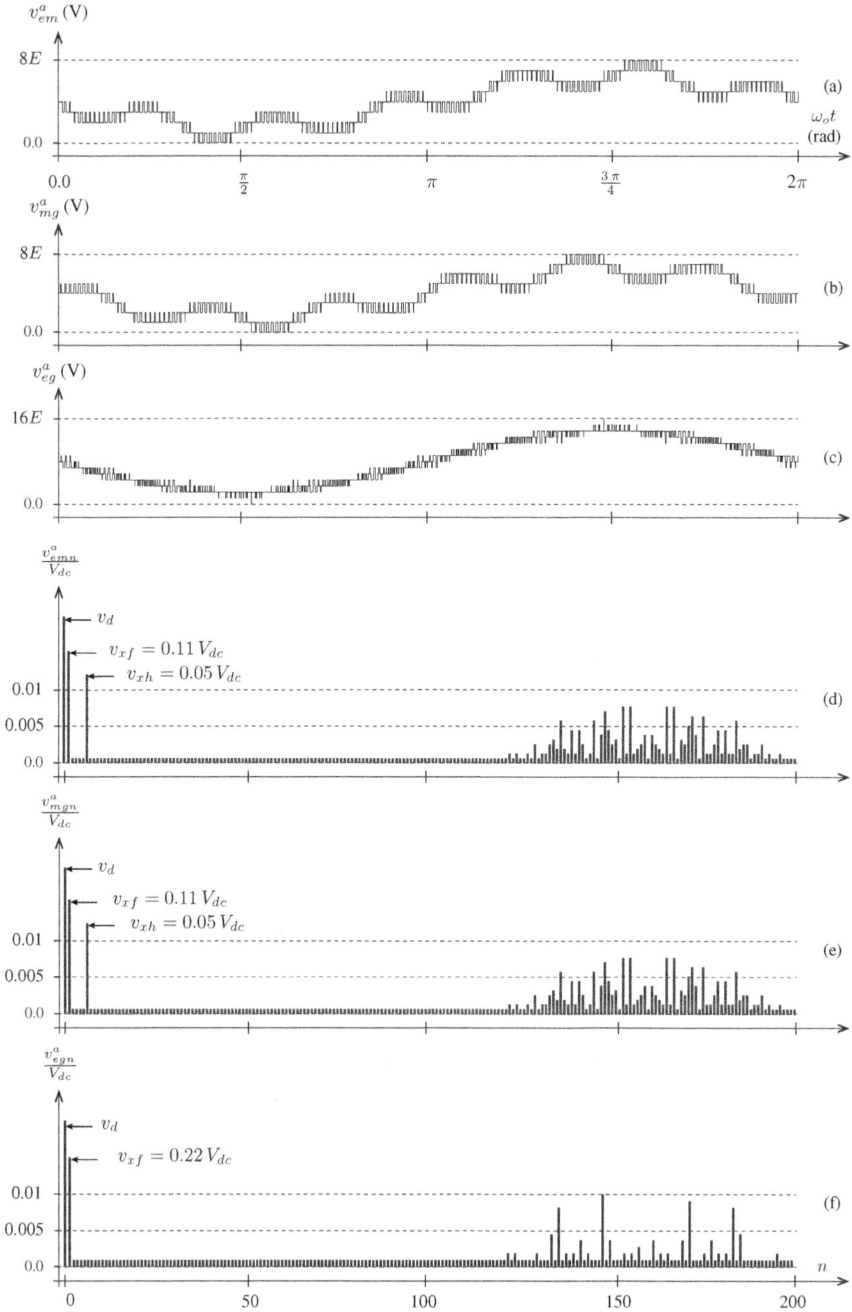

Figure 5.4 Performance of the PC-MMC with a PSC-PWM.

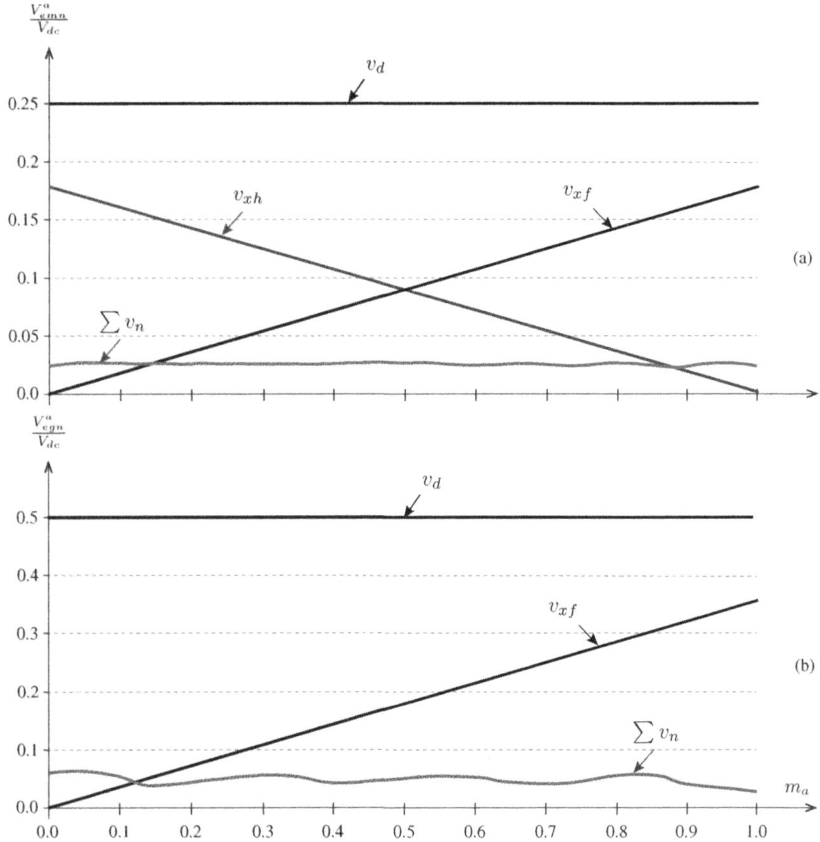

Figure 5.5 Analysis of sub-arm and arm voltages at different modulation indices.

■

5.4 LOW/ZERO FREQUENCY OPERATION OF PC-MMC

In conventional MMC, the submodule capacitors voltage has a very high ripple at low/zero frequency operation. These ripples increase the voltage stress on semiconductor devices and submodule capacitors, and affect the reliability and efficiency of conventional MMC. These issues limit the application of conventional modular multilevel converter for variable speed motor drives. For minimizing the capacitors voltage ripple, a high-frequency common-mode voltage (CMV) is externally injected into the system. The injected CMV increases the voltage stress on the motor winding insulation, and increases the shaft voltage which could lead to bearing and shaft failures. This issue can be resolved by using an input transformer. However, the transformer increases the cost and size of the drive. On the other hand, the proposed PC-MMC is capable of operating at low/zero frequency without increasing the size of submodule capacitors and motor neutral voltage. The cross-connected capacitor in PC-MMC helps to balance the power between the upper and lower arms and provides a path for the injected high-frequency current component. The design procedure of cross-connected capacitor and its impact on SM capacitors voltage and CMV at low/zero frequency operation are analyzed in the next section.

5.4.1 Equivalent Circuit

The equivalent circuit of a three-phase PC-MMC is shown in Figure 5.6, where the submodules in each sub-arm are modeled as a controlled voltage source. The controlled voltage source consists of DC (v_d), fundamental frequency (v_{xf}), and high-frequency (v_{xh}) voltage components.

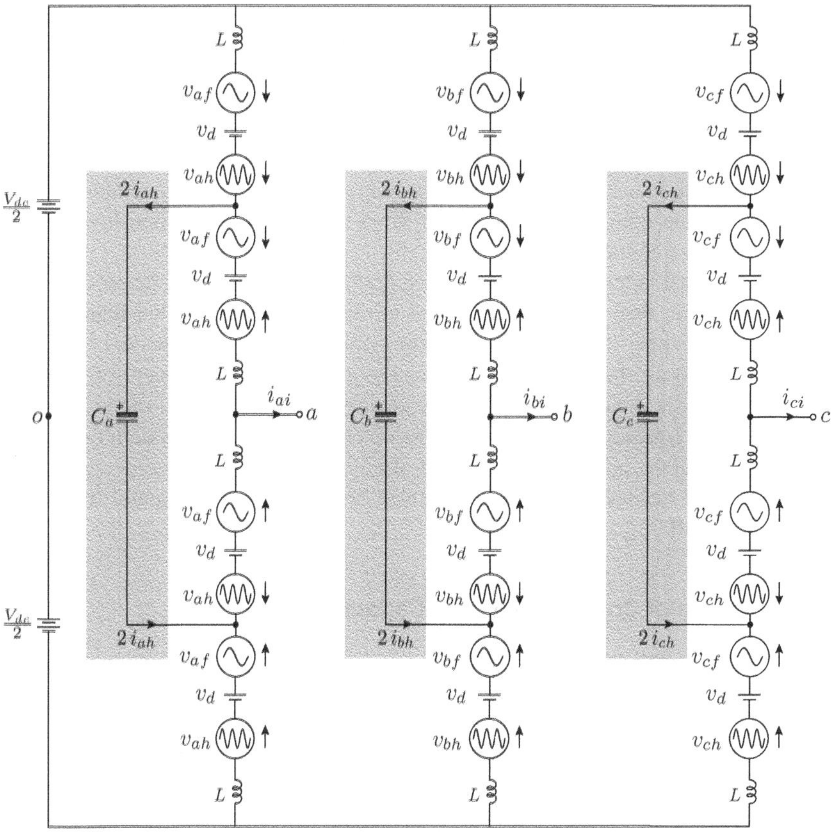

Figure 5.6 Equivalent circuit of a three-phase PC-MMC.

From equations (5.7), (5.8), and (5.9), the magnitudes of these voltage components are given by

$$\begin{aligned} v_d &= \frac{1}{4} V_{dc} \\ v_{xf} &= \frac{m_a}{4} V_{dc} \sin\left(\omega_o\, t + \theta_x\right) \\ v_{xh} &= \frac{1 - m_a}{4} V_{dc} \sin\left(\omega_h\, t\right) \end{aligned} \tag{5.10}$$

The DC voltage component in each sub-arm is controlled to support the DC-bus voltage V_{dc} while the fundamental AC voltage components in the upper and lower arms are controlled to generate a voltage component v_{xi} at the AC output terminals. The high-frequency voltage component is introduced to minimize the capacitors voltage ripple and to achieve the power balance between the upper and lower arms. The high-frequency voltage component in the upper and lower sub-arms are in opposite phase and cancel each other. Therefore, the injected high-frequency voltage component do not have any effect on the DC and AC system of PC-MMC.

The fundamental frequency voltage v_{xf} generates a current component i_{xi} in the arm. The interaction between the fundamental (low) frequency voltage and corresponding current components in the arm causes a ripple in submodule capacitors voltage. These voltage ripples increase the charging and discharging periods of submodule capacitors, which leads to an energy imbalance between the upper and lower arms. The injected high-frequency voltage component v_{xh} generates a high-frequency current component i_{xh}, which flows among the upper and lower arms through the cross-connected capacitor C_x.

When the total submodule capacitor voltage of the upper arm is larger than that of the lower arm, then the high-frequency current i_{xh} (usually higher than 60 Hz) flows from the upper arm to the lower arm through the cross-connected capacitor C_x. In other words, the upper arm releases the power while the lower arm absorbs it. Similarly, if the total capacitor voltage of upper arm is less than the lower arm, then the high-frequency current flows in the reverse direction to maintain the power balance between the arms. The same philosophy is used to maintain the capacitors voltage balance between the upper and lower arms at low fundamental frequency operation as well.

5.4.2 Design of Cross-Connected Capacitor

The cross-connected capacitor C_x plays a key role in the power balancing between the upper and lower arms of a PC-MMC at low/zero fundamental frequency. The cross-connected capacitor is designed such that a smooth power transfer between the upper and lower arms can be obtained without affecting the operation of PC-MMC. The single-phase equivalent circuit of PC-MMC corresponding to the high-frequency voltage is shown in Figure 5.7(a). The equivalent circuit is further simplified by canceling the opposite phase high-frequency voltage in the upper and lower arms as shown in Figure 5.7(b). The transfer function and resonant frequency (f_r) of the equivalent circuit is given by

$$
\begin{aligned}
G_{LC} &= s\,L + \frac{1}{s\,C_x} \\
f_r &= \frac{1}{2\pi\sqrt{L\,C_x}}
\end{aligned}
\tag{5.11}
$$

The high-frequency voltage component is injected at a frequency equal to the resonant frequency formed by the cross-connected capacitor and the arm inductor. Therefore, the equivalent impedance of the cross-connected power channel is significantly reduced. According to this criteria, the frequency of injected voltage is given by

$$f_h \approx f_r \tag{5.12}$$

From equations (5.11) and (5.12), the value of cross-connected capacitor C_x is given by

$$C_x = \frac{1}{4\,\pi^2\,f_h^2\,L} \tag{5.13}$$

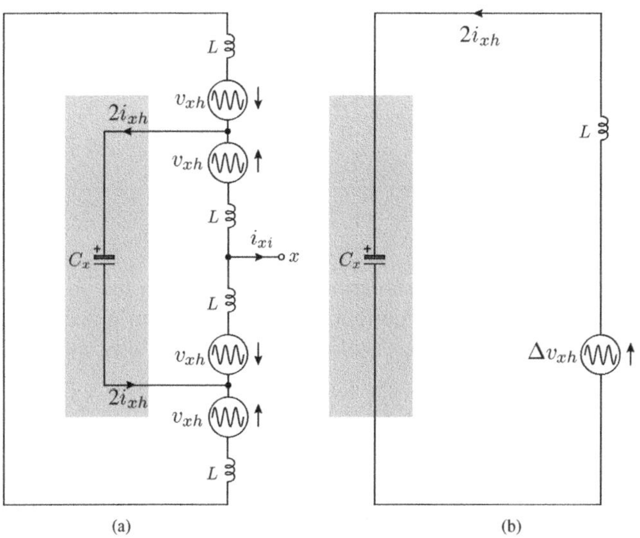

Figure 5.7 High-frequency equivalent circuit of PC-MMC.

5.4.3 Submodule Capacitor Voltage Ripple

From the equivalent circuit shown in Figure 5.6, the current flowing through each sub-arm is given by

$$\begin{aligned} i_{xu}^{m} &= i_{xd} + \frac{1}{2} i_{xi} + i_{xh} \\ i_{xu} &= i_{xd} + \frac{1}{2} i_{xi} - i_{xh} \\ i_{xl} &= i_{xd} - \frac{1}{2} i_{xi} - i_{xh} \\ i_{xl}^{m} &= i_{xd} - \frac{1}{2} i_{xi} + i_{xh} \end{aligned} \tag{5.14}$$

The sub-arm current consists of DC current component (i_{xd}) , fundamental frequency current component (i_{xi}), and high-frequency current component (i_{xh}). For the analysis of capacitors voltage ripple, these current components are defined as

$$\begin{aligned} i_{xd} &= I_{dcx} \\ i_{xh} &= I_h \sin(\omega_h t) \\ i_{xi} &= I \sin(\omega_o t + \theta_x + \varphi) \end{aligned} \tag{5.15}$$

where, I_h and I are the peak value of high-frequency and fundamental frequency current components, respectively.

From equations (5.7), (5.8), and (5.10), the sub-arm voltages are given by

$$\begin{aligned} v_{em}^{x} &= \frac{V_{dc}}{4} - v_{xf} - v_{xh} \\ v_{mg}^{x} &= \frac{V_{dc}}{4} - v_{xf} + v_{xh} \\ v_{rw}^{x} &= \frac{V_{dc}}{4} + v_{xf} - v_{xh} \\ v_{ws}^{x} &= \frac{V_{dc}}{4} + v_{xf} + v_{xh} \\ v_{xf} &= \frac{v_{xi}}{2} \end{aligned} \tag{5.16}$$

The AC side power of each sub-arm is calculated from the product of sub-arm voltage and corresponding arm current. The AC side arm power is equal to the summation of two sub-arm voltages belonging to the same arm, and is given by

$$\begin{aligned} p_{eg}^{x} &= v_{em}^{x} \times i_{xu}^{m} + v_{mg}^{x} \times i_{xu} \\ p_{rs}^{x} &= v_{rw}^{x} \times i_{xl} + v_{ws}^{x} \times i_{xl}^{m} \end{aligned} \tag{5.17}$$

The upper and lower arm powers are obtained by substituting equations (5.14) and (5.16) in (5.17). The resultant arm powers are given by

$$\begin{aligned} p_{eg}^{x} &= \frac{V_{dc}\, I_{dcx}}{2} - \frac{v_{xi}\, i_{xi}}{2} + \frac{V_{dc}\, i_{xi}}{4} - v_{xi}\, I_{dcx} - 2\, v_{xh}\, i_{xh} \\ p_{rs}^{x} &= \frac{V_{dc}\, I_{dcx}}{2} - \frac{v_{xi}\, i_{xi}}{2} - \frac{V_{dc}\, i_{xi}}{4} + v_{xi}\, I_{dcx} + 2\, v_{xh}\, i_{xh} \end{aligned} \tag{5.18}$$

In equation (5.18), the first two terms on the right-hand side have the same polarity in the upper and lower arms. These terms represent the total power of the converter leg. The last three terms are in opposite polarity and affect the power imbalance between the upper and lower arms. From equation (5.18), the following conditions are depicted to maintain the average leg power zero, and to cancel the deviation of the upper and lower arm powers at steady state. The stead-state operating conditions are summarized as

$$\begin{aligned} \frac{V_{dc}\, I_{dcx}}{2} - \frac{v_{xi}\, i_{xi}}{2} &= 0 \\ \frac{V_{dc}\, i_{xi}}{4} - v_{xi}\, I_{dcx} - V_h\, I_h &= 0 \end{aligned} \tag{5.19}$$

From equation (5.19), the magnitude of DC and high-frequency current components are resolved as

$$\begin{aligned} I_{dcx} &= \frac{v_{xi}\, i_{xi}}{V_{dc}} \\ I_h &= \frac{0.25\, V_{dc}\, i_{xi} - v_{xi}\, I_{dcx}}{V_h} \end{aligned} \tag{5.20}$$

The submodule capacitor voltage ripple is depicted from the relationship between submodule instantaneous power and DC-side voltage by considering the conditions given in equation (5.19). During this analysis, it is assumed that each sub-arm power is equally distributed among the submodules. The individual submodule power for SM-1 in the top-half upper arm is given by

$$p^{x}_{em1} = \frac{2}{N}\, [v^{x}_{em}\, i^{m}_{xu}] \tag{5.21}$$

From equation (5.14), (5.16), and (5.20), the components presented in the individual submodule power is given by

$$\begin{aligned} p^{x}_{em1} = \frac{2}{N}\, [&\frac{V_{dc}\, i_{xh}}{4} - \frac{v_{xi}\, i_{xh}}{2} - v_{xh}\, I_{dcx} - \frac{v_{xh}\, i_{xi}}{2} \\ &+ \frac{V_h\, I_h}{2} \sin(2\,\omega_h\, t)] \end{aligned} \tag{5.22}$$

From the definition of voltage and current components given in equations (5.8), (5.10), and (5.15), the instantaneous submodule power is given by

$$\begin{aligned} p^{x}_{em1} = \frac{2}{N}\, [&\frac{V_{dc}\, I_h}{4} \sin(\omega_h\, t) - \frac{V_f\, I_h}{2} \sin(\omega_o\, t + \theta_x) \sin(\omega_h\, t) \\ &- V_h\, I_{dcx} \sin(\omega_h\, t) - \frac{V_h\, I}{2} \sin(\omega_o\, t + \theta_x + \varphi) \sin(\omega_h\, t) \\ &+ \frac{V_h\, I_h}{2} \sin(2\,\omega_h\, t)] \end{aligned} \tag{5.23}$$

The submodule capacitor voltage in terms of the power absorbed by the submodule is given by

$$v^{x}_{Cy1} = \sqrt{\frac{2}{C}\left[\frac{1}{2} C\, V_C^2 + \int p^{x}_{em1}\, dt\right]} \tag{5.24}$$

Equations (5.23) and (5.24) show that the instantaneous value of submodule power consists of a high-frequency term of "sin $(\omega_h t)$," which reduces the charging and discharging periods of submodule capacitors. Hence, the voltage ripple is greatly minimized.

CASE STUDY 5.2 Voltage Ripple Reduction at Low-Frequency Operation

Objective: In this case study, the effect of injected high-frequency voltage on submodule capacitor voltage ripple is studied using MATLAB simulation. The impact of high-frequency voltage injection is also studied.

Parameters: The system parameters of the PC-MMC are as follows: rated DC-bus voltage $V_{dc} = 7$ kV, number of SMs per arm $N = 8$, SM capacitor voltage $v_C = 875$ V, SM capacitance $C = 4700$ μF, cross-connected capacitor $C_x = 400$ μF, load inductance $L_o = 20$ mH, load resistance $r_o = 34$ Ω, arm inductor $L = 2$ mH, and carrier frequency $f_c = 1000$ Hz.

Analysis:
In this case study, the DC side of PC-MMC is modeled as a stiff DC voltage source V_{dc}, and the AC side is connected to a three-phase passive load formed by a resistor r_o and an inductor L_o. The three-phase currents flowing through the load are controlled by a close-loop current control, which will be discussed in Section 5.5. The amplitude and frequency of the three-phase reference currents are set to 50 A and 1 Hz, respectively. The simulation results with and without injection of the high-frequency component as shown in Figure 5.8. Under these conditions, the PC-MMC is operating at an amplitude modulation index (m_a) of 0.5, corresponding to the current magnitude is 50 A at steady state.

Figure 5.8(a) shows the submodule capacitors voltage of a PC-MMC without the injection of high-frequency voltage and current components. Without high-frequency voltage, there is no power transfer between the upper and lower arms through a cross-connected capacitor. Hence, the PC-MMC is simply operating like a conventional MMC. The PC-MMC generates an output voltage at 1 Hz frequency corresponding to the 50 A current. The low-frequency load current component is equally distributed among the upper and lower arms and interacts with the DC voltage components in the arms. This interaction causing a fluctuating power component in the arm is shown in equation (5.18) (third term on the right-hand side). The upper and lower arms also carry a DC-bus current component corresponding to the power demanded by the load. The DC-bus current interacts with the low-frequency AC output voltage component in the arm and results in another fluctuating power component in the arm power as stated in equation (5.18) (fourth term on right-hand side). These two terms have opposite polarity in the upper and lower arm powers. Therefore, the fluctuations in the upper and lower arm submodule capacitors voltage are in opposite phase as well (i.e., during the charging period of the upper arm submodule capacitors, the lower arm submodule capacitors will discharge and vice versa). The peak-to-peak ripple in submodule capacitors voltage is as high as 850 V for a duration of 1 s, and the voltage ripple is 48.6% with respect to the nominal submodule capacitors voltage of 875 V.

Figure 5.8(b) shows the simulation results with the injection of high-frequency voltage component. The injected high-frequency voltage generates a controllable high-frequency current component, which flows through the cross-connected capacitor in a PC-MMC. The cross-connected capacitor helps to reduce the voltage ripple in submodule capacitors voltage and maintains the power balance between the upper and lower arms. Figure 5.8(b) shows that the voltage fluctuation is minimized as low as 25 V for a duration of 1/180 s. The ripple in submodule capacitors voltage is around 1.4% of its nominal voltage. Therefore, the size of submodule capacitor in a PC-MMC is significantly reduced due to the smaller ripple compared to the conventional MMC.

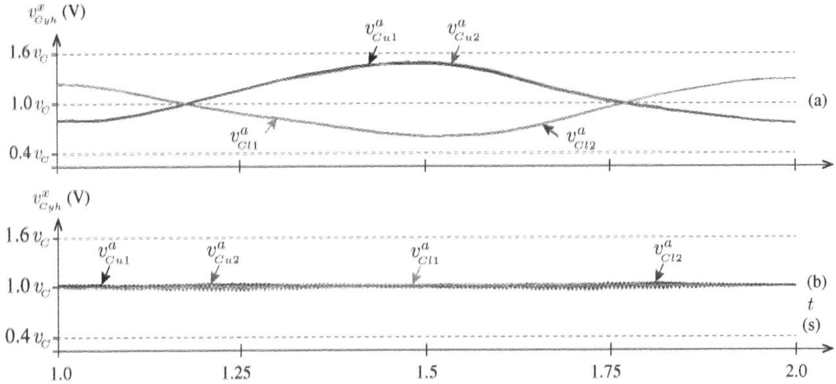

Figure 5.8 Submodule capacitor voltage (a) without high-frequency voltage, and (b) with high-frequency voltage at low-frequency operation.

■

5.4.4 Common-Mode Voltage

In conventional MMC, the high-frequency voltage component is injected into the system to minimize the voltage ripple at a low frequency of operation. The injected voltage appears in the common-mode voltage (CMV) and increases its peak value. The increased CMV increases the voltage stress on motor winding insulation, and generates shaft voltage in a motor drive system. In the case of PC-MMC, the injected high-frequency voltage does not appear in CMV and effectively reduces the voltage ripple at a low frequency of operation. The CMV generated by PC-MMC is given by

$$v_{cmi} = \frac{1}{3}\left[v_{ai} + v_{bi} + v_{ci}\right] \tag{5.25}$$

where v_{ai}, v_{bi}, and v_{ci} represents the phase-to-neutral voltage of the three phases, and v_{cmi} is the inverter common-mode voltage. From equations (5.2), (5.3), and (5.4), the output voltage is given by

$$v_{xi} = \frac{1}{2}(v_{rs}^{x} - v_{eg}^{x}) - 2\,L\,\frac{d\,i_{xi}}{dt} \tag{5.26}$$

From equations (5.25) and (5.26), the mathematical representation of CMV is given under the assumption that the three-phase currents are balanced and their summation become zero (i.e., $i_{ai} + i_{bi} + i_{ci} = 0$). The CMV in terms of arm voltage is given by

$$v_{cmi} = \frac{1}{6}\left[v_{rs}^{a} + v_{rs}^{b} + v_{rs}^{c} - v_{eg}^{a} - v_{eg}^{b} - v_{eg}^{c}\right] \tag{5.27}$$

The arm voltages are generated by controlling the submodules using a PWM modulator. The PWM schemes introduce a small amount of switching ripple v_{sw} besides the dominant components in the reference arm voltage. The switching ripple v_{sw} is added to the reference arm voltage given in equation (5.6) and results in

$$\begin{aligned} v_{eg}^{x} &= \frac{V_{dc}}{2} - V_f \sin(\omega_o t + \theta_x) + v_{sw} \\ v_{rs}^{x} &= \frac{V_{dc}}{2} + V_f \sin(\omega_o t + \theta_x) + \overline{v}_{sw} \end{aligned} \tag{5.28}$$

Equation (5.28) is substituted in equation (5.27) and the CMV produced by the PC-MMC is given by

$$v_{cmi} = \frac{1}{2}(v_{sw} + \overline{v}_{sw}) \tag{5.29}$$

Equation (5.29) shows that the injected high-frequency voltage is completely eliminated from the CMV produced by the PC-MMC, and it only consists of a small switching ripple generated by the PWM scheme, which makes the PC-MMC a potential candidate for motor drive applications, especially where the low-speed, high-torque is required.

CASE STUDY 5.3 Effect of High-Frequency Voltage in a PC-MMC

Objective: In this case study, the effect of injected high-frequency voltage on CMV in a PC-MMC is studied through MATLAB simulation.

Parameters: The system parameters of the PC-MMC are as follows: rated DC-bus voltage $V_{dc} = 7$ kV, number of SMs per arm $N = 8$, SM capacitor voltage $v_C = E$ V, SM capacitor $C = 4700$ μF, cross-connected capacitor $C_x = 400$ μF, load inductance $L_o = 20$ mH, load resistance $r_o = 34$ Ω, arm inductor $L = 2$ mH, and carrier frequency $f_c = 600$ Hz.

Analysis:
This case study investigates the CMV generated by the PC-MMC. Each arm consists of $N = 8$ submodules and each submodule capacitor voltage is set to E. The modulation signals of four sub-arms in each leg are synthesized using equation (5.9). The four submodules in each sub-arm are controlled using a single modulation signal. The PSC-PWM scheme requires 16 triangular carrier signals to control the 16 submodules in each arm, which are phase-shifted by $360°/16$ with each other. In this simulation, the fundamental frequency of the modulation signal is steeply changed from 1 Hz to 10 Hz and 30 Hz, and their amplitude modulation index is set to 0.7. The voltage waveform of the six arms is shown in Figure 5.9.

Figure 5.9(a) shows the voltage waveforms of the upper and lower arms in phase-a. The high-frequency voltage components are arranged in opposite phase in the two half-arms. The sub-arm voltages consist of a DC component ($8E$) and a fundamental voltage component along with the switching ripples. The AC components in the upper and lower arm voltages are in opposite phase. As the fundamental frequency of the modulation signals is steeply changed at a time instants of $t = 1$ s and $t = 1.5$ s, and the time period of arm voltages is also varied accordingly. Figures 5.9(b) and 5.9(c) show the upper and lower arm voltages in phase-b and -c, respectively. These voltages have similar components as that of the phase-a, but the fundamental components are phase-shifted by 120° and 240° with respect to the phase-a, respectively.

According to equation (5.27), the common-mode voltage generated by PC-MMC can be obtained from the arm voltages as shown in Figure 5.9(d). The waveform shows that the CMV contains only a small amount of switching ripples of 0.5 E, which is around 3% of the nominal DC-bus voltage V_{dc}. Also, the injected high-frequency voltage (f_h = 180 Hz) does not appear in the CMV.

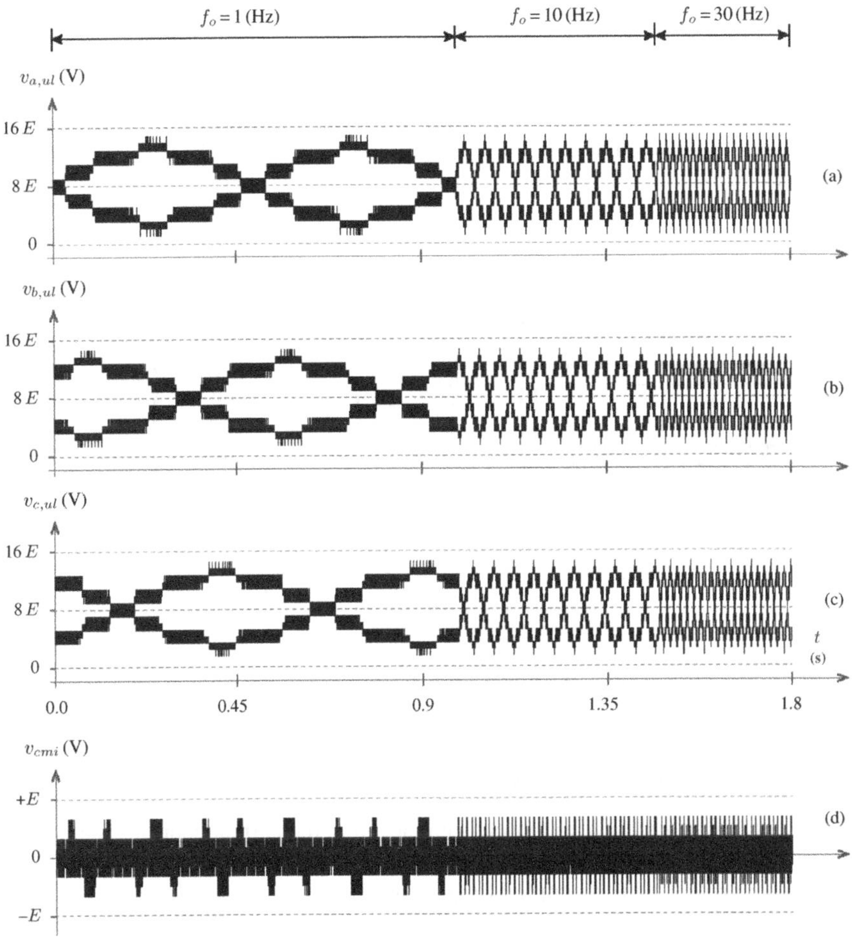

Figure 5.9 Simulated waveforms of CMV in a PC-MMC.

■

5.5 CLASSICAL CONTROL OF PC-MMC

For the stable operation of PC-MMC, the AC output current and submodule capacitors voltage must be controlled using a closed-loop control method. Figure 5.10

shows the control block diagram of PC-MMC, where the AC output current control is implemented at the system level, which gives three-phase reference modulation signals. The AC output current is generated by controlling the submodules in each arm. These submodules have floating capacitors, and their voltages must be controlled to achieve the perfect current tracking. The submodule capacitors voltage control is implemented in the local controller, which generates a compensating signal. The compensating signal and modulation signals are added together and result in final modulation signal for each sub-arm. These modulation signals are applied to the phase-shifted carrier modulator (PSC-PWM). The comparison of modulation and triangular carrier signals gives the gating signals for each submodule. The AC output current and submodule capacitor voltage control require feedback signals. These feedback signals are calculated from the measured arm currents and submodule capacitors voltage.

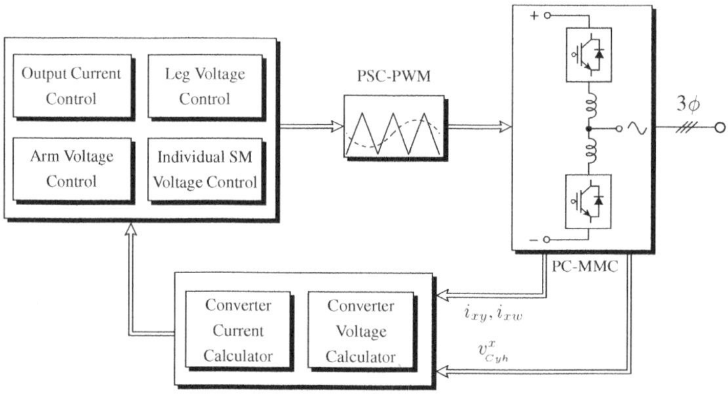

Figure 5.10 Block diagram of the PC-MMC control system.

5.5.1 Output Current Control

The synchronous-dq frame approach is employed in the current control. The three-phase time-varying currents are converted into two-phase DC current components and are controlled using a simple PI-regulator. The synchronous-dq frame requires two PI-regulators in comparison to three PI-regulators in the stationary-abc frame control.

Figure 5.11 shows the structure of output current control in the synchronous-dq reference frame. The three-phase output currents ($\mathbf{i}_{\mathbf{i,abc}}$) are calculated from the measured upper and lower arm currents. These currents are converted into a synchronous-dq reference frame rotating at the output frequency (θ). The three-phase currents become two DC currents i_{di} and i_{qi} in the synchronous-dq frame. The feedback signals i_{di} and i_{qi} are compared with the reference dq-currents of i^*_{di} and i^*_{qi}, respectively. The current errors are passed through PI-regulators and generate a reference voltage commands of v^*_{di} and v^*_{qi}, respectively. The reference voltages in the dq-frame are converted back to abc-frame using $\mathbf{T}_{dq \Rightarrow abc}$ transformation matrix given in Chapter 3. The reference voltages in abc-frame are used with PWM scheme to generate the gating signals for the submodules. The v_{di}^* and v_{qi}^* are used to obtain the amplitude modulation index of PC-MMC, which is used in the calculation of high-frequency voltage component magnitude.

The generation of reference dq-currents i_{di}^* and i_{qi}^* depends on the specific application. For example, in HVDC applications, the i_{di}^* and i_{qi}^* are generated according to the active and reactive power demands, respectively. In motor drive application, the reference currents are generated using a field-oriented control (FOC) or other advanced drive control schemes.

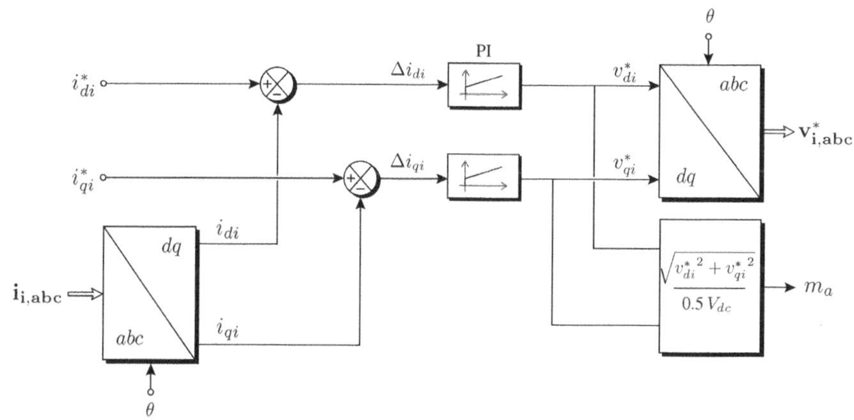

Figure 5.11 Output current control in synchronous dq-frame.

CASE STUDY 5.4 Performance of dq-Frame Current Control for PC-MMC

Objective: In this case study, the dynamic performance of output current control in dq-frame for PC-MMC is studied through MATLAB simulation.

Parameters: The system parameters of the PC-MMC are as follows: rated DC-bus voltage $V_{dc} = 7$ kV, number of SMs per arm $N = 8$, SM capacitor voltage $v_C = E = 875$ V, SM capacitor $C = 4700\ \mu$F, cross-connected capacitor $C_x = 400\ \mu$F, load inductance $L_o = 20$ mH, load resistance $r_o = 34\ \Omega$, arm inductor $L = 2$ mH, fundamental frequency $f_o = 10$ Hz, high-frequency $f_h = 180$ Hz, and carrier frequency $f_c = 1000$ Hz.

Analysis:

In this case study, the DC-bus of PC-MMC is modeled as a constant DC voltage source rated at 7000 V, and the AC system is configured to use a passive load with a resistor r_o and an inductor L_o. The dynamic performance of output current control at a fundamental frequency of $f_o = 10$ Hz is shown in Figure 5.12. In this study, the reference dq-currents i_{di}^* and i_{qi}^* are defined by the user. The magnitude of d-axis current component is proportional to the active power demanded by the load. Similarly, the q-axis current component is proportional to the reactive power demand (i.e., power factor). Initially, i_{di}^* is set to 100 A and i_{qi}^* is set to zero. At $t = 0.15$ s, the i_{di}^* is changed from 100 A to 200 A while i_{qi}^* is maintained at zero as shown in Figure 5.12(a). The actual currents are perfectly following the reference currents. The currents in the dq-frame are transformed to the stationary-abc frame as shown in Figure 5.12(b). The three-phase currents in abc-frame have a peak value equal to the magnitude of i_{di}^*.

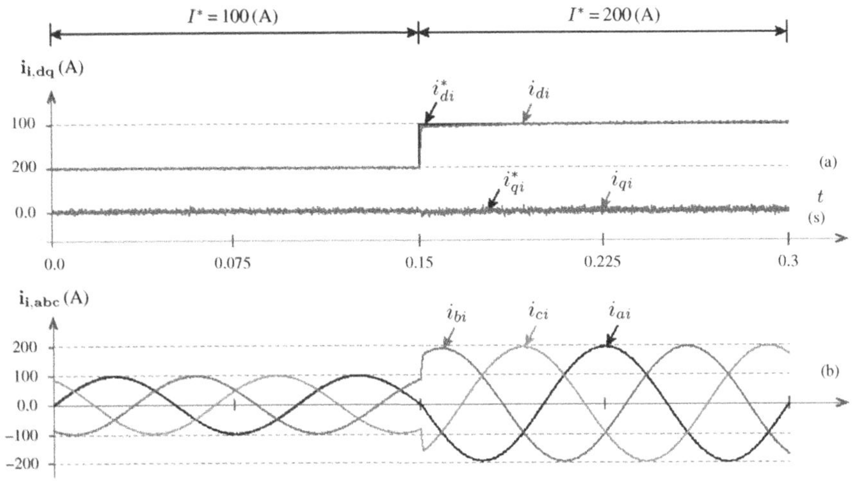

Figure 5.12 Dynamic performance of output current control in the synchronous-dq frame. ■

5.5.2 Submodule Capacitor Voltage Control

The overall submodule capacitor voltage control is divided into 1) leg capacitor voltage control, 2) arm capacitor voltage control, and 3) individual submodule capacitor voltage control. This section provides the implementation and working principle of each control method.

(1) Leg Capacitor Voltage Control: The main goal of the leg capacitor voltage control is to maintain the average DC voltage of each leg at the given reference value V_C^*. This objective is achieved by controlling the DC current component flowing through each converter leg. The structure of leg capacitor voltage control is shown in Figure 5.13. The leg capacitor voltage control consists of an outer voltage loop and an inner current control loop. In the outer voltage loop, the reference average DC voltage is compared with the actual average DC voltage v_{Cx}. The average DC voltage of each leg is given by

$$v_{Cx} = \frac{1}{2N}\left[\sum_{h=1}^{N} v_{Cuh}^{x} + \sum_{h=1}^{N} v_{Clh}^{x}\right] \tag{5.30}$$

The voltage error is controlled using a PI-regulator. The output of the PI-regulator is added to the feed-forward term given in equation (5.20) and results in a reference DC current component flowing through the converter leg. The feed-forward term compensates the error between the input (DC) and output (AC) active powers. The reference DC-bus current is compared with the actual DC-bus current flowing through each leg, and its error is applied to the PI-regulator. The current controller minimizes the current error and gives a reference voltage command. The reference voltage command is added to the modulation signals, which are obtained from the output current control.

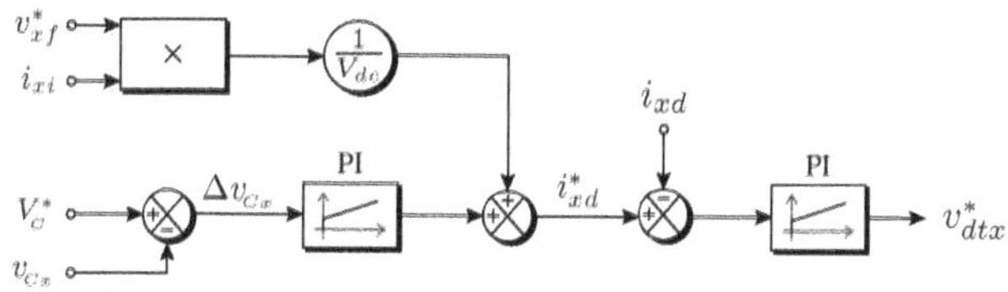

Figure 5.13 Block diagram of leg capacitor voltage control.

(2) Arm Capacitor Voltage Control: The main objective of the arm capacitor voltage control is to maintain the voltage balancing between the upper and lower arm capacitor voltages. The arm voltage balancing is achieved by controlling the high-frequency current flowing through the arms. The block diagram of the arm capacitor voltage control is shown in Figure 5.14. The arm capacitor voltage control consists of two independent control loops. The first control loop generates a high-frequency voltage component proportional to the modulation index m_a, which is obtained from the output current control. The magnitude of the injected high-frequency voltage in the upper and lower arms are adjusted by using a closed-loop controller. This controller has an outer voltage loop, which minimizes the error between the upper and lower arm average capacitor voltage and is given by

$$\begin{aligned} v_{Cu} &= \frac{1}{N}\sum_{h=1}^{N} v_{Cuh}^x \\ v_{Cl} &= \frac{1}{N}\sum_{h=1}^{N} v_{Clh}^x \end{aligned} \tag{5.31}$$

The voltage loop gives the high-frequency reference current and is added to the feed-forward term given in equation (5.20). The final output is multiplied with the unity magnitude sinusoidal signal generated at a frequency of ω_h. This signal is compared with the actual high-frequency current component obtained from the feedback signals (measurements), and its error is minimized using a PI-regulator. The current controller gives a compensating high-frequency voltage command Δv_{xh}^*. The compensating voltage command is added to the upper and lower arm modulation signals.

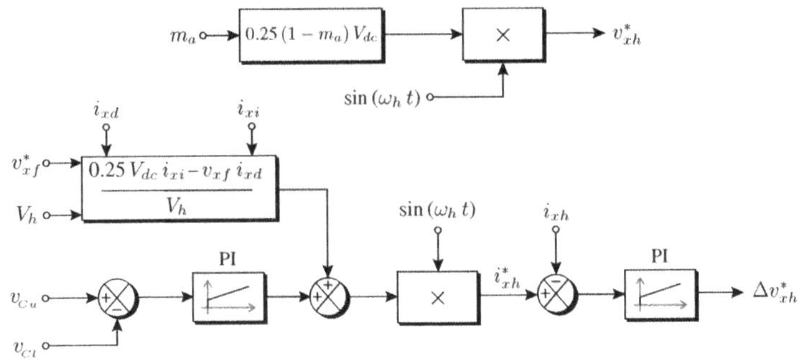

Figure 5.14 Block diagram of arm capacitor voltage control.

(3) Individual Submodule Capacitor Voltage Control: The individual submodule capacitor voltage control maintains each submodule capacitor voltage at the given reference value V_C^*. Figure 5.15 shows the individual submodule capacitor voltage control, where the measured submodule capacitor voltage is compared with the reference voltage V_C^*. The voltage error is minimized using a PI-regulator. The controller output is multiplied with the arm current to obtain the reference voltage component Δv^*_{xyh}. The reference voltage component is used to control the charging and discharging of each submodule capacitor. The current flowing through the first upper sub-arm and second lower sub-arm are given by

$$\begin{aligned} i^m_{xu} &= i_{xu} + \frac{1}{2}\, i_{xw} \\ i^m_{xl} &= i_{xl} + \frac{1}{2}\, i_{xw} \end{aligned} \tag{5.32}$$

where i_{xu}, i_{xw}, and i_{xl} are the measured currents of second upper sub-arm, cross-connected capacitor, and first lower sub-arm currents.

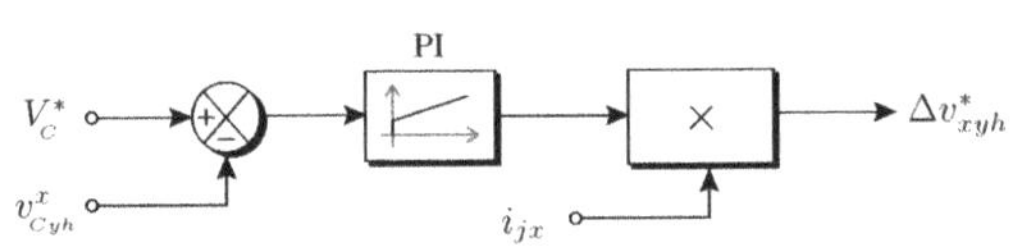

Figure 5.15 Block diagram of individual SM capacitor voltage control.

CASE STUDY 5.5 Performance of Capacitor Voltage Control for PC-MMC

Objective: In this case study, the dynamic performance of submodule capacitor voltage control for PC-MMC is studied through MATLAB simulation.

Parameters: The system parameters of the PC-MMC are as follows: rated DC-bus voltage $V_{dc} = 7$ kV, number of SMs per arm $N = 8$, SM capacitor voltage $v_C = E = 875$ V, SM capacitor

$C = 4700\ \mu$F, cross-connected capacitor $C_x = 400\ \mu$F, load inductance $L_o = 20$ mH, load resistance $r_o = 8.5\ \Omega$, arm inductor $L = 2$ mH, fundamental frequency $f_o = 10$ Hz, high-frequency $f_h = 180$ Hz, and carrier frequency $f_c = 1000$ Hz.

Analysis:
The dynamic performance of submodule capacitor voltage control for PC-MMC at a fundamental frequency of $f_o = 10$ Hz is shown in Figure 5.16. The total DC-bus voltage is set to 7000 V. Each arm has a total of 8 submodules, and the total DC-bus voltage is equally distributed among the submodules. The rated submodule capacitor voltage is equal to 7000/8 = 875 V.

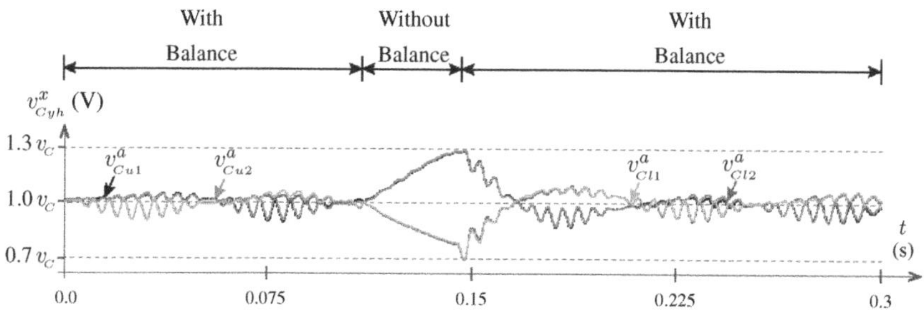

Figure 5.16 Dynamic performance of SM capacitor voltage control.

Initially, the converter is controlled to supply an output current of $I = 200$ A. The submodule capacitors in the upper and lower arms are perfectly maintained at their nominal value. At $t = 0.11$ s, the capacitor voltage control is disabled. The submodule capacitors voltage diverges from its nominal value, which will affect the reliability and operation of PC-MMC. At $t = 0.14$ s, the voltage control is activated, and the submodule capacitors voltage is converges toward its rated value. ■

5.5.3 Synthesis of Modulation Signals

The output current control and various stages of submodule capacitor voltage control are discussed in Sections 5.5.1 and 5.5.2. These controllers generate various voltage commands such as the DC (V_{dc}), fundamental (v^*_{xf}), and high-frequency voltage components (v^*_{xh}), and their compensating voltage components v^*_{dtx}, Δv^*_{xh}, and Δv^*_{xyh}, respectively. These voltage commands are used to synthesize the modulation signals for each submodule in the upper and lower sub-arms as

$$\begin{aligned} v^*_{xuh} &= \frac{V_{dc}}{2N} - \frac{v^*_{xf}}{N} \mp \frac{2\,v^*_{xh}}{N} - v^*_{dtx} \mp \Delta v^*_{xh} + \Delta v^*_{xyh} \\ v^*_{xlh} &= \frac{V_{dc}}{2N} + \frac{v^*_{xf}}{N} \mp \frac{2\,v^*_{xh}}{N} - v^*_{dtx} \pm \Delta v^*_{xh} + \Delta v^*_{xyh} \end{aligned} \tag{5.33}$$

where the symbols $\mp$ and $\pm$ are used to indicate the top half and bottom half of the upper and lower arms, respectively.

The voltage components V_{dc}, v^*_{xf}, and v^*_{xh} are equally shared by the submodules in each leg, arm, and sub-arms, respectively. The DC compensating voltage v^*_{dtx} is assigned the minus sign in the entire leg due to the same polarity of DC current component in the entire leg. The compensating high-frequency voltage Δv^*_{xh} in each sub-arm has a reverse sign. This voltage component has a positive relationship with AC circulating current. Therefore, the compensating component Δv^*_{xh} is assigned the same sign as that of the high-frequency voltage component v^*_{xh}. The individual capacitor compensating voltage Δv^*_{xyh} is assigned the positive sign because of its positive relationship with submodule power resulting from the same polarity of the sub-arm current.

CASE STUDY 5.6 Steady-State Performance Analysis of a PC-MMC

Objective: In this case study, the steady state performance of the PC-MMC is studied through MATLAB simulation.

Parameters: The system parameters of PC-MMC are as follows: rated DC-bus voltage $V_{dc} = 7$ kV, number of SMs per arm $N = 8$, SM capacitor voltage $v_C = E = 875$ V, SM capacitor $C = 4700\ \mu$F, cross-connected capacitor $C_x = 400\ \mu$F, load inductance $L_o = 20$ mH, load resistance $r_o = 8.5\ \Omega$, arm inductor $L = 2$ mH, fundamental frequency $f_o = 10$ Hz, high-frequency $f_h = 180$ Hz, and carrier frequency $f_c = 1000$ Hz.

Analysis:
The steady-state performance of the PC-MMC is investigated in this case study. The DC-bus is connected to a stiff DC source rated at 7000 V, and the AC terminals of PC-MMC are connected to a passive load. The reference dq-currents of output current control are set to $i^*_{di} = 200$ A and $i^*_{qi} = 0$ A. The reference submodule capacitor voltage is set to 875 V. The converter voltage and current waveforms are shown in Figures 5.17 and 5.18.

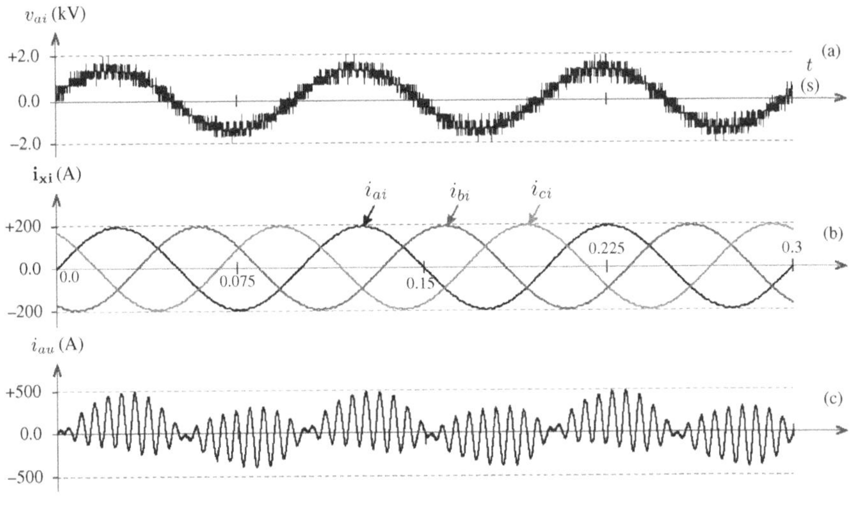

Figure 5.17 Steady-state performance of the PC-MMC.

The converter generates a voltage waveform with a fundamental voltage of 1750 V corresponding to the modulation index (m_a) of 0.5 as shown in Figure 5.17(a). The generated AC voltage forces the required magnitude of the output current of 200 A at a fundamental frequency of 10 Hz through the load as shown in Figure 5.17(b). In addition, the high-frequency voltage, and current components are generated proportional to the modulation index. The high-frequency current component is observed in the arm current as shown in 5.17(c), which maintains the power balance between the upper and lower arm submodule capacitors voltage. The arm current also contains DC and fundamental current components. The DC component balances the input and output power and maintains the submodule capacitors voltage at their rated value while the fundamental current is supplied to the load. Figure 5.18(a) shows the current flowing through the cross-connected capacitor, which consists of high-frequency current components only. Figure 5.18(b) shows the current flowing through the top-half lower arm. This current contains similar current components as that of the upper arm current except for the polarity of DC and fundamental components. The common-mode voltage generated by PC-MMC consists of switching ripple with a peak value of 400 V as shown in Figure 5.18(c). The submodule capacitors in the upper and lower arms are perfectly regulated at their nominal value of 875 V as shown in Figure 5.18(d).

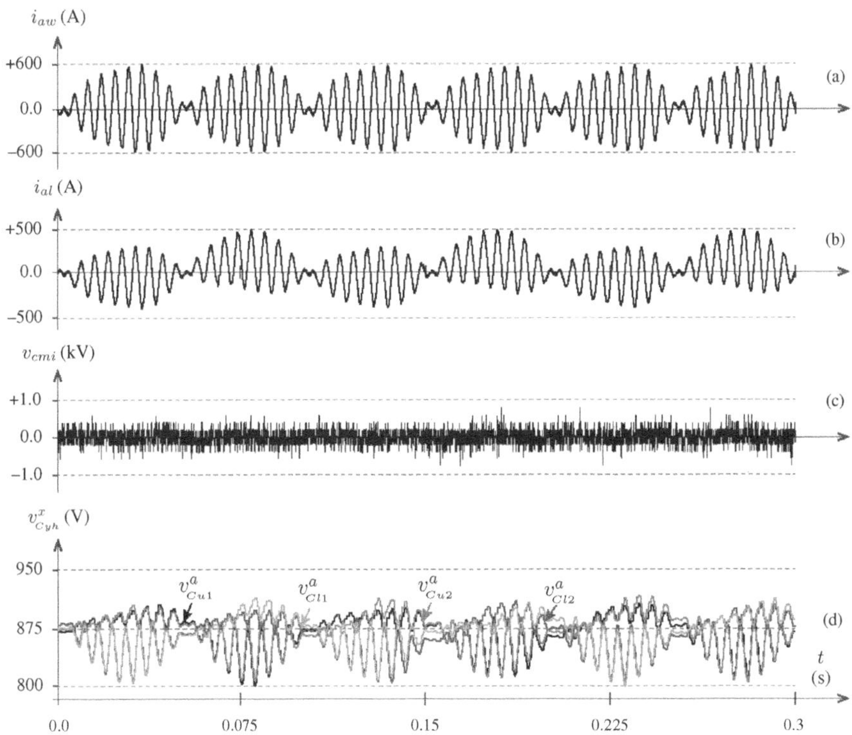

Figure 5.18 Steady-state performance of the PC-MMC.

5.6 SUMMARY

The conventional MMC has a large ripple in submodule capacitor voltages when operated at low/zero fundamental frequency. The ripple in capacitors voltage can be minimized by injecting high-frequency voltage component into the system. The injected voltage component appears in common-mode voltage and results in increased motor neutral point and shaft voltage in motor drive system. To deal with this problem, a modified configuration of MMC, referred to as a PC-MMC is introduced in this chapter. The operation and performance of the PC-MMC are analyzed with PSC-PWM scheme. An equivalent circuit is further developed to analyze the submodule capacitors voltage ripple and the common-mode voltage generated by PC-MMC at low/zero frequency of operation.

In PC-MMC, the mid-point of upper and lower arms are connected through a capacitor. The cross-connected capacitor carries a current component corresponding to the injected high-frequency voltage. The high-frequency voltage component greatly minimizes the submodule capacitors voltage ripple at low/zero frequency of operation. The high-frequency voltage components are opposition in-phase in the top-half and bottom-half arm voltages and cancel each other, and will not appear in the CMV. Also, the cross-connected capacitor in a PC-MMC provides a path for the power transfer between the upper and lower arms in each leg to achieve the voltage balance between the arms. The systematic design procedure of cross-connected capacitor and selection of the frequency of injected voltage component is presented with the case studies.

For the stable and reliable operation of PC-MMC, a closed-loop controller for the control of output currents and submodule capacitors voltage including leg, arm, and individual submodule capacitor voltage control is developed. The steady-state and dynamic performances of these control methods is analyzed through the case studies.

REFERENCES

1. Siemens, "The smart way hvdc plus-one step ahead." [Online]. Available: https://www.energy.siemens.com/us/en/power-transmission/hvdc/hvdcplus/
2. ABB, "Hvdc light gen 4." [Online]. Available: https://www.abb.com/systems/hvdc/
3. Alstom, "Hvdc maxsine." [Online]. Available: https://www.alstom.com/grid/products-and-services/engineered-energy-solutions/hvdc-transmission-systems/HVDC_Maxsine/
4. EPRI, "Hvdc flexible." [Online]. Available: https://www.epri/sgcc.com.cn/prgc/english/Product_Solution/HVDC_-Flexible/
5. M. Hagiwara, K. Nishimura, and H. Akagi, "A medium-voltage motor drive with a modular multilevel pwm inverter," *IEEE Trans. Power Electron.*, vol. 25, no. 7, pp. 1786–1799, Jul 2010.
6. J. A. Ferreira, "The multilevel modular dc converter," *IEEE Trans. Power Electron.*, vol. 28, no. 10, pp. 4460–4465, Oct 2013.
7. G. J. Kish, M. Ranjram, and P. W. Lehn, "A modular multilevel dc/dc converter with fault blocking capability for hvdc interconnects," *IEEE Trans. Power Electron.*, vol. 30, no. 1, pp. 148–162, Jan 2015.
8. S. Debnath, J. Qin, and M. Saeedifard, "Control and stability analysis of modular multilevel converter under low-frequency operation," *IEEE Trans. Ind. Electron.*, vol. 62, no. 9, pp. 5329–5339, Sep 2015.
9. M. Hagiwara, I. Hasegawa, and H. Akagi, "Start-up and low-speed operation of an electric motor driven by a modular multilevel cascade inverter," *IEEE Trans. Ind. Appl.*, vol. 49, no. 4, pp. 1556–1565, Jul 2013.

10. R. Picas, S. Ceballos, J. Pou, J. Zaragoza, G. Konstantinou, and V. Agelidis, "Closed-loop discontinuous modulation technique for capacitor voltage ripples and switching losses reduction in modular multilevel converters," *IEEE Trans. Power Electron.*, vol. 30, no. 9, pp. 4714–4725, Sep 2015.

11. A. Antonopoulos, L. Angquist, S. Norrga, K. Ilves, L. Harnefors, and H.-P. Nee, "Modular multilevel converter ac motor drives with constant torque from zero to nominal speed," *IEEE Trans. Ind. Appl.*, vol. 50, no. 3, pp. 1982–1993, May 2014.

12. J. Pou, S. Ceballos, G. Konstantinou, V. Agelidis, R. Picas, and J. Zaragoza, "Circulating current injection methods based on instantaneous information for the modular multilevel converter," *IEEE Trans. Ind. Electron.*, vol. 62, no. 2, pp. 777–788, Feb 2015.

13. Y. Okazaki, M. Hagiwara, and H. Akagi, "A speed-sensorless start-up method of an induction motor driven by a modular multilevel cascade inverter (mmci-dscc)," *IEEE Trans. Ind. Appl.*, vol. 50, no. 4, pp. 2671–2680, Jul 2014.

14. K. Wang, Y. Li, Z. Zheng, and L. Xu, "Voltage balancing and fluctuation-suppression methods of floating capacitors in a new modular multilevel converter," *IEEE Trans. Ind. Electron.*, vol. 60, no. 5, pp. 1943–1954, May 2013.

15. A. Nami, J. Liang, F. Dijkhuizen, and G. Demetriades, "Modular multilevel converters for hvdc applications: Review on converter cells and functionalities," *IEEE Trans. Power Electron.*, vol. 30, no. 1, pp. 18–36, Jan 2015.

CHAPTER 6

ACTIVE CROSS-CONNECTED MODULAR MULTILEVEL CONVERTERS

6.1 INTRODUCTION

The voltage ripple in the submodule capacitors is one of the major issues in conventional modular multilevel converter based motor drive systems. These ripples are quite high at low-speed and constant torque operation and increase the voltage and current stress on the semiconductor devices [1, 2]. To address the issue of capacitors voltage ripple, a passive cross-connected modular multilevel converter (PC-MMC) is presented in Chapter 5. In PC-MMC, the upper and lower arm of each leg is cross-connected through a capacitor [3]. The cross-connected capacitor provides a path for the high-frequency current component to mitigate the capacitors voltage ripple.

Although, it is possible to use different high-frequency current components such as sinusoidal, square wave, and third-harmonic signals to minimize the capacitor voltage ripple [4, 5]. But, the cross-connected capacitor in PC-MMC cannot shape the high-frequency current component. In this chapter, the cross-connected capacitor is replaced with an active branch, named "active cross-connected modular multilevel converter (AC-MMC)." In AC-MMC, the voltage rating of the cross-connected branch can be increased by adding the active devices in cascade unlike PC-MMC. Therefore, the AC-MMC can be used in high-voltage, high-power applications such

Modular Multilevel Converters: Analysis, Control, and Applications, First Edition.
Sixing Du, Apparao Dekka, Bin Wu, and Navid Zargari.

as medium- to high-voltage motor drives and high-voltage DC–DC conversion systems [6, 7].

In this chapter, the operation and control of AC-MMC-based motor drive systems are presented. The circuit configuration and switching states are presented in Section 6.2. The modeling and operation of AC-MMC with phase-shifted carrier modulation is presented in Section 6.3. The performance of PC-MMC at low-frequency is presented in Section 6.4. The generalized control scheme is presented in Section 6.5, and a summary is presented in Section 6.6.

6.2 ACTIVE CROSS-CONNECTED MMC

The new configuration of a modular multilevel converter is referred to as an active cross-connected modular multilevel converter (AC-MMC), in which the active submodules are connected across the mid-point of the upper and lower arm of each leg. The active submodules are controlled to limit the current flowing through the cross-connected branch and provide a path for the power transfer between the upper and lower arms. Hence, the cross-connected branch helps to minimize the ripple in submodule capacitors voltage during low/zero frequency operation and limits the current stress in the semiconductor devices. The circuit configuration and operation of AC-MMC are presented in the following subsections.

6.2.1 Circuit Configuration of AC-MMC

A three-phase AC-MMC with N submodules per arm is shown in Figure 6.1, where each leg is composed of two arms (upper and lower arms). The upper and lower arms form an AC output terminal "x," where $x \in \{a, b, c\}$ represents the output phase. Each arm is further divided into two halves referred to as a sub-arm. The mid-point of upper and lower arm of each leg is connected through a group of submodules. The number of submodules in the cross-connected branch is equal to the number of submodules in the arm. Hence, the submodules with identical voltage and current rating can be used in the arms and cross-connected branch. The submodules in the cross-connected branch are controlled to generate a high-frequency current component. This high-frequency current component interacts with the high-frequency arm voltage component and reduces the submodule capacitor voltage fluctuations.

Similar to other modular multilevel converters, the AC-MMC can be realized by using half-bridge (HB) submodule. In AC-MMC, the arm inductor ("L") is divided into two halves and connected at both ends of each arm as shown in Figure 6.1. This inductor is designed to mitigate the switching ripples in the arm current.

6.2.2 Switching States and Output Voltage

The configuration of converter leg in an AC-MMC system is shown in Figure 6.2. Each converter leg is divided into two arms, and each arm consists of four half-bridge (HB) submodules and two arm inductors. The half-bridge submodule has a simple

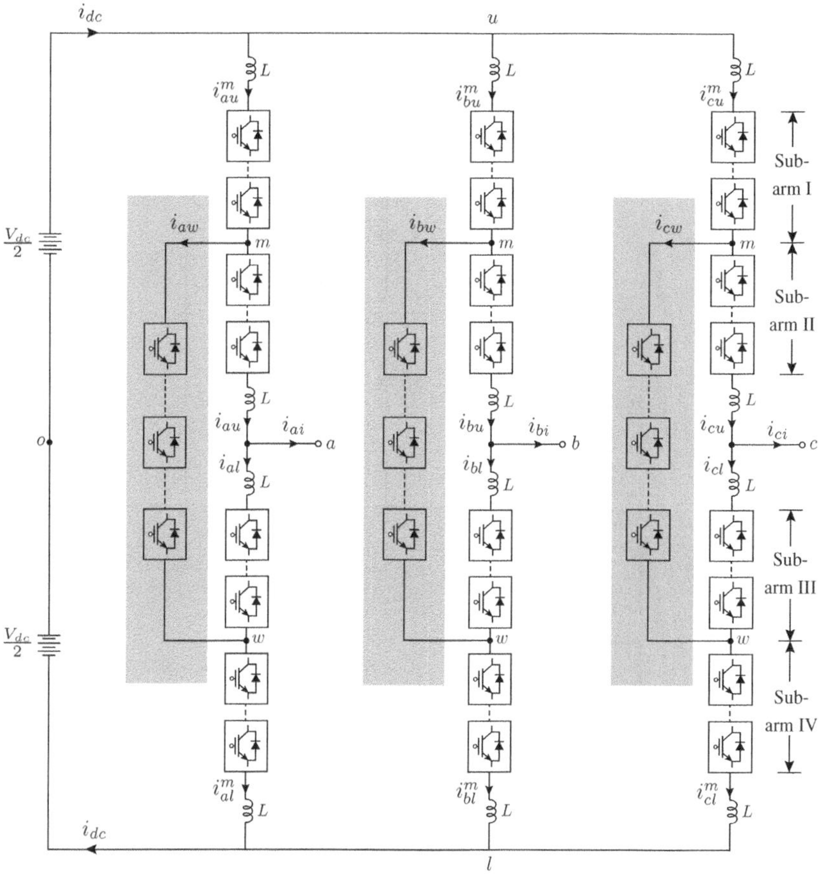

Figure 6.1 Power circuit configuration of AC-MMC.

construction and requires one independent gating signal to control the devices. With four submodules per arm, the AC-MMC generates five voltage levels across the arm terminals of "e" and "g" in the upper arm and "r" and "s" in the lower arm. It should be noted that the cross-connected branch also consists of four half-bridge submodules but does not contribute to the generation of AC output voltage.

The switching states and output voltage levels of the upper arm are shown in Table 6.1. When switches "S_1," "S_2," "S_3," and "S_4" are turned ON, each half-bridge submodule generates a voltage of "v_C", and the resultant arm voltage is $v_{eg}^x = 4\, v_C$. Similarly, when switches "S_1," "S_2," "S_3," and "S_4" are turned OFF, the voltage across the arm is $v_{eg}^x = 0$. There are several redundant switching states to generate the remaining voltage levels of "v_C," "$2\, v_C$," and "$3\, v_C$" as shown in Table 6.1. The redundant switching states are used to achieve the voltage balancing in a modular multilevel converter.

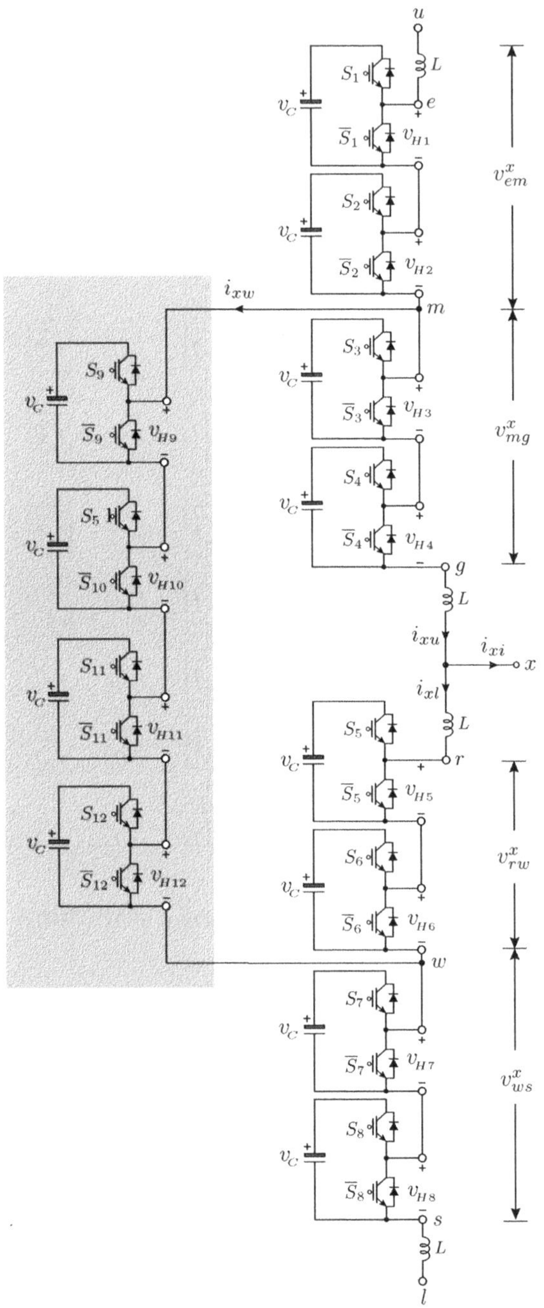

Figure 6.2 Configuration of converter leg in an AC-MMC.

Table 6.1 Switching states and output voltage levels of an arm

S_1	S_2	S_3	S_4	v_{H1}	v_{H2}	v_{H3}	v_{H4}	v^x_{em}	v^x_{mg}	v^x_{eg}	Voltage Level
0	0	0	0	0	0	0	0	0	0	0	0
1	0	0	0	v_C	0	0	0	v_C	0	v_C	
0	1	0	0	0	v_C	0	0	v_C	0	v_C	1
0	0	1	0	0	0	v_C	0	0	v_C	v_C	
0	0	0	1	0	0	0	v_C	0	v_C	v_C	
1	1	0	0	v_C	v_C	0	0	$2v_C$	0	$2v_C$	
1	0	1	0	v_C	0	v_C	0	v_C	v_C	$2v_C$	
1	0	0	1	v_C	0	0	v_C	v_C	v_C	$2v_C$	2
0	1	1	0	0	v_C	v_C	0	v_C	v_C	$2v_C$	
0	1	0	1	0	v_C	0	v_C	v_C	v_C	$2v_C$	
0	0	1	1	0	0	v_C	v_C	0	$2v_C$	$2v_C$	
1	1	1	0	v_C	v_C	v_C	0	$2v_C$	v_C	$3v_C$	
1	1	0	1	v_C	v_C	0	v_C	$2v_C$	v_C	$3v_C$	3
1	0	1	1	v_C	0	v_C	v_C	v_C	$2v_C$	$3v_C$	
0	1	1	1	0	v_C	v_C	v_C	v_C	$2v_C$	$3v_C$	
1	1	1	1	v_C	v_C	v_C	v_C	$2v_C$	$2v_C$	$4v_C$	4

The entire arm voltage v^x_{eg} is uniformly distributed between the sub-arms (v^x_{em} and v^x_{mg}). For example, the arm voltage $v^x_{eg} = 2\,v_C$ can be obtained by using the sub-arm voltage combinations of ($2\,v_C$, 0), (v_C, v_C), and (0, $2\,v_C$). The sub-arm voltage $v^x_{em} = v_C$ or $v^x_{mg} = v_C$ can be generated using the redundant switching states. Typically, the AC-MMC is designed with an even number of submodules in each arm. Therefore, it is easy to form a mid-point to cross-connect the upper and lower arms through a group of submodules. The number of voltage levels generated by the entire arm is given by

$$v_{Lev} = N + 1 \tag{6.1}$$

where v_{Lev} represents the arm voltage level and N represents the number of submodules per arm.

6.3 PRINCIPLES OF OPERATION

The submodules in the upper and lower arm are controlled to generate the required AC output voltage waveform while maintaining the balance between the converter input and output power. On the other hand, the submodules in the cross-connected arm are controlled to minimize the fluctuations in the arm power while maintaining the balance between the upper and lower arm powers. According to this criterion, the modulation signals of the upper, lower, and cross-connected arms are defined. These modulation signals are used with the phase-shifted carrier modulation scheme

(PSC-PWM) to generate the switching signals for submodules. The performance of the AC-MMC with a PSC-PWM is presented in the following subsections.

6.3.1 Modeling of AC-MMC

Each leg of the AC-MMC has an additional cross-connected arm along with the upper and lower arms. The upper and lower arms are further divided into sub-arms, and each sub-arm has an inductor and two half-bridge submodules. The cross-connected arm consists of four half-bridge submodules only. The total DC-bus voltage V_{dc} is equally distributed between the sub-arms. The submodules in each sub-arm are controlled to produce a phase voltage v_{xi}. From Figure 6.1, the voltage across the upper and lower arms is given by

$$\begin{aligned} v_{em}^{x} + v_{mg}^{x} &= \frac{V_{dc}}{2} - v_{xi} - L\frac{d\left(i_{xu} + i_{xw}\right)}{dt} - L\frac{d\,i_{xu}}{dt} \\ v_{rw}^{x} + v_{ws}^{x} &= \frac{V_{dc}}{2} + v_{xi} - L\frac{d\left(i_{xl} + i_{xw}\right)}{dt} - L\frac{d\,i_{xl}}{dt} \end{aligned} \tag{6.2}$$

where v_{em}^{x}, v_{mg}^{x} represents the voltage across the upper sub-arms of phase-x and v_{rw}^{x}, v_{ws}^{x} represents the voltage across the lower sub-arms of phase-x.

The sub-arm current consists of DC, high-frequency, and fundamental frequency current components. These current components are obtained from the measured currents of i_{xw}, i_{xu}, and i_{xl}, and are given by

$$\begin{aligned} i_{xd} &= \frac{1}{2}\left(i_{xu} + i_{xl} + i_{xw}\right) \\ i_{xh} &= \frac{1}{2}i_{xw} \end{aligned} \tag{6.3}$$

where $i_{xd} \in \{i_{ad}, i_{bd}, i_{cd}\}$ and $i_{xh} \in \{i_{ah}, i_{bh}, i_{ch}\}$ are the DC and high-frequency components in the arm current, $i_{xu} \in \{i_{au}, i_{bu}, i_{cu}\}$ is the upper arm current, $i_{xl} \in \{i_{al}, i_{bl}, i_{cl}\}$ is the lower arm current, and $i_{xw} \in \{i_{aw}, i_{bw}, i_{cw}\}$ is the cross-connected arm current.

From Figure 6.2, the fundamental frequency current component is given by

$$i_{xi} = i_{xu} - i_{xl} \tag{6.4}$$

where $i_{xi} \in \{i_{ai}, i_{bi}, i_{ci}\}$ represents the output current flowing through the load.

The arm voltage is generated by controlling the submodules in the sub-arms. The arm voltage is equally distributed between the four sub-arms. A high-frequency voltage (v_h) is introduced in the sub-arm voltages as an additional control degree of freedom to control the submodule capacitors voltage ripple during the low-frequency operation. The magnitude of high-frequency voltage component is set to zero ($v_h = 0$) at high-frequency (high-speed) operation. The resultant sub-arm voltages of phase-x

are given by

$$\begin{aligned}
v_{em}^{x} &= \frac{V_{dc}}{4} - \frac{v_{xi}}{2} - L\,\frac{d\,i_{xd}}{dt} - \frac{L}{2}\,\frac{d\,i_{xi}}{dt} - v_h \\
v_{mg}^{x} &= \frac{V_{dc}}{4} - \frac{v_{xi}}{2} - L\,\frac{d\,i_{xd}}{dt} - \frac{L}{2}\,\frac{d\,i_{xi}}{dt} + v_h \\
v_{rw}^{x} &= \frac{V_{dc}}{4} + \frac{v_{xi}}{2} - L\,\frac{d\,i_{xd}}{dt} + \frac{L}{2}\,\frac{d\,i_{xi}}{dt} - v_h \\
v_{ws}^{x} &= \frac{V_{dc}}{4} + \frac{v_{xi}}{2} - L\,\frac{d\,i_{xd}}{dt} + \frac{L}{2}\,\frac{d\,i_{xi}}{dt} + v_h
\end{aligned} \tag{6.5}$$

From Figure 6.2, it is observed that the voltage across the cross-connected arm is equal to the summation of sub-arm II and sub-arm III voltages (v_{mg}^{x} and v_{rw}^{x}) along with the voltage drop across arm inductor "L." The cross-connected arm voltage is given by

$$v_{mw}^{x} = v_{mg}^{x} + v_{rw}^{x} + L\,\frac{d\,i_{xu}}{dt} + L\,\frac{d\,i_{xl}}{dt} \tag{6.6}$$

The output voltage of cross-connected arm given in equation (6.6) is further simplified by using equations (6.3) and (6.5), and results in

$$v_{mw}^{x} = \frac{V_{dc}}{2} - 2L\,\frac{d\,i_{xh}}{dt} \tag{6.7}$$

From the above equation, it is observed that the cross-connected arm voltage consists of a DC component along with a small high-frequency component corresponding to the inductive drop. The DC component is utilized to cancel the counterpart voltage produced by the sub-arms. The high-frequency voltage component is used to shape the high-frequency current into a desirable waveform.

6.3.2 Phase-Shifted Carrier Modulation for AC-MMC

The AC-MMC requires N-carrier signals to control the submodules in each leg. The carrier signals are arranged with a phase angle shift of $360°/N$ while a phase angle shift of 0° is maintained between the upper and lower arm carrier signals [8]. The submodules in the cross-connected arm are modulated with identical carrier signals as that of the upper and lower arms. The upper and lower arm modulation signals given in equation (6.5) consists of fundamental (v_{xi}) and high-frequency voltage components (v_h). These voltage components are defined as

$$\begin{aligned}
v_{xi} &= V_f \sin\left(\omega_o\, t + \theta_x\right) \\
v_h &= V_h\, S_h
\end{aligned} \tag{6.8}$$

where V_f represents the peak value of fundamental voltage component, ω_o represents the fundamental angular frequency (rad), θ_x represents the initial angle of fundamental voltage component, V_h represents the peak value of high-frequency voltage, and S_h represents the wave shape of high-frequency voltage.

It is possible to use sinusoidal, square, trapezoidal, and third-harmonic wave shapes as a high-frequency voltage [4]. In this chapter, the sinusoidal and square waveforms are used to investigate their impact on the converter performance and device current stress. The square wave shape of high-frequency voltage is given by

$$S_h = \begin{cases} 1 & 0 \leq t < \frac{\pi}{\omega_h} \\ -1 & \frac{\pi}{\omega_h} \leq t < \frac{2\pi}{\omega_h} \end{cases} \tag{6.9}$$

The sinusoidal high-frequency voltage is given by

$$S_h = \sin(\omega_h t) \tag{6.10}$$

where ω_h represents the angular frequency of high-frequency voltage component.

The peak value of fundamental (V_f) and high-frequency (V_h) voltage components are selected such that the converter operates in the linear modulation index (m_a) range of 0–1. Accordingly, the magnitudes are defined as

$$\begin{aligned} V_f &= \frac{m_a V_{dc}}{2} \\ V_h &= \frac{(1 - m_a) V_{dc}}{4} \end{aligned} \tag{6.11}$$

The resultant sub-arm modulation signals are obtained by substituting equations (6.8), (6.9), and (6.11) in equation (6.5). The resultant modulation signals are normalized with respect to the half of the DC-bus voltage $V_{dc}/2$ and become as

$$\begin{aligned} m_{u1}^x &= \frac{1}{2} - \frac{m_a}{2} \sin(\omega_o t + \theta_x) - \frac{1 - m_a}{2} S_h \\ m_{u2}^x &= \frac{1}{2} - \frac{m_a}{2} \sin(\omega_o t + \theta_x) + \frac{1 - m_a}{2} S_h \\ m_{l1}^x &= \frac{1}{2} + \frac{m_a}{2} \sin(\omega_o t + \theta_x) - \frac{1 - m_a}{2} S_h \\ m_{l2}^x &= \frac{1}{2} + \frac{m_a}{2} \sin(\omega_o t + \theta_x) + \frac{1 - m_a}{2} S_h \end{aligned} \tag{6.12}$$

where the voltage drop across the arm inductor is neglected at steady state.

According to equation (6.7), the modulation signal of cross-connected arm depends on the derivative of the high-frequency current component (i_{xh}). Furthermore, the magnitude and shape of the high-frequency current component vary with the high-frequency voltage (v_h). For example, the sinusoidal high-frequency voltage generates a sinusoidal high-frequency current, and its derivative will also be a sinusoidal signal. On the other hand, the square wave high-frequency voltage generates a square wave high-frequency current and its derivative will be a pulse with finite magnitude. In this study, the square wave high-frequency voltage is added to the modulating signals. Accordingly, the pulse (S_p) is defined as

$$S_p = \frac{2L}{0.5 V_{dc}} \frac{d i_{xh}}{dt} \tag{6.13}$$

Equation (6.13) is substituted in equation (6.7) and the resultant equation is normalized with respect to the DC-bus voltage V_{dc}. The normalized modulation signal of cross-connected arm (m_c^x) is given by

$$m_c^x = \frac{1}{2} - \frac{1}{2} S_p \tag{6.14}$$

By using equations (6.12) and (6.14), the modulation signals of upper, lower, and cross-connected arm are generated. These modulation signals are compared with the corresponding arm carrier signals. The resultant comparison gives the gating signals and are applied to the submodules.

CASE STUDY 6.1 Analysis of AC-MMC with a PSC-PWM Scheme

Objective: In this case study, the implementation and performance of an AC-MMC with a PSC-PWM scheme is studied through MATLAB simulation.

Parameters: The system parameters of the AC-MMC are as follows: rated DC-bus voltage $V_{dc} = 7$ kV, fundamental frequency $f_o = 60$ Hz, high-frequency $f_h = 180$ Hz, number of SMs per arm $N = 8$, arm inductor $L = 2$ mH, and carrier frequency $f_c = 180$ Hz.

Analysis:
In this study, each arm of the AC-MMC is designed with eight half-bridge submodules. Each submodule capacitor is replaced with an ideal DC-voltage source of $E = 875$ V. The operation of nine-level AC-MMC with PSC-PWM scheme is shown in Figure 6.3. The modulation signals m_{u1}^a and m_{u2}^a of phase-a, sub-arms I and II are generated with an amplitude modulation index $m_a = 0.55$, fundamental frequency $f_o = 60$ Hz, and zero initial phase angle (θ_x). In addition, a square wave high-frequency voltage component (S_h) with a frequency of $f_h = 180$ Hz is added to the modulation signals as shown in Figure 6.3(a). The modulation signals are given by

$$\begin{aligned} m_{u1}^a &= \frac{1}{2} - \frac{0.55}{2} \sin(120\,\pi\, t) - \frac{1-0.55}{2} S_h \\ m_{u2}^a &= \frac{1}{2} - \frac{0.55}{2} \sin(120\,\pi\, t) + \frac{1-0.55}{2} S_h \end{aligned} \tag{6.15}$$

Similarly, the modulation signals of sub-arms III and IV are generated as per the equations given in (6.12). The nine-level AC-MMC required eight triangular signals, which are generated with a carrier frequency of $f_c = 180$ Hz as shown in Figure 6.3(a). The carrier signals are arranged with an initial phase angle of 45°. The modulation signal m_{u1}^a is compared with the carrier signals T_{u1}, T_{u3}, T_{u5}, and T_{u7}, and generates the gating signal of submodules in sub-arm I as shown in Figure 6.3(b)–(e). These four carrier signals are arranged with a phase angle shift of 90°. The gating signals are applied to the submodules in the sub-arm I. Each submodule generates two voltage levels of "E" and "0." The output voltages of submodules in the sub-arm I are added together and result in a five-level voltage waveform (v_{em}^a) as shown in Figure 6.3(f). The sub-arm I voltage has a total five voltage levels with a maximum value of "$4E$" and a minimum value of "0."

Similarly, the modulation signal m_{u2}^a is compared with the carrier signals T_{u2}, T_{u4}, T_{u6} and T_{u8}, and yields the gating signal of submodules in sub-arm II as shown in Figure 6.3(g)–(j). The output voltage of submodules in sub-arm II is added together, which forms a sub-arm voltage (v_{mg}^a) as shown in Figure 6.3(k).

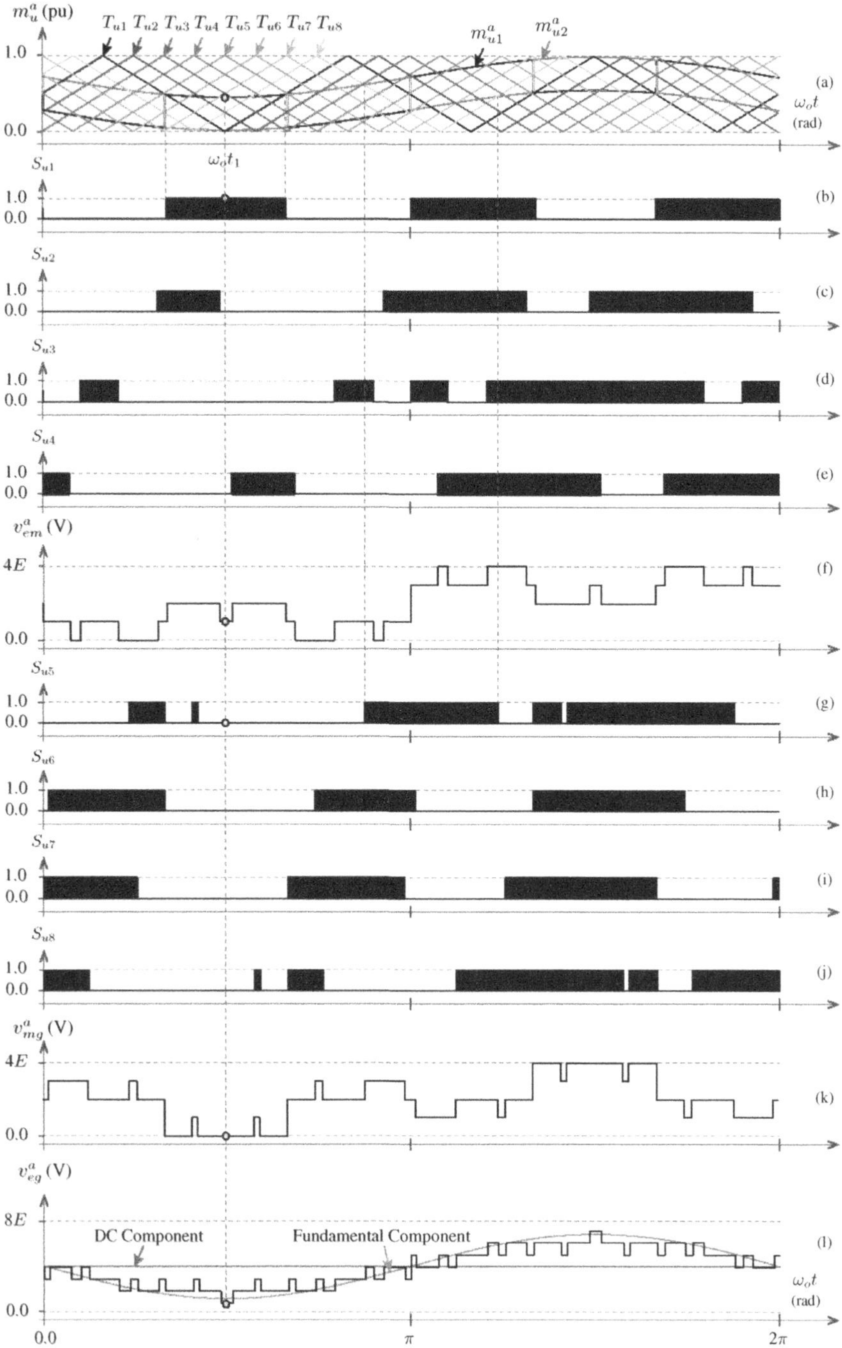

Figure 6.3 Implementation of PSC-PWM for an AC-MMC.

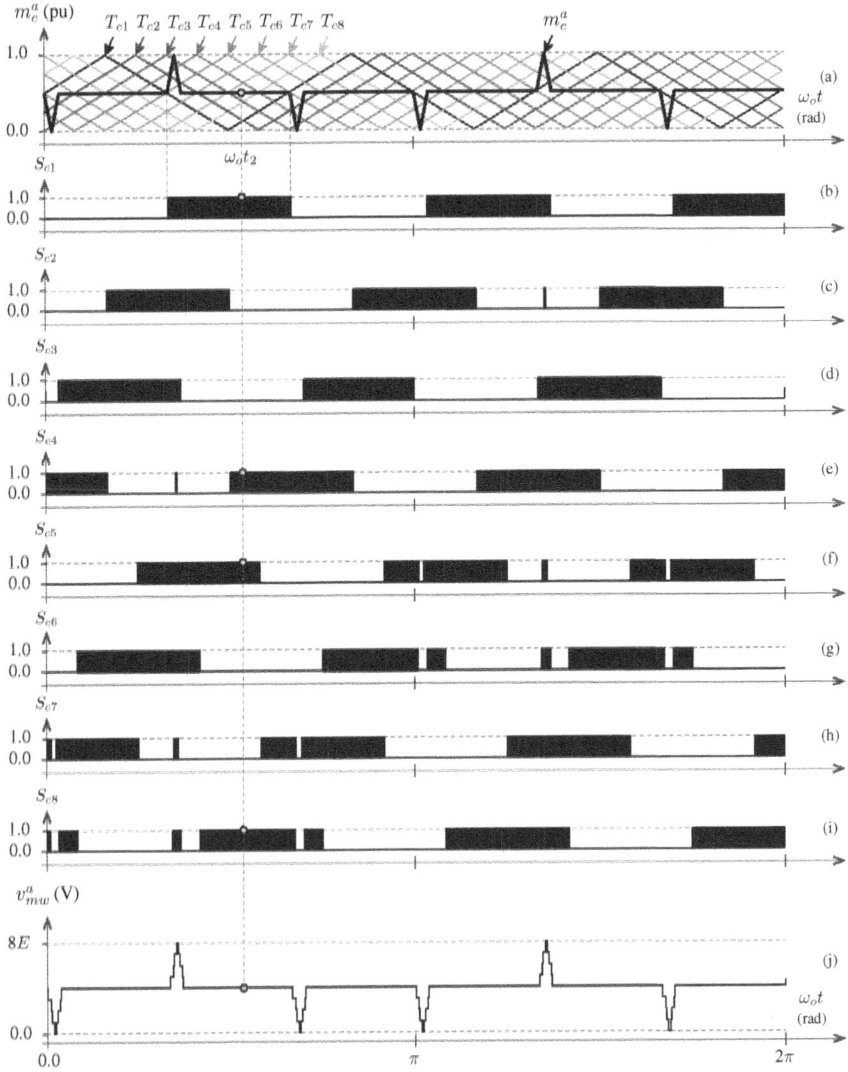

Figure 6.4 Implementation of PSC-PWM for an AC-MMC.

The output voltages of sub-arms I and II are added, and result in the upper arm voltage (v_{eg}^a) as shown in Figure 6.3(l). This upper arm voltage is a seven-level voltage waveform corresponding to the modulation index of $m_a = 0.55$. The output voltage has a maximum value of "$7E$" and a minimum value of "E." Each voltage step is equal to the rated value of submodule capacitor voltage "E." At $m_a = 1$, the AC-MMC can generate a total of nine voltage levels across the arm. The above principle can be applied to the lower arm submodules as well.

For example, at $\omega_o t_1$, SM-1 is turned ON while SM-2, SM-3, and SM-4 are turned OFF in sub-arm I. Hence, the sub-arm I voltage is equal to $v^a_{em} = E$. In the case of sub-arm II, SM-5, SM-6, SM-7, and SM-8 are turned OFF. Hence, sub-arm II voltage is equal to $v^a_{mg} = 0$. Sub-arm I and II voltages are added together, resulting in the arm voltage $v^a_{eg} = E$ as shown in Figure 6.3(l).

According to equation (6.14), the cross-connected arm requires a single modulation signal m^a_c, which consists of a small pulse with DC-offset as shown in Figure 6.4(a). The modulation signal is compared with the carrier signals $T_{u1} - T_{u8}$, which are similar to the upper and lower arm carrier signals. The comparison produces the gating signals for the cross-connected arm submodules as shown in Figure 6.4(b)–(i). Each submodule generates a voltage level "E," when it is turned ON. Similarly, the submodule generates a voltage level "0," when it is turned OFF. The resultant voltage across the cross-connected arm is shown in Figure 6.4(j), which is similar to the reference modulation signal. This voltage is used to shape the high-frequency current component flowing through the cross-connected arm. For example, at $\omega_o t_2$, the SM-1, SM-4, SM-5, and SM-8 are turned ON while the SM-2, SM-3, SM-6, and SM-7 are turned OFF in the cross-connected arm. Hence, the output voltage of cross-connected arm is equal to $v^a_{mw} = 4E$ as shown in Figure 6.4(j).

■

6.4 LOW-FREQUENCY OPERATION OF AC-MMC

To minimize the capacitor voltage ripple during low-frequency operation, a sinusoidal high-frequency voltage is added to the modulation signals. The sinusoidal high-frequency voltage effectively minimizes the voltage ripple, but it increases the magnitude of peak circulating current flowing in the arm. On the other hand, the square wave component is injected into the system, which significantly minimizes the peak circulating currents by 50% compared with the sinusoidal voltage [4]. The PC-MMC topology consists of a simple capacitor in the cross-connected arm, and cannot shape the high-frequency component [3, 9, 10]. Hence, the sinusoidal high-frequency voltage is employed in the minimization of voltage ripple in the PC-MMC. On the other hand, the AC-MMC can shape the high-frequency component and uses square wave as a high-frequency voltage component. The detailed analysis and operation of AC-MMC during low-frequency is presented in the following subsections.

6.4.1 Equivalent Circuit

An equivalent model is required to understand and analyze the behavior of AC-MMC during low-frequency operation. The equivalent model of a three-phase AC-MMC is shown in Figure 6.5, in which the submodules in each sub-arm are modeled as a controlled voltage source. The controlled voltage source consists of DC component (v_d), fundamental frequency (v_{xf}), and high-frequency (v_{xh}) voltage components. The submodules in the cross-connected arm are modeled as a constant DC source (v_d) and a high-frequency pulse (v_{xp}). From equations (6.12) and (6.14), the magnitude of these voltage components is given by

$$\begin{aligned}
v_d &= \frac{1}{4} V_{dc} \\
v_{xf} &= \frac{m_a}{4} V_{dc} \sin(\omega_o t + \theta_x) \\
v_{xh} &= \frac{1 - m_a}{4} V_{dc} S_h \\
v_{xp} &= \frac{1}{2} V_{dc} S_p
\end{aligned} \tag{6.16}$$

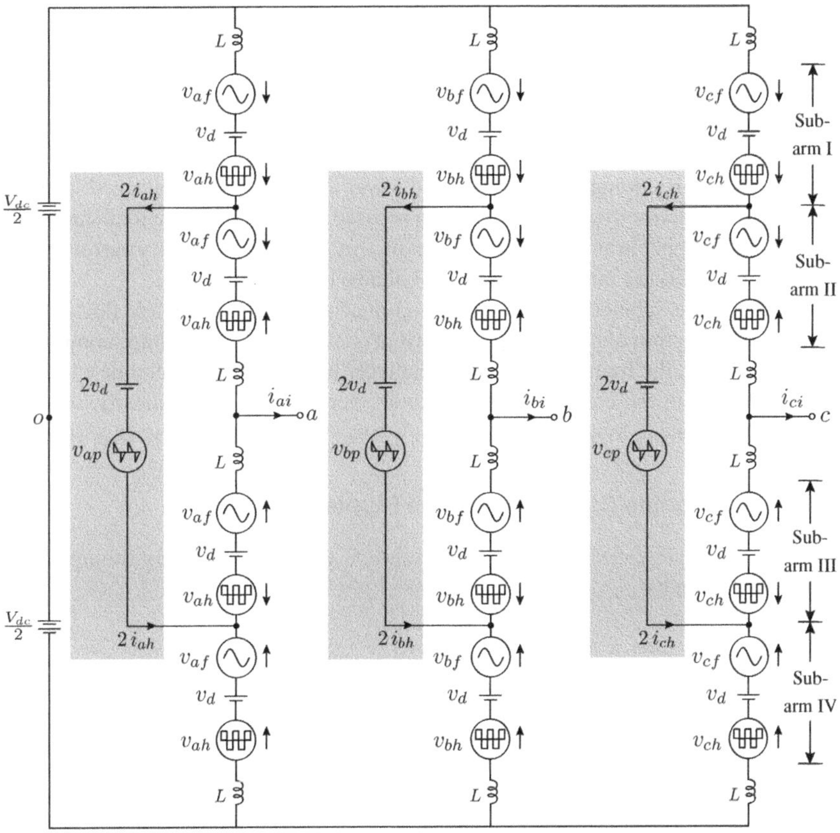

Figure 6.5 Equivalent circuit of a three-phase AC-MMC.

The DC voltage component in each sub-arm is controlled to support the DC-bus voltage V_{dc} while the fundamental AC voltage components in the upper and lower arms are controlled to generate a voltage component v_{xi} at the AC output terminals. The high-frequency voltage component is introduced to minimize the capacitors voltage ripple and to achieve the power balance between the upper and lower arms. The high-frequency voltage component in the upper and lower sub-arms are in opposite phase and cancel each other. Therefore, the injected high-frequency voltage component does not have any effect on the DC and AC system of AC-MMC. The cross-connected arm generates DC and pulse voltage components. The DC voltage ($2v_d$) is generated to cancel the counterpart DC voltages in sub-arms II and III. The pulse voltage (v_{xp}) is generated to compensate the voltage drop across the arm inductors due to the square-wave high-frequency current (i_{xh}).

The fundamental frequency voltage v_{xf} generates a current component i_{xi} in the arm. The interaction between the fundamental (low) frequency current and DC voltage component in the arm leads to larger charging and discharging periods of submodule capacitors. The voltage fluctuation becomes worse when the DC current component (i_{xd}) interacts with the low-frequency voltage component (v_{xf}) in the submodule. The cross-connected arm helps to mitigate the large fluctuation of submodule capacitor voltage. When the total energy of the upper arm is larger than that of the lower arm, the high-frequency current component (i_{xh}) flows from the upper arm to the lower arm such that the upper arm releases the power, and the lower arm absorbs it. By doing so, the upper and lower arm powers are balanced, and the submodule capacitors voltage ripple is suppressed. Similarly, if the total capacitor voltage of the upper arm is less than the lower arm, the high-frequency current flows in reverse direction to maintain the power balance between the arms.

Similarly, the submodules in the cross-connected arm have voltage fluctuations due to the interaction of high-frequency current (i_{xh}) with the DC voltage component ($2v_d$). However, the frequency of i_{xh} is relatively high and the submodule capacitor voltage ripples in the cross-connected arm are very small. Consequently, the AC-MMC can operate at low-frequency region with a smaller capacitor voltage ripples.

6.4.2 Submodule Capacitor Voltage Ripple

From the equivalent circuit shown in Figure 6.5, the current flowing through each sub-arm is given by

$$\begin{aligned} i_{xu}^{m} &= i_{xd} + \frac{1}{2} i_{xi} + i_{xh} \\ i_{xu} &= i_{xd} + \frac{1}{2} i_{xi} - i_{xh} \\ i_{xl} &= i_{xd} - \frac{1}{2} i_{xi} - i_{xh} \\ i_{xl}^{m} &= i_{xd} - \frac{1}{2} i_{xi} + i_{xh} \end{aligned} \tag{6.17}$$

The sub-arm current consists of DC current component (i_{xd}) , fundamental frequency current component (i_{xi}), and high-frequency current component (i_{xh}). For the analysis of capacitors voltage ripple, these current components are defined as

$$\begin{aligned} i_{xd} &= I_{dcx} \\ i_{xh} &= I_h\, S_h \\ i_{xi} &= I \sin\left(\omega_o t + \theta_x + \varphi\right) \end{aligned} \tag{6.18}$$

where I_h and I are the peak values of high-frequency and fundamental frequency current components, respectively.

From equation (6.5), the sub-arm voltages are given by

$$\begin{aligned} v_{em}^x &= \frac{V_{dc}}{4} - \frac{v_{xi}}{2} - v_{xh} \\ v_{mg}^x &= \frac{V_{dc}}{4} - \frac{v_{xi}}{2} + v_{xh} \\ v_{rw}^x &= \frac{V_{dc}}{4} + \frac{v_{xi}}{2} - v_{xh} \\ v_{ws}^x &= \frac{V_{dc}}{4} + \frac{v_{xi}}{2} + v_{xh} \end{aligned} \tag{6.19}$$

The submodule capacitor voltage ripple is obtained from the relationship between submodule instantaneous power and DC-side voltage. During this analysis, it is assumed that each sub-arm power is equally distributed among the submodules. Each submodule power in sub-arm I is given by

$$p_{em1}^x = \frac{2}{N}\,[v_{em}^x\, i_{xu}^m] \tag{6.20}$$

From equation (6.17), (6.18), and (6.19), the components presented in the individual submodule power are given by

$$\begin{aligned} p_{em1}^x = {} & \frac{2}{N}\,[(\frac{V_{dc}\,I_{dcx}}{4} - \frac{v_{xi}\,i_{xi}}{4}) + (\frac{V_{dc}\,i_{xi}}{8} - \frac{v_{xi}\,I_{dcx}}{2} - 2\,V_h\,I_h\,S_{hw}) \\ & + (\frac{V_{dc}\,i_{xh}}{4} - \frac{v_{xi}\,i_{xh}}{2} - \frac{v_{xh}\,i_{xi}}{2} - v_{xh}\,I_{dcx})] \end{aligned} \tag{6.21}$$

where S_{hw} represents the root mean square (RMS) value of the high-frequency voltage, which is given by

$$S_{hw} = \begin{cases} 1, & \text{Square wave} \\ 0.5, & \text{Sinusoidal wave} \end{cases} \tag{6.22}$$

At steady state, the average submodule power given in equation (6.21) should be equal to zero. Therefore, from equation (6.21), the following steady-state operating conditions are obtained as

$$\begin{aligned} \frac{V_{dc}\,I_{dcx}}{4} - \frac{v_{xi}\,i_{xi}}{4} &= 0 \\ \frac{V_{dc}\,i_{xi}}{8} - \frac{v_{xi}\,I_{dcx}}{2} - V_h\,I_h\,S_{hw} &= 0 \end{aligned} \tag{6.23}$$

From equation (6.23), the magnitudes of DC and high-frequency current components are

$$\begin{aligned} I_{dcx} &= \frac{v_{xi}\,i_{xi}}{V_{dc}} \\ I_h &= \frac{0.25\,V_{dc}\,i_{xi} - v_{xi}\,I_{dcx}}{2\,V_h\,S_{hw}} \end{aligned} \tag{6.24}$$

From equations (6.21) and (6.24), the instantaneous value of submodule power is given by

$$p^x_{em1} = \frac{2}{N}\Big[\frac{V_{dc}\,I_h}{4}\,S_h - \frac{V_f\,I_h}{2}\sin(\omega_o\,t + \theta_x)\,S_h - V_h\,I_{dcx}\,S_h - \frac{V_h\,I}{2}\sin(\omega_o\,t + \theta_x + \varphi)\,S_h\Big] \tag{6.25}$$

The submodule AC output power is stored in the DC capacitor. By considering the submodule AC output power and DC-side energy, the submodule capacitor voltage is given by

$$v^x_{Cy1} = \sqrt{\frac{2}{C}\left[\frac{1}{2}C\,V_C^2 + \int p^x_{em1}\,dt\right]} \tag{6.26}$$

Equations (6.25) and (6.26) show that the instantaneous value of submodule power consists of a high-frequency term of S_h, which reduces the charging and discharging periods of submodule capacitors and its voltage ripple.

CASE STUDY 6.2 Voltage Ripple Reduction at Low-Frequency Operation

Objective: In this case study, the effect of injected high-frequency voltage on submodule capacitor voltage ripple is studied using MATLAB simulation.

Parameters: The system parameters of the AC-MMC are as follows: rated DC-bus voltage $V_{dc} = 7$ kV, number of SMs per arm $N = 8$, SM capacitor voltage $v_C = 875$ V, SM capacitor $C = 4700\ \mu$F, load current $I_o = 200$ A, power factor angle $\varphi = 0°$, arm inductor $L = 2$ mH, and carrier frequency $f_c = 1000$ Hz.

Analysis:
In this case study, the DC side of AC-MMC is modeled as a stiff DC voltage source V_{dc}, and the AC side is connected to a three-phase load. The three-phase currents flowing through the load are controlled by a close-loop current control, which will be discussed in Section 6.5. The amplitude and frequency of the three-phase reference currents are set to 200 A and 10 Hz, respectively. The simulation results with and without injection of the high-frequency component is shown in Figure 6.6. Under these conditions, the AC-MMC is operating at an amplitude modulation index (m_a) of 0.2, corresponding to the current magnitude of 200 A at steady state. The phase-a, modulation signal of sub-arm I without high-frequency voltage ($v_h = 0$) is given by

$$v^x_{em} = \frac{7000}{4} - \frac{0.2 \times 7000}{4}\sin(20\pi\,t) \tag{6.27}$$

Figure 6.6(a) shows the submodule capacitors voltage of AC-MMC without the injection of high-frequency voltage and current components. Without high-frequency voltage, there is no power transfer between the upper and lower arms through a cross-connected arm. Hence, the AC-MMC is simply operating as a conventional MMC. The AC-MMC generates an output voltage at 10 Hz frequency corresponding to the 200 A current. The load current component is equally distributed among the upper and lower arms and interacts with the DC voltage component in the arms. This interaction causes a fluctuation in the upper and lower arm submodule capacitors voltage. The peak-to-peak ripple in submodule capacitors voltage is as high as 386 V, and the voltage ripple is 44.10% with respect to the nominal submodule

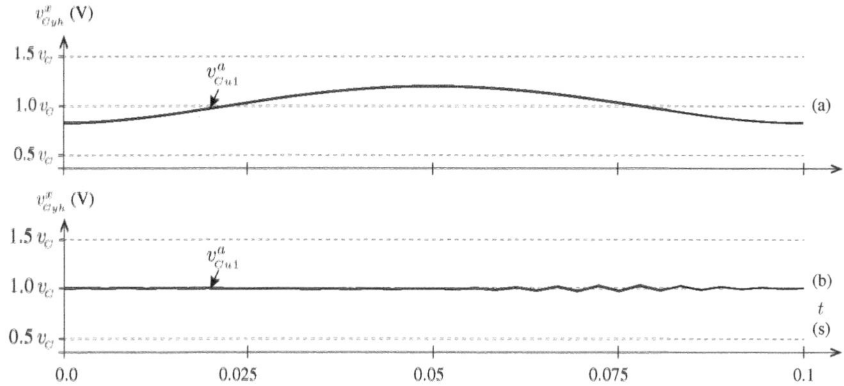

Figure 6.6 Submodule capacitors voltage: (a) without high-frequency voltage, and (b) with high-frequency voltage at low-frequency operation.

capacitor voltage of 875 V. The ripple becomes larger as the output frequency decreases and increases the voltage stress on submodule capacitors and devices.

Figure 6.6(b) shows the simulation results with the injection of square wave high-frequency voltage component. The phase-a, modulation signal of sub-arm I with square wave high-frequency voltage is given by

$$v^x_{em} = \frac{7000}{4} - \frac{0.2 \times 7000}{4}\sin(20\pi\, t) - \frac{(1-0.2)\times 7000}{4} S_h \tag{6.28}$$

The injected high-frequency voltage generates a controllable high-frequency current component, which flows through the cross-connected arm in an AC-MMC. The submodules in the cross-connected arm helps to reduce the voltage ripple in submodule capacitors voltage and maintains the power balance between the upper and lower arms. Figure 6.6(b) shows that the voltage fluctuation is minimized as low as 15 V. The ripple in submodule capacitors voltage is around 1.7% of its nominal voltage. Therefore, the AC-MMC require smaller submodule capacitors compared to the conventional MMC. ■

6.4.3 Common-Mode Voltage

In the conventional MMC, the externally injected high-frequency voltage component appears in the motor neutral voltage, also known as common-mode voltage (CMV). The externally injected voltage increases the magnitude of motor neutral voltage and increases the voltage stress on the motor winding insulation, and generates shaft voltages in a motor drive system. However, in the case of AC-MMC, the injected high-frequency voltage does not appear in motor neutral voltage. The CMV generated by AC-MMC is given by

$$v_{cmi} = \frac{1}{3}\left[v_{ai} + v_{bi} + v_{ci}\right] \tag{6.29}$$

where v_{ai}, v_{bi}, and v_{ci} represents the phase-to-neutral voltage of three phases. From equations (6.2), (6.3), and (6.4), the output voltage with respect to the mid-point of DC-bus o is given by

$$v_{xi} = \frac{1}{2}(v_{rs}^{x} - v_{eg}^{x}) - 2\,L\,\frac{di_{xi}}{dt} \tag{6.30}$$

From equations (6.29) and (6.30), the CMV in terms of arm voltage is given by

$$v_{cmi} = \frac{1}{6}\left[v_{rs}^{a} + v_{rs}^{b} + v_{rs}^{c} - v_{eg}^{a} - v_{eg}^{b} - v_{eg}^{c}\right] \tag{6.31}$$

The arm voltage consists of fundamental voltage component along with switching ripples (v_{sw}) generated by the pulse width modulation scheme. The switching ripple v_{sw} is added to the arm voltage given in equation (6.6) and results in

$$\begin{aligned} v_{eg}^{x} &= \frac{V_{dc}}{2} - V_f \sin\left(\omega_o t + \theta_x\right) + v_{sw} \\ v_{rs}^{x} &= \frac{V_{dc}}{2} + V_f \sin\left(\omega_o t + \theta_x\right) + \overline{v}_{sw} \end{aligned} \tag{6.32}$$

Equation (6.32) is substituted in equation (6.31) and the CMV generated by the AC-MMC is given by

$$v_{cmi} = \frac{1}{2}(v_{sw} + \overline{v}_{sw}) \tag{6.33}$$

Equation (6.33) shows the injected high-frequency voltage is completely eliminated from the CMV produced by the AC-MMC, and it only consists of small switching ripples generated by the PWM scheme, which makes the AC-MMC a potential candidate for low-speed, high-torque motor drive applications.

CASE STUDY 6.3 Effect of High-Frequency Voltage in an AC-MMC

Objective: In this case study, the effect of injected high-frequency voltage on CMV in an AC-MMC is studied through MATLAB simulation.

Parameters: The system parameters of the AC-MMC are as follows: rated DC-bus voltage $V_{dc} = 7$ kV, number of SMs per arm $N = 8$, SM capacitor voltage $v_C = E$ V, SM capacitor $C = 4700\ \mu$F, load inductance $L_o = 20$ mH, load resistance $r_o = 34\ \Omega$, arm inductor $L = 2$ mH, and carrier frequency $f_c = 600$ Hz.

Analysis:

This case study investigates the CMV generated by the AC-MMC. Each arm consists of $N = 8$ submodules and each submodule capacitor voltage is set to E. The modulation signal of sub-arms I to IV in each leg are synthesized using equation (6.12). The frequency of square wave voltage component is set to 180 Hz.

The four submodules in each sub-arm are controlled using a single modulation signal. The PSC-PWM scheme requires eight-triangular carrier signals to control the 8 submodules in each arm, which is phase-shifted by $360°/8$ from each other. In this simulation, the fundamental frequency of the modulation signal is steeply changed from 1 Hz to 5 Hz and 10 Hz, and their amplitude modulation index is set to 0.85.

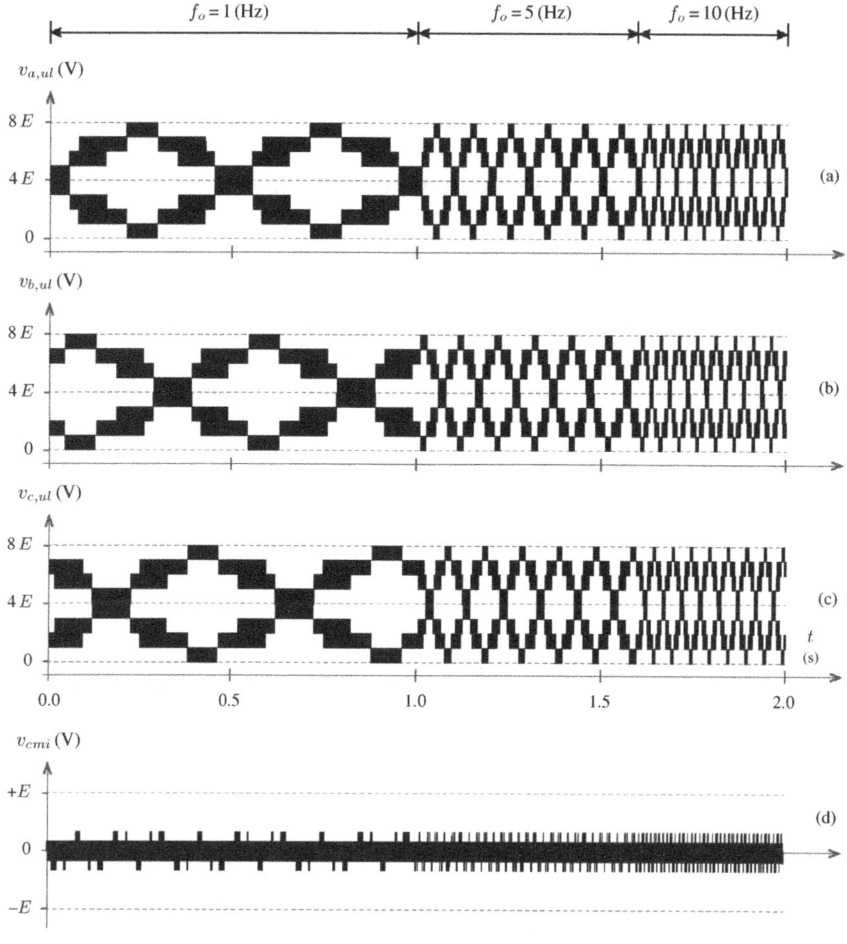

Figure 6.7 Simulated waveforms of CMV in an AC-MMC.

The voltage waveform of the six arms is shown in Figure 6.7. Figure 6.7(a) shows the voltage waveforms of the upper and lower arms in phase-a. The high-frequency voltage components are arranged in opposite phase in the two half-arms. The arm voltages consist of a DC component ($4E$) and a fundamental voltage component along with the switching ripples. The AC components in the upper and lower arm voltages are in opposite phase. As the fundamental frequency of the modulation signals is steeply changed at time instants of $t = 1$ s and 1.6 s, the time period of arm voltages is also varied accordingly. Figures 6.7(b) and 6.7(c) show the upper and lower arm voltages in phase-b and -c, respectively. These voltages have similar components as that of the phase-a, but the fundamental components are phase-shifted by 120° and 240° with respect to the phase-a voltage.

According to equation (6.33), the common-mode voltage generated by AC-MMC can be obtained from the arm voltages as shown in Figure 6.7(d). The waveform shows that the

CMV contains only small switching ripple of 0.5 E, which is around 6% of the nominal DC-bus voltage V_{dc}. Also, the injected high-frequency voltage ($f_h = 180$ Hz) does not appear in the CMV. ■

6.4.4 Current Stress on Semiconductor Devices

The arm current consists of DC, fundamental, and high-frequency current components. The externally injected high-frequency current component appear in the circulating current and increases the peak value of the arm current. The increment varies with the type of injected waveform. In the case of sinusoidal wave injection, the peak value of arm current can reach up to three times the rated current during low-frequency, constant torque operation. With square wave injection, the peak value of arm current can be 50% smaller than the sinusoidal wave injection [4]. The peak value of sub-arm and cross-connected arm currents in an AC-MMC is given by

$$\begin{aligned} i_{xu}^{m} &= \frac{v_{xi}\, i_x}{V_{dc}} + \frac{i_{xi}}{2} + \frac{V_{dc}^2\, i_{xi} - 4\, v_{xi}^2\, i_{xi}}{8\, V_h\, V_{dc} S_{hw}} S_h \\ i_{xw} &= \frac{V_{dc}^2\, i_{xi} - 4\, v_{xi}^2\, i_{xi}}{4\, V_h\, V_{dc} S_{hw}} S_h \end{aligned} \tag{6.34}$$

where S_h represents the type of injected waveform, $S_{hw} = 1$ represents the RMS value of injected square waveform and $S_{hw} = 0.5$ represents the RMS value of injected sinusoidal waveform.

The device current stress in the AC-MMC is analyzed with the square and sinusoidal high-frequency waveforms by using MATLAB simulations. In this study, the AC-MMC is designed with the following parameters: rated DC-bus voltage $V_{dc} = 7$ kV, number of SMs per arm $N = 8$, SM capacitor $C = 4700\ \mu$F, load resistance $r_o = 3.5\ \Omega$, arm inductor $L = 2$ mH, and carrier frequency $f_c = 600$ Hz. The output current controller is designed to control a magnitude of $I = 200$ A, and generates a modulation signal with an amplitude modulation index (m_a) of 0.2, and fundamental frequency (f_o) is 10 Hz. The mathematical expression of output voltage and output current of phase-a is given by

$$\begin{aligned} v_{ai} &= 0.5 \times 0.2 \times 7000 \times \sin(20\,\pi t) \\ i_{ai} &= 200 \times \sin(20\,\pi t) \end{aligned} \tag{6.35}$$

The sub-arm and cross-connected arm currents of AC-MMC with square and sinusoidal injected waveforms are shown in Figure 6.8. With square wave injection, the peak value of sub-arm current is 1.2 pu with respect to the rated output current as shown in Figure 6.8(a). Figure 6.8(b) shows the waveform of the cross-connected arm current, which consists of high-frequency square wave component only. The peak value of the cross-connected arm current is also 1.2 pu. Therefore, the AC-MMC can be designed with identical rated power semiconductors.

The performance of AC-MMC with sinusoidal wave injection is shown in Figure 6.8(c). The peak value of sub-arm current is 1.8 pu, resulting in a current stress

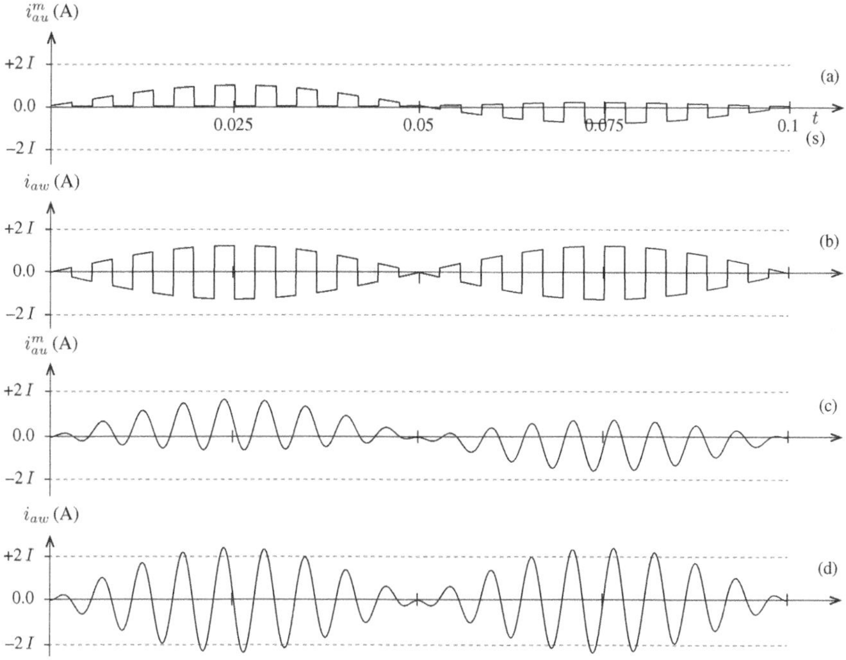

Figure 6.8 Performance comparison between square and sinusoidal high-frequency voltage in an AC-MMC.

80% higher than the rated current stress. The current flowing through the cross-connected arm is shown in Figure 6.8(d). It is observed that the peak value of current is 2.5 pu. Thus, the current stress on the devices in the cross-connected arm is 150% higher than the rated current stress. With sinusoidal wave injection, the AC-MMC requires devices with different ratings to construct the submodules in the sub-arms and cross-connected arms. Overall, the square wave injection reduces the sub-arm device current stress by 50% compared to the sinusoidal wave injection.

6.5 CLASSICAL CONTROL OF AC-MMC

The AC-MMC requires an additional voltage controller to control the submodules in the cross-connected arm. The control block diagram of AC-MMC is shown in Figure 6.9. The capacitor voltage control is divided into leg voltage, arm voltage, and individual capacitor voltage controllers. Typically, the current control is implemented in a central controller and generates a modulation signal at a fundamental frequency. The capacitor voltage control is implemented in a local controller and generates a compensation signal, which is added to the fundamental frequency modulation signal. The modulation signals are applied to the phase-shifted carrier modulator (PSC-

PWM), and compared with the carrier signals and generated the gating signals to the submodules. To implement the current and voltage controllers, the control system requires feedback signals, which are measured using voltage and current sensors.

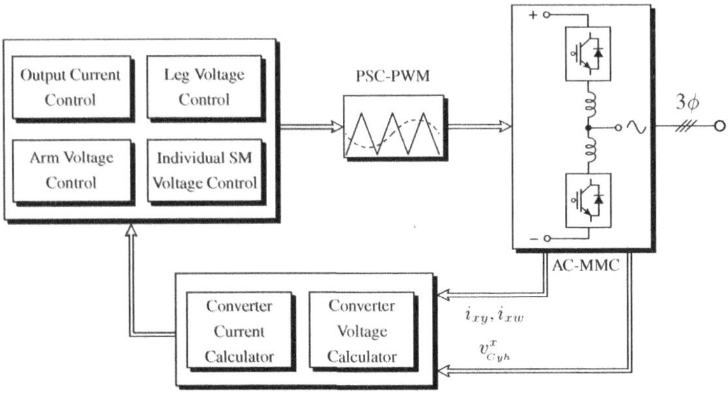

Figure 6.9 Block diagram of the AC-MMC control system.

6.5.1 Output Current Control

In motor drive applications, the current control is implemented in the rotor field reference frame rotating at the speed of rotor field flux [11]. The current control requires reference and feedback signals in the rotor-dq frame. The reference d-axis current component is obtained from the rotor flux controller while the speed/torque controller generates the reference q-axis current component. The feedback currents are measured and transformed into rotor-dq frame using a transformation matrix. The dq-axis reference and feedback signals are controlled using PI-controllers and generate reference voltage commands of v^*_{di} and v^*_{qi}, respectively. The reference voltages in dq-frame are converted back to abc-frame using $\mathbf{T}_{dq\Rightarrow abc}$ transformation matrix given in Chapter 3. The reference voltages in abc-frame are used with PWM scheme to generate the gating signals for the submodules. The details of current control are discussed in Chapter 3.

6.5.2 Submodule Capacitor Voltage Control

The submodule capacitor voltage control of AC-MMC is divided into 1) leg capacitor voltage control, 2) arm capacitor voltage control, and 3) individual submodule capacitor voltage control. The implementation and working principle of each control method is discussed in the following subsections.

(1) Leg Capacitor Voltage Control: The leg voltage control maintains the average DC voltage of each leg at a given reference value of V^*_C. To achieve this objective, the DC component in the arm current is used as a control variable. The control block diagram of the leg voltage control is shown in Figure 6.10 in which, the actual DC voltage is compared with the reference voltage command, and their difference is minimized using a PI-controller. The average DC voltage (v_{C_x}) of each leg is given by

$$v_{C_x} = \frac{1}{2N}\left[\sum_{h=1}^{N} v^x_{C_{uh}} + \sum_{h=1}^{N} v^x_{C_{lh}}\right] \tag{6.36}$$

The output of PI-controller is added to a feed-forward term corresponding to the active power delivered to load. The resultant output becomes the reference DC current component, which is controlled using a PI-controller and gives a compensation signal. The compensation signals are added to the modulation signal obtained from the AC output current control.

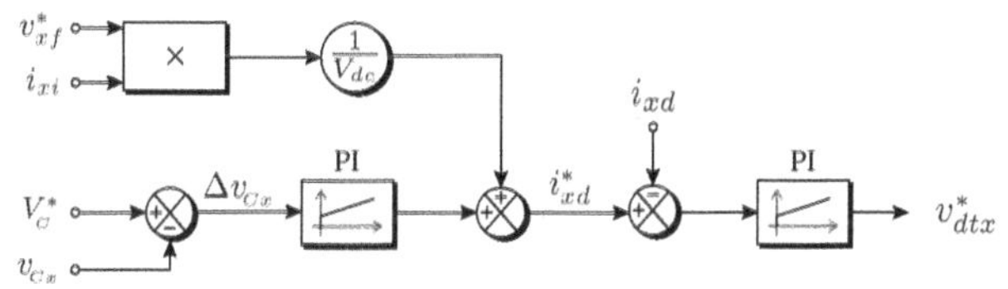

Figure 6.10 Block diagram of leg capacitor voltage control.

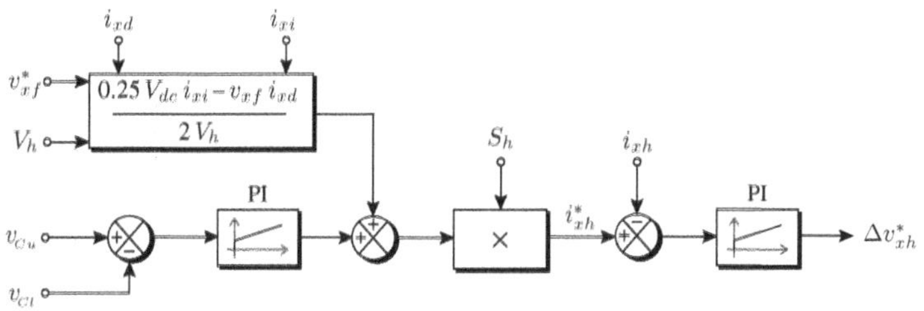

Figure 6.11 Block diagram of arm capacitor voltage control.

(2) Arm Capacitor Voltage Control: The arm capacitor voltage control aims to balance the average submodule voltages between the upper and lower arms. The arm voltage balancing is achieved by controlling the high-frequency current flowing through the arms. The block diagram of arm capacitor voltage control is shown in Figure 6.11. It consists of an outer voltage loop and an inner current control loop. The voltage loop regulates the average capacitors voltage of the upper and lower arms at an identical value. The average voltage of the upper and lower arms is given by

$$\begin{aligned} v_{Cu} &= \frac{1}{N}\sum_{h=1}^{N} v^x_{Cuh} \\ v_{Cl} &= \frac{1}{N}\sum_{h=1}^{N} v^x_{Clh} \end{aligned} \tag{6.37}$$

The voltage loop gives the high-frequency reference current and is added to the feed-forward term as shown in Figure 6.11. The final output is multiplied with a unity magnitude high-frequency current component. The high-frequency component is controlled to achieve the voltage balance between the upper and lower arms. The current controller gives a compensation voltage signal Δv^*_{xh}, which is added to the modulation signals.

(3) Individual Submodule Capacitor Voltage Control: The individual submodule capacitor voltage control aims to regulate each submodule capacitor voltage at a given reference value V^*_C. The control block diagram of individual submodule capacitor voltage control is shown in Figure 6.12, where the measured submodule capacitor voltage is compared with the reference voltage V^*_C and the error is minimized using a PI-regulator. The PI-controller output is multiplied with the direction of arm current to obtain the reference voltage component Δv^*_{xyh}, which is added to the modulation signals.

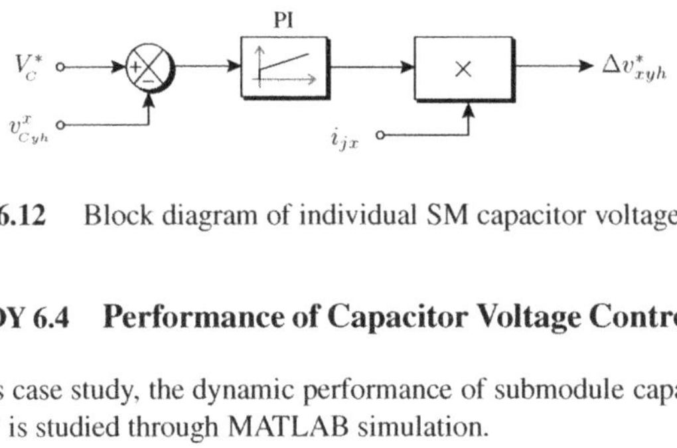

Figure 6.12 Block diagram of individual SM capacitor voltage control.

CASE STUDY 6.4 Performance of Capacitor Voltage Control for AC-MMC

Objective: In this case study, the dynamic performance of submodule capacitor voltage control for AC-MMC is studied through MATLAB simulation.

Parameters: The system parameters of the AC-MMC are as follows: rated DC-bus voltage $V_{dc} = 7$ kV, number of SMs per arm $N = 8$, SM capacitor voltage $v_C = E = 875$ V, SM capacitor $C = 4700$ μF, load inductance $L_o = 20$ mH, load resistance $r_o = 5$ Ω, arm inductor $L = 2$ mH, fundamental frequency $f_o = 10$ Hz, high-frequency $f_h = 100$ Hz, and carrier frequency $f_c = 1000$ Hz.

Analysis:
The dynamic performance of submodule capacitor voltage control for AC-MMC at a fundamental frequency $f_o = 10$ Hz is shown in Figure 6.13. The total DC-bus voltage is set to 7000 V. Each arm has a total of 8 submodules, and the total DC-bus voltage is equally distributed among the submodules. The rated submodule capacitor voltage is equal to 7000/8 = 875 V.

Initially, the converter is controlled to supply an output current of $I = 200$ A. The submodule capacitors in the upper and lower arms are perfectly maintained at their nominal value. At $t = 1.11$ s, the capacitor voltage control is disabled. The submodule capacitors voltage is diverging from their nominal value, which affects the reliability and operation of AC-MMC. At $t = 1.14$ s, the voltage control is activated, and the submodule capacitors voltage is converging towards their rated value.

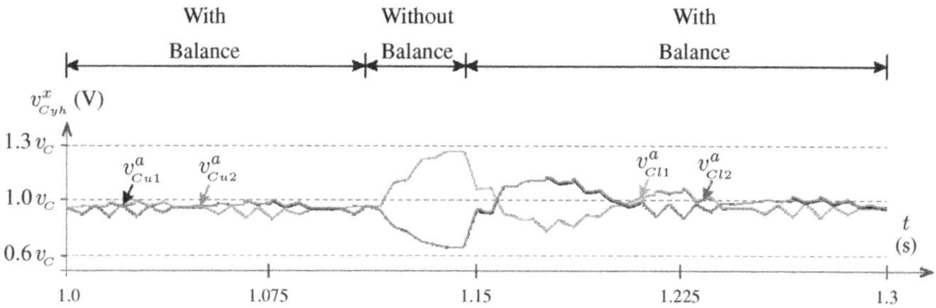

Figure 6.13 Dynamic performance of SM capacitor voltage control.

■

6.5.3 Synthesis of Modulation Signals

The output current control and various stages of submodule capacitor voltage control are discussed in Sections 6.5.1 and 6.5.2. These controllers generate various voltage commands such as the DC (V_{dc}), fundamental (v^*_{xf}), and high-frequency voltage components (v^*_{xh}), and their compensation voltage components v^*_{dtx}, Δv^*_{xh}, and Δv^*_{xyh}, respectively. These voltage commands are used to synthesize the modulation signals for each submodule in the upper, lower sub-arms, and cross-connected arm as

$$\begin{aligned}
m^*_{xuh} &= \frac{1}{V_C}\left[\frac{V_{dc}}{2N} - \frac{v^*_{xf}}{N} \mp \frac{2\,v^*_{xh}}{N} - v^*_{dtx} + \Delta v^*_{xyh}\right] \\
m^*_{xlh} &= \frac{1}{V_C}\left[\frac{V_{dc}}{2N} + \frac{v^*_{xf}}{N} \mp \frac{2\,v^*_{xh}}{N} - v^*_{dtx} + \Delta v^*_{xyh}\right] \\
m^*_{xch} &= \frac{1}{V_C}\left[\frac{V_{dc}}{2N} - \frac{\Delta v^*_{xh}}{N} + \Delta v^*_{xyh}\right]
\end{aligned} \tag{6.38}$$

where the symbols $\mp$ is used to distinguish the top-half and bottom-half of the upper and lower arms, $-$ is assigned to the top-half sub-arms, and $+$ is assigned to the bottom-half sub-arms.

The voltage components V_{dc}, v^*_{xf}, and v^*_{xh} are equally shared by the submodules in each leg, arm, and sub-arms, respectively. The DC compensating voltage v^*_{dtx} is assigned with the minus sign in the entire leg since the DC current component has the same polarity in the entire leg. The compensating high-frequency voltage Δv^*_{xh} is arranged with an opposite sign in each sub-arm, since the submodule power fluctuations are in phase opposition. This compensating voltage component Δv_{xh}^* is assigned with the same sign as that of the high-frequency voltage component v_{xh}^*. The individual capacitor compensation voltage Δv_{xyh}^* is assigned with the positive sign in the arms. The actual sign of compensating voltage varies with the current direction. On the other hand, the modulation signal of cross-connected arm consists of V_{dc}, Δv_{xh}^*, and Δv_{xyh}^*. The Δv_{xh}^* term has negative sign to compensate the voltage drop due to the high-frequency current component flowing through the cross-connected arm. The Δv_{xyh}^* term is added to modulation signals to control the submodule capacitors voltage.

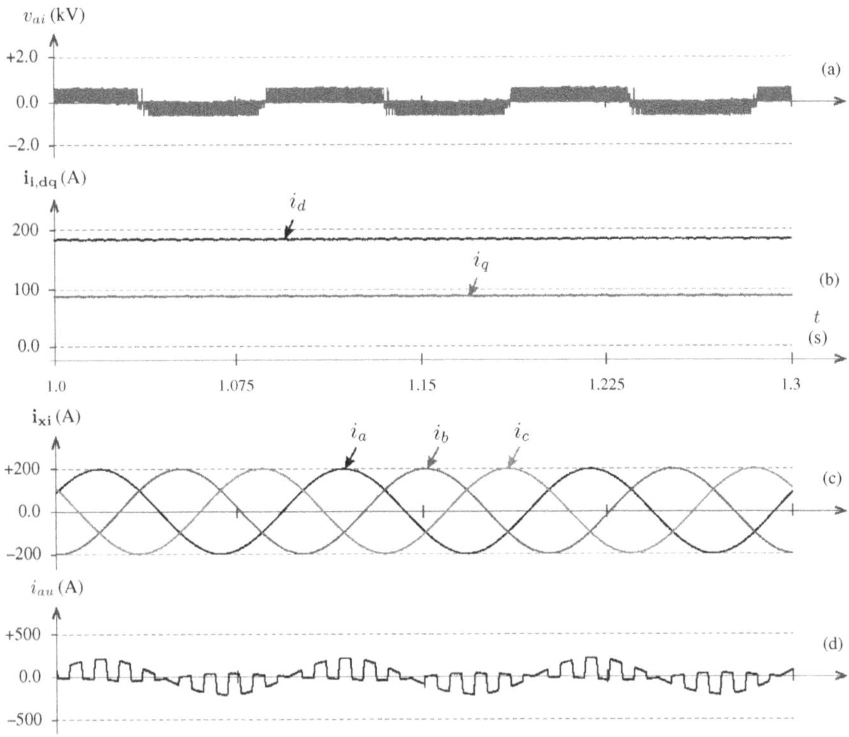

Figure 6.14 Steady-state performance of the AC-MMC.

CASE STUDY 6.5 Steady-State Performance Analysis of an AC-MMC

Objective: In this case study, the steady-state performance of an AC-MMC is studied through MATLAB simulation.

Parameters: The system parameters of the AC-MMC are as follows: rated DC-bus voltage $V_{dc} = 7$ kV, number of SMs per arm $N = 8$, SM capacitor voltage $v_C = E = 875$ V, SM capacitor $C = 4700$ μF, load inductance $L_o = 20$ mH, load resistance $r_o = 5$ Ω, arm inductor $L = 2$

mH, fundamental frequency $f_o = 10$ Hz, high-frequency $f_h = 100$ Hz, and carrier frequency $f_c = 1000$ Hz.

Analysis:

The DC-bus of AC-MMC is connected to a stiff DC source rated to 7000 V, and the AC terminals of AC-MMC are connected to a passive load. The reference dq-currents of output current control are set to $i_{di}^* = 200$ A and $i_{qi}^* = 0$ A. The reference submodule capacitor voltage is set to 875 V. The converter voltage and current waveforms are shown in Figure 6.14 and 6.15.

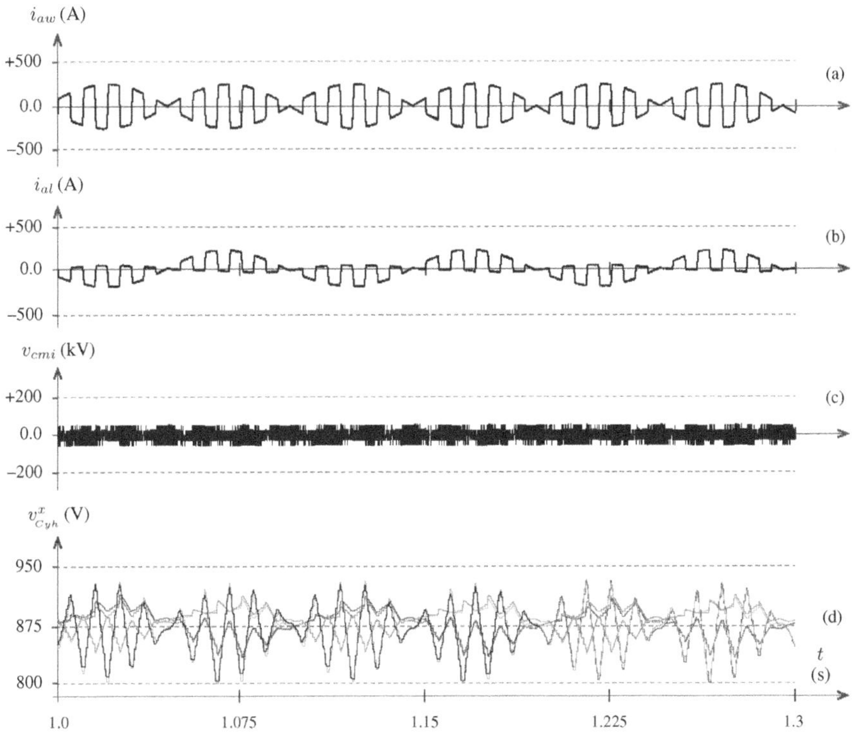

Figure 6.15 Steady-state performance of the AC-MMC.

The converter generates an output voltage with a fundamental voltage of 500 V corresponding to the modulation index (m_a) of 0.167 as shown in Figure 6.14(a). The d- and q-axis current components are perfectly regulated at 200 A and 0 A, respectively, as shown in Figure 6.14(b). The dq-axis current components are transformed to abc-frame, which have a peak value of 200 A at a fundamental frequency of 10 Hz as shown in Figure 6.14(c). The modulation signal consists of DC and square wave high-frequency components, which generates corresponding current components flowing through the sub-arms as shown in Figure 6.14(d). The square wave high-frequency current component in the arm current is used to maintain the power balance between the upper and lower arm submodule capacitors voltage.

Figure 6.15(a) shows the waveform of cross-connected arm current, which includes the high-frequency square wave current component. This current component is controlled to minimize the voltage ripple during low-frequency operation. The current flowing through the sub-arm I is shown in Figure 6.15(b), which is exactly opposite in-phase with the sub-arm II current shown in Figure 6.14(d). This current contains similar current components as that of the upper arm current except for the polarity of DC and fundamental components. The common-mode voltage generated by AC-MMC consists of switching ripple with a peak value of 50 V as shown in Figure 6.15(c). The submodule capacitors in the upper and lower arms are perfectly regulated at their nominal value of 875 V as shown in Figure 6.15(d). ■

6.6 SUMMARY

In this chapter, a novel modular multilevel converter named active cross-connected modular multilevel converter (AC-MMC) is presented. The AC-MMC has an additional arm, interconnecting the upper and lower arm of each leg. The operation and performance of the AC-MMC are analyzed with PSC-PWM scheme. An equivalent circuit is further developed to analyze the submodule voltage ripple and the common-mode voltage generated by AC-MMC at low/zero frequency operation. Also, the device current stress is analyzed with square and sinusoidal high-frequency voltage components. It is observed that the square wave injection reduces the peak arm current by 50% compared to the sinusoidal wave. All the devices in the upper, lower, and cross-connected arms have uniform current stress with square wave high-frequency voltage.

Finally, a closed-loop control method is presented, which includes an output current control and submodule voltage control. The steady-state and dynamic performance of output current and capacitor voltage control schemes are analyzed through case studies.

REFERENCES

1. M. Spichartz, V. Staudt, and A. Steimel, "Analysis of the module-voltage fluctuations of the modular multilevel converter at variable speed drive applications," in *Optimization of Electrical and Electronic Equipment (OPTIM), 2012 13th International Conference on*, May 2012, pp. 751–758.
2. S. Debnath, J. Qin, and M. Saeedifard, "Control and stability analysis of modular multilevel converter under low-frequency operation," *IEEE Trans. Ind. Electron.*, vol. 62, no. 9, pp. 5329–5339, Sep 2015.
3. S. Du, B. Wu, N. Zargari, and Z. Cheng, "A flying-capacitor modular multilevel converter (fc-mmc) for medium-voltage motor drive," *IEEE Trans. Power Electron.*, vol. 32, no. 3, pp. 2081–2089, Mar 2017.
4. M. Hagiwara, I. Hasegawa, and H. Akagi, "Start-up and low-speed operation of an electric motor driven by a modular multilevel cascade inverter," *IEEE Trans. Ind. Appl.*, vol. 49, no. 4, pp. 1556–1565, Jul 2013.
5. M. Perez and S. Bernet, "Capacitor voltage ripple minimization in modular multilevel converters," in *Industrial Technology (ICIT), 2015 IEEE International Conference on*, Mar 2015, pp. 3022–3027.
6. S. Du, B. Wu, D. Xu, and N. R. Zargari, "A transformerless bipolar multistring dc-dc converter based on series-connected modules," *IEEE Trans. Power Electron.*, vol. 32, no. 2, pp. 1006–1017, Feb 2017.
7. S. Du, B. Wu, K. Tian, D. Xu, and N. R. Zargari, "A novel medium-voltage modular multilevel dc-dc converter," *IEEE Trans. Ind. Electron.*, vol. 63, no. 12, pp. 7939–7949, Dec 2016.

8. M. Hagiwara, K. Nishimura, and H. Akagi, "A medium-voltage motor drive with a modular multilevel pwm inverter," *IEEE Trans. Power Electron.*, vol. 25, no. 7, pp. 1786–1799, Jul 2010.

9. L. He, K. Zhang, J. Xiong, S. Fan, and Y. Xue, "Low-frequency ripple suppression for medium-voltage drives using modular multilevel converter with full-bridge submodules," *IEEE Trans. Emerg. Sel. Topics Power Electron.*, vol. 4, no. 2, pp. 657–667, Jun 2016.

10. B. Li, Y. Zhang, G. Wang, W. Sun, D. Xu, and W. Wang, "A modified modular multilevel converter with reduced capacitor voltage fluctuation," *IEEE Trans. Ind. Electron.*, vol. 62, no. 10, pp. 6108–6119, Oct 2015.

11. J.-J. Jung, H.-J. Lee, and S.-K. Sul, "Control of the modular multilevel converter for variable-speed drives," in *Power Electronics, Drives and Energy Systems (PEDES), 2012 IEEE International Conference on*, 2012, pp. 1–6.

CHAPTER 7

STAR AND DELTA-CHANNEL MODULAR MULTILEVEL CONVERTERS

7.1 INTRODUCTION

In earlier chapters, the passive (PC-MMC) and active (AC-MMC) cross-connected modular multilevel converters were introduced to overcome the technical issues in the conventional modular multilevel converter (MMC) during low/zero-frequency operation. These converters use capacitors or additional submodules to interconnect the upper and lower arms of each leg [1, 2]. The cross-connected arm provides an alternative path to exchange the power between the upper and lower arms. The power flowing through the cross-connected arm mainly consists of high-frequency components, which are used to minimize the capacitor voltage ripple during low-frequency operation [3, 4].

Another possibility is to interconnect the converter legs rather than the arms belonging to the same leg using additional submodules in either star or delta configuration. These topologies are referred to as star- or delta-channel modular multilevel converters, depending on the type of interconnection between the legs. This interconnection allows the power transfer from the higher energy leg to the lower energy leg, and balance the power fluctuations, which are phase-shifted by 120°. The star- and delta-channel modular multilevel converters have the capability to operate at

Modular Multilevel Converters: Analysis, Control, and Applications, First Edition.
Sixing Du, Apparao Dekka, Bin Wu, and Navid Zargari.

low/zero-frequency region with smaller capacitor voltage ripple and motor neutral voltage.

In this chapter, the operation and control of star-channel MMC are presented. The circuit configuration and switching states are presented in Section 7.2. The modeling and operation of star-channel MMC with phase-shifted carrier modulation is discussed in Section 7.3. The voltage ripple and common-mode voltage of star-channel MMC during low-frequency operation are analyzed in Section 7.4. The generalized control scheme to control the output currents and submodule capacitor voltages is presented in Section 7.5. The operation and control of delta-channel MMC are quite similar to the star-channel MMC, and are briefly introduced in Section 7.6. Finally, a comparison of advanced modular multilevel converters including PC-MMC, AC-MMC, star-channel MMC, and delta-channel MMC is presented in Section 7.7, and a summary is given in Section 7.8.

7.2 STAR-CHANNEL MODULAR MULTILEVEL CONVERTER

The conventional modular multilevel converter is modified into a star-channel modular multilevel converter by interconnecting the converter legs using additional submodules in star configuration. The star-channel provides a path for high-frequency current components flowing in the system and helps to minimize the capacitor voltage ripple during low-frequency operation.

7.2.1 Circuit Configuration of Star-Channel MMC

The circuit configuration of a three-phase star-channel MMC with N submodules per arm is shown in Figure 7.1, where each leg is composed of two arms named as upper and lower. The upper and lower arms form an AC output terminal "x," where $x \in \{a, b, c\}$ represents the output phase. Each arm is further divided into two halves and referred to as a sub-arm. The sub-arms I and II of three converter legs are interconnected using additional submodules in star configuration (nu). Similarly, the sub-arms III and IV of three converter legs are connected together in star configuration (nl).

The star-channels carry bipolar high-frequency current components. To control these currents, the star-channels are designed with full-bridge submodules. These submodules can generate both positive and negative voltage levels corresponding to the bipolar current. On the other hand, half-bridge submodules are used in the sub-arms, which have low power losses and are suitable to achieve a higher efficiency. The required number of submodules in the sub-arms and star-channels depends on the converter voltage rating. Generally, the star-channel requires only $\frac{1}{4}$ of the submodules in the main arms. In star-channel MMC, each sub-arm is equipped with an arm inductor ("L"), which mitigates the switching ripples and the inrush current. The star-channel MMC requires twice the buffer inductors, but the equivalent (total) inductance remains same as that of the conventional MMC.

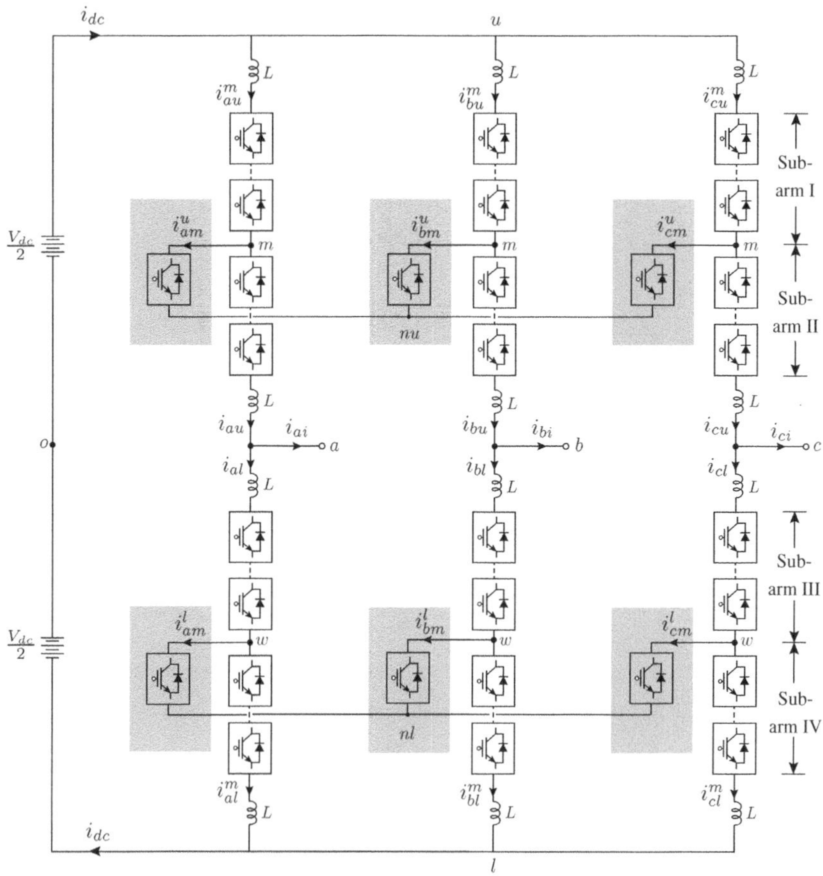

Figure 7.1 Power circuit configuration of star-channel MMC.

7.2.2 Switching States and Output Voltage

The configuration of a converter leg in star-channel MMC is shown in Figure 7.2. Each converter leg is divided into two arms, and each arm consists of four half-bridge (HB) submodules and two arm inductors. To simplify the analysis, the half-bridge submodule capacitor voltage is assumed to be constant at v_C. Each half-bridge submodule requires one independent gating signal to control the devices and generate a voltage of either v_C or 0. With four submodules per arm, the star-channel MMC generates five voltage levels across the arm terminals of "e" and "g" in the upper arm and "r" and "s" in the lower arm.

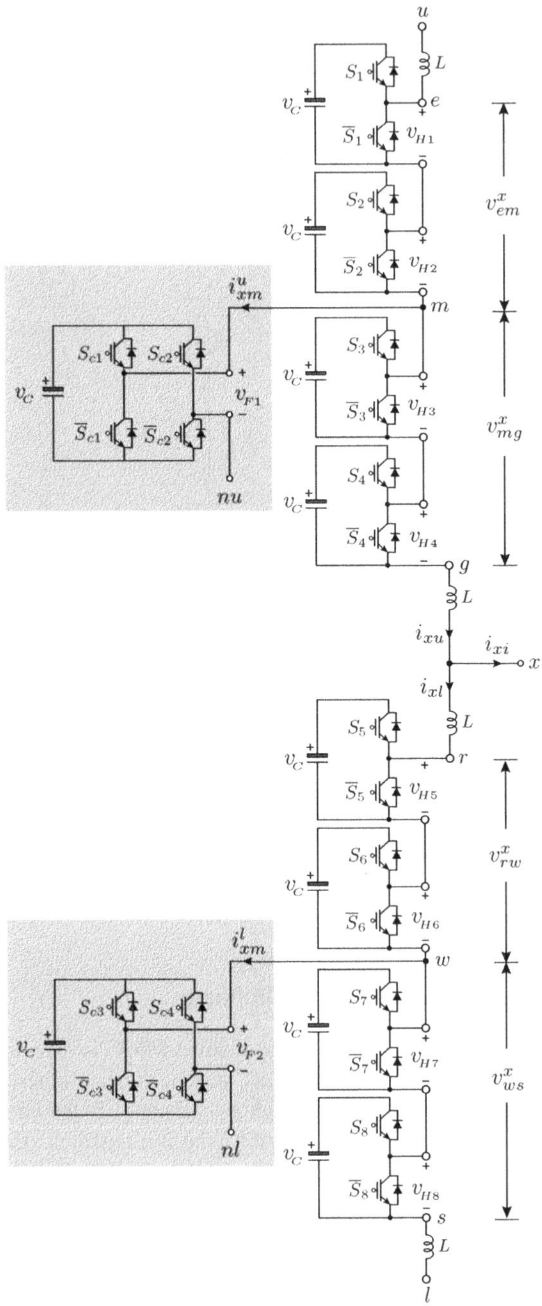

Figure 7.2 Configuration of converter leg in star-channel MMC.

On the other hand, the star-channel arm consists of full-bridge submodules to control the bidirectional high-frequency current components flowing amongst the converter legs. These submodule capacitor voltages are maintained at a voltage of v_C. The switching states and output voltage levels of the upper arm are shown in Table 7.1. When switches "S_1," "S_2," "S_3," and "S_4" are turned ON, each half-bridge submodule generates a voltage of "v_C," and the resultant arm voltage is $v_{eg}^x = 4\,v_C$. Similarly, the zero voltage level ($v_{eg}^x = 0$) can be generated by turning OFF the switches "S_1," "S_2," "S_3," and "S_4." Several redundant switching states are available to generate the remaining voltage levels of "v_C," "$2\,v_C$," and "$3\,v_C$" as shown in Table 7.1. These switching states are used to achieve the voltage balancing in a star-channel modular multilevel converter.

Table 7.1 Switching states and output voltage levels of an arm

S_1	S_2	S_3	S_4	v_{H1}	v_{H2}	v_{H3}	v_{H4}	v_{em}^x	v_{mg}^x	v_{eg}^x	**Voltage Level**
0	0	0	0	0	0	0	0	0	0	0	0
1	0	0	0	v_C	0	0	0	v_C	0	v_C	
0	1	0	0	0	v_C	0	0	v_C	0	v_C	1
0	0	1	0	0	0	v_C	0	0	v_C	v_C	
0	0	0	1	0	0	0	v_C	0	v_C	v_C	
1	1	0	0	v_C	v_C	0	0	$2v_C$	0	$2v_C$	
1	0	1	0	v_C	0	v_C	0	v_C	v_C	$2v_C$	
1	0	0	1	v_C	0	0	v_C	v_C	v_C	$2v_C$	2
0	1	1	0	0	v_C	v_C	0	v_C	v_C	$2v_C$	
0	1	0	1	0	v_C	0	v_C	v_C	v_C	$2v_C$	
0	0	1	1	0	0	v_C	v_C	0	$2v_C$	$2v_C$	
1	1	1	0	v_C	v_C	v_C	0	$2v_C$	v_C	$3v_C$	
1	1	0	1	v_C	v_C	0	v_C	$2v_C$	v_C	$3v_C$	3
1	0	1	1	v_C	0	v_C	v_C	v_C	$2v_C$	$3v_C$	
0	1	1	1	0	v_C	v_C	v_C	v_C	$2v_C$	$3v_C$	
1	1	1	1	v_C	v_C	v_C	v_C	$2v_C$	$2v_C$	$4v_C$	4

The entire upper arm voltage v_{eg}^x is uniformly distributed between the sub-arms I and II (v_{em}^x and v_{mg}^x). For example, the arm voltage $v_{eg}^x = 3\,v_C$ can be obtained by using the sub-arm voltage combinations of ($2\,v_C$, v_C) and (v_C, $2\,v_C$). The sub-arm voltage $v_{em}^x = 2\,v_C$ or $v_{mg}^x = v_C$ can be generated using the redundant switching states. Typically, the star-channel MMC is designed with an even number of submodules in each arm. Therefore, it is easy to form a mid-point to interconnect the three converter legs through a group of submodules.

In Figure 7.2, the star-channel is designed with a single full-bridge submodule corresponding to the $\frac{1}{4}$ of the submodules in the main arm. The full-bridge submodule requires two independent gating signals to control the devices. The switching states and possible voltage combinations across the submodules with respect to the upper star-channel neutral point nu are shown in Table 7.2. The zero voltage level can be generated by turning ON the devices S_{c1} and S_{c2} or turning them OFF. Other voltage levels $+1$ and -1 are generated by turning ON either S_{c1} or S_{c2}; then the output voltage is equal to either $v_{mnu} = +v_C$ or $v_{mnu} = -v_C$.

Table 7.2 Switching states and output voltage levels of a star-channel

S_{c1}	S_{c2}	v_{mnu}	**Voltage Level**
0	0	0	0
1	1	0	
1	0	$+v_C$	+1
0	1	$-v_C$	-1

7.3 PRINCIPLES OF OPERATION

The converter AC output voltage is generated by controlling the AC output current components in the upper and lower arms. At the same time, the submodule voltage balancing is necessary to ensure the AC output voltage has symmetrical voltage steps. On the other hand, the star-channel full-bridge submodules are controlled based on the high-frequency current components in the upper and lower arms. The high-frequency current component is used to minimize the ripple in the submodule capacitor voltage. Considering the above control objectives, the modulation signals are generated for star-channel MMC. The modulation signals are then compared with the triangular carrier signals to generate the gating signals for the half-bridge and full-bridge submodules in each converter leg. The mathematical modeling and generation of gating signals using PSC-PWM are discussed in the following subsections.

7.3.1 Modeling of Star-Channel MMC

The star-channel MMC has half-bridge submodules in the main arms and full-bridge submodules in the star-channel arms. The submodules in the main arm are controlled to maintain the power balance between the input (DC) and output (AC) sides of the converter. Whereas the submodules in the star-channel arm are controlled to minimize the submodule voltage ripple while maintaining the power balance amongst the converter legs. To control the submodules in the main and star-channel arms, a set of equations are required. In this section, the modeling of main and star-channel arms is presented.

(1) Modeling of Main Converter Arms: The converter DC input power is transferred to the output load through the submodules in the main arms. During this process, the DC input voltage V_{dc} is equally distributed between the sub-arms I and II in the upper arm (sub-arms III and IV in the lower arm). The submodules generate an AC voltage and form a sub-arm voltage. The summation of sub-arm I and II voltage gives the upper arm voltage. Similarly, the lower arm voltage is obtained by adding the sub-arm III and IV voltages. The difference between the lower and upper arm voltage appears across the AC output load terminals. From Figure 7.1, the voltage across the upper and lower arms is given by

$$\begin{aligned} v_{em}^{x} + v_{mg}^{x} &= \frac{V_{dc}}{2} - v_{xi} - L\frac{di_{xu}^{m}}{dt} - L\frac{di_{xu}}{dt} \\ v_{rw}^{x} + v_{ws}^{x} &= \frac{V_{dc}}{2} + v_{xi} - L\frac{di_{xl}^{m}}{dt} - L\frac{di_{xl}}{dt} \end{aligned} \tag{7.1}$$

where v_{em}^{x}, v_{mg}^{x} represent the voltages across the upper sub-arms of phase-x; v_{rw}^{x}, v_{ws}^{x} represent the voltages across the lower sub-arms of phase-x; i_{xu}^{m} and i_{xu} represent the current flowing through the sub-arms I and II, respectively, and i_{xl}^{m} and i_{xl} represent the current flowing through the sub-arms III and IV, respectively.

Each sub-arm current consists of DC, high-frequency, and fundamental frequency current components. These current components are given by

$$\begin{aligned} i_{xd} &= \frac{1}{2}\left(i_{xu}^{m} + i_{xu} + i_{xl}^{m} + i_{xl}\right) \\ i_{xh} &= \frac{1}{2}\left(i_{xu}^{m} - i_{xu}\right) \text{ (or) } \frac{1}{2}\left(i_{xl}^{m} - i_{xl}\right) \\ i_{xi} &= i_{xu} - i_{xl} \end{aligned} \tag{7.2}$$

where $x \in \{a, b, c\}$ represents the phase, i_{xh} represents the high-frequency current component, and i_{xi} represents the fundamental frequency current component flowing through the load.

The arm voltage is generated by controlling the submodules in the sub-arms. The arm voltage is equally distributed between the two sub-arms. A high-frequency voltage (v_{xh}, where $v_{ah} = v_{bh} = v_{ch} = v_h$) is introduced in the sub-arm voltages as an additional control degree of freedom to minimize the submodule capacitors voltage ripple during the low-frequency operation. The magnitude of high-frequency voltage component is set to zero ($v_h = 0$) at high-frequency (high-speed) operation. The resultant sub-arms voltage of phase-x are given by

$$\begin{aligned} v_{em}^{x} &= \frac{V_{dc}}{4} - \frac{v_{xi}}{2} - L\frac{di_{xd}}{dt} - \frac{L}{2}\frac{di_{xi}}{dt} - v_h \\ v_{mg}^{x} &= \frac{V_{dc}}{4} - \frac{v_{xi}}{2} - L\frac{di_{xd}}{dt} - \frac{L}{2}\frac{di_{xi}}{dt} + v_h \\ v_{rw}^{x} &= \frac{V_{dc}}{4} + \frac{v_{xi}}{2} - L\frac{di_{xd}}{dt} + \frac{L}{2}\frac{di_{xi}}{dt} - v_h \\ v_{ws}^{x} &= \frac{V_{dc}}{4} + \frac{v_{xi}}{2} - L\frac{di_{xd}}{dt} + \frac{L}{2}\frac{di_{xi}}{dt} + v_h \end{aligned} \tag{7.3}$$

The above equations show that the main arm voltages consist of DC, fundamental, and high-frequency voltage components. Each voltage component is used to control various current components flowing in the main arms.

(2) Modeling of Star-Channel Arms: The star-channel arm forms an internal loop to exchange the power between the converter legs. From Figure 7.1, the voltage across the upper star-channel submodules with respect to the DC-bus mid-point (o) is given by

$$v_{nuo}^{x} = \frac{V_{dc}}{2} - v_{em}^{x} - v_{mnu}^{x} - L\frac{d\,i_{xu}^{m}}{dt} \tag{7.4}$$

where v_{nuo}^{x} represents the per-phase voltage across the neutral point of star-channel and DC-bus mid-point.

Equation (7.4) is further simplified by substituting equations (7.2) and (7.3) as

$$v_{nuo}^{x} = \frac{V_{dc}}{4} + \frac{v_{xi}}{2} + v_h - v_{mnu}^{x} - L\frac{d\,i_{xh}}{dt} \tag{7.5}$$

The three-phase voltage of v_{nuo} is expressed in terms of the average of three-phase voltages as

$$\begin{aligned} v_{nuo} &= \frac{1}{3}\left(v_{nuo}^{a} + v_{nuo}^{b} + v_{nuo}^{c}\right) \\ &= \frac{V_{dc}}{4} + \frac{1}{6}\left(v_{ai} + v_{bi} + v_{ci}\right) + v_h - \frac{1}{3}\left(v_{mnu}^{a} + v_{mnu}^{b} + v_{mnu}^{c}\right) \\ &\quad - \frac{L}{3}\frac{d\left(i_{ah} + i_{bh} + i_{ch}\right)}{dt} \end{aligned} \tag{7.6}$$

Under balanced operation, the summation of three-phase voltages and currents become zero, and is given by

$$\begin{aligned} v_{ai} + v_{bi} + v_{ci} &= 0 \\ v_{mnu}^{a} + v_{mnu}^{b} + v_{mnu}^{c} &= 0 \\ i_{ah} + i_{bh} + i_{ch} &= 0 \end{aligned} \tag{7.7}$$

Equation (7.7) is substituted in equation (7.6) and results in

$$v_{nuo} = \frac{V_{dc}}{4} + v_h \tag{7.8}$$

The voltage across the star-channel is obtained by subtracting equation (7.8) from (7.5), and results in

$$v_{mnu}^{x} = \frac{v_{xi}}{2} - L\frac{d\,i_{xh}}{dt} \tag{7.9}$$

Equation (7.9) shows that the star-channel arm voltage consists of fundamental voltage component v_{xi} and a voltage drop due to the high-frequency current component. The fundamental voltage component is bipolar in nature and capable of controlling the bidirectional current flowing through the star-channel arms. According to equation (7.8), the externally injected high-frequency voltage component v_h appears in the voltage across the star-channel neutral point and DC-bus mid-point, and there is no effect on the motor neutral voltage.

7.3.2 Phase-Shifted Carrier Modulation for Star-Channel MMC

The phase-shifted carrier modulation scheme is widely used to control the modular multilevel converters. The pulse width modulator requires carrier signals and modulation signals. The main arm carrier signals are horizontally phase-shifted by an angle of $360°/N$ and star-channel carrier signals are phase-shifted by $360°/2N_c$, where N represents the number of submodules in the main arm and N_c represents the number of submodules in the star-channel arm [5]. The modulation signals of main and star-channel arms are generated by using the mathematical models given in equations (7.3) and (7.9). According to these equations, the modulation signals consist of fundamental (v_{xi}) and high-frequency (v_h) voltage components, which are defined as

$$\begin{aligned} v_{xi} &= V_f \sin(\omega_o t + \theta_x) \\ v_h &= V_h S_h \end{aligned} \tag{7.10}$$

where V_f represents the peak value of fundamental voltage component, ω_o represents the fundamental angular frequency (rad), θ_x represents the initial angle of fundamental frequency voltage component, V_h represents the peak value of high-frequency voltage, and S_h represents the wave shape of high-frequency voltage.

Several wave shapes such as sinusoidal, square, trapezoidal, and third-harmonic can be used as a high-frequency voltage components [6]. The square wave high-frequency voltage helps to minimize the voltage ripple during low-frequency operation and reduces the peak value of current flowing through the main arms by 50% compared to the sinusoidal high-frequency voltage [6]. Considering the above advantages, the square wave is considered as a high-frequency voltage, and defined as

$$S_h = \begin{cases} 1 & 0 \le t < \frac{\pi}{\omega_h} \\ -1 & \frac{\pi}{\omega_h} \le t < \frac{2\pi}{\omega_h} \end{cases} \tag{7.11}$$

where ω_h represents the angular frequency of high-frequency voltage component.

The peak value of fundamental (V_f) and high-frequency (V_h) voltage components are defined as

$$\begin{aligned} V_f &= \frac{m_a V_{dc}}{2} \\ V_h &= \frac{(1 - m_a) V_{dc}}{4} \end{aligned} \tag{7.12}$$

where m_a represents the amplitude modulation index of 0–1.

The sub-arm modulation signals are obtained by substituting equations (7.10) and (7.12) in equations (7.3) and (7.9). The resultant modulation signals are normalized with respect to the half of the DC-bus voltage $V_{dc}/2$ and results in

$$
\begin{aligned}
m_{u1}^{x} &= \frac{1}{2} - \frac{m_a}{2}\sin(\omega_o t + \theta_x) - \frac{1-m_a}{2} S_h \\
m_{u2}^{x} &= \frac{1}{2} - \frac{m_a}{2}\sin(\omega_o t + \theta_x) + \frac{1-m_a}{2} S_h \\
m_{l1}^{x} &= \frac{1}{2} + \frac{m_a}{2}\sin(\omega_o t + \theta_x) - \frac{1-m_a}{2} S_h \\
m_{l2}^{x} &= \frac{1}{2} + \frac{m_a}{2}\sin(\omega_o t + \theta_x) + \frac{1-m_a}{2} S_h \\
m_{c}^{x} &= m_a \sin(\omega_o t + \theta_x)
\end{aligned}
\tag{7.13}
$$

where the voltage drop across the inductor is neglected at steady state, m_{u1}^{x} and m_{u2}^{x} represent the normalized modulation signals of sub-arms I and II, respectively, m_{l1}^{x} and m_{l2}^{x} represent the normalized modulation signals of sub-arms III and IV, respectively, and m_c^x represents the modulation signal of star-channel arm in phase-x. The modulation signals are compared with the corresponding arm carrier signals, and yield the gating signals that are then applied to the submodules.

CASE STUDY 7.1 Analysis of Star-Channel MMC with a PSC-PWM scheme

Objective: In this case study, the implementation and performance of star-channel MMC with a PSC-PWM scheme is studied through MATLAB simulation.

Parameters: The system parameters of the star-channel MMC are as follows: rated DC-bus voltage $V_{dc} = 7$ kV, fundamental frequency $f_o = 55$ Hz, high-frequency $f_h = 165$ Hz, number of SMs per arm $N = 8$, number of star-channel submodules $N_c = 2$, arm inductor $L = 2$ mH, and carrier frequency $f_c = 165$ Hz.

Analysis:
In this study, the main arm of star-channel MMC is designed with eight half-bridge submodules. Each submodule capacitor is replaced with an ideal DC-voltage source of $E = 875$ V. On the other hand, the star-channel arm is designed with two full-bridge submodules, which is equal to the $1/4$ of the submodules in the main arm. With eight submodules in the main arm, the star-channel MMC generates a total of nine voltage levels at AC output terminals. The operation of nine-level star-channel MMC with PSC-PWM scheme is shown in Figure 7.3. The modulation signals m_{u1}^{a} and m_{u2}^{a} of phase-a, sub-arms I and II are generated with an amplitude modulation index $m_a = 0.9$, fundamental frequency $f_o = 55$ Hz, and zero initial phase angle (θ_x). In addition, a square wave high-frequency voltage component (S_h) with a frequency of $f_h = 165$ Hz is added to the modulation signals as shown in Figure 7.3(a). The modulation signals are given by

$$
\begin{aligned}
m_{u1}^{a} &= \frac{1}{2} - \frac{0.9}{2}\sin(110\,\pi\, t) - \frac{1-0.9}{2} S_h \\
m_{u2}^{a} &= \frac{1}{2} - \frac{0.9}{2}\sin(110\,\pi\, t) + \frac{1-0.9}{2} S_h
\end{aligned}
\tag{7.14}
$$

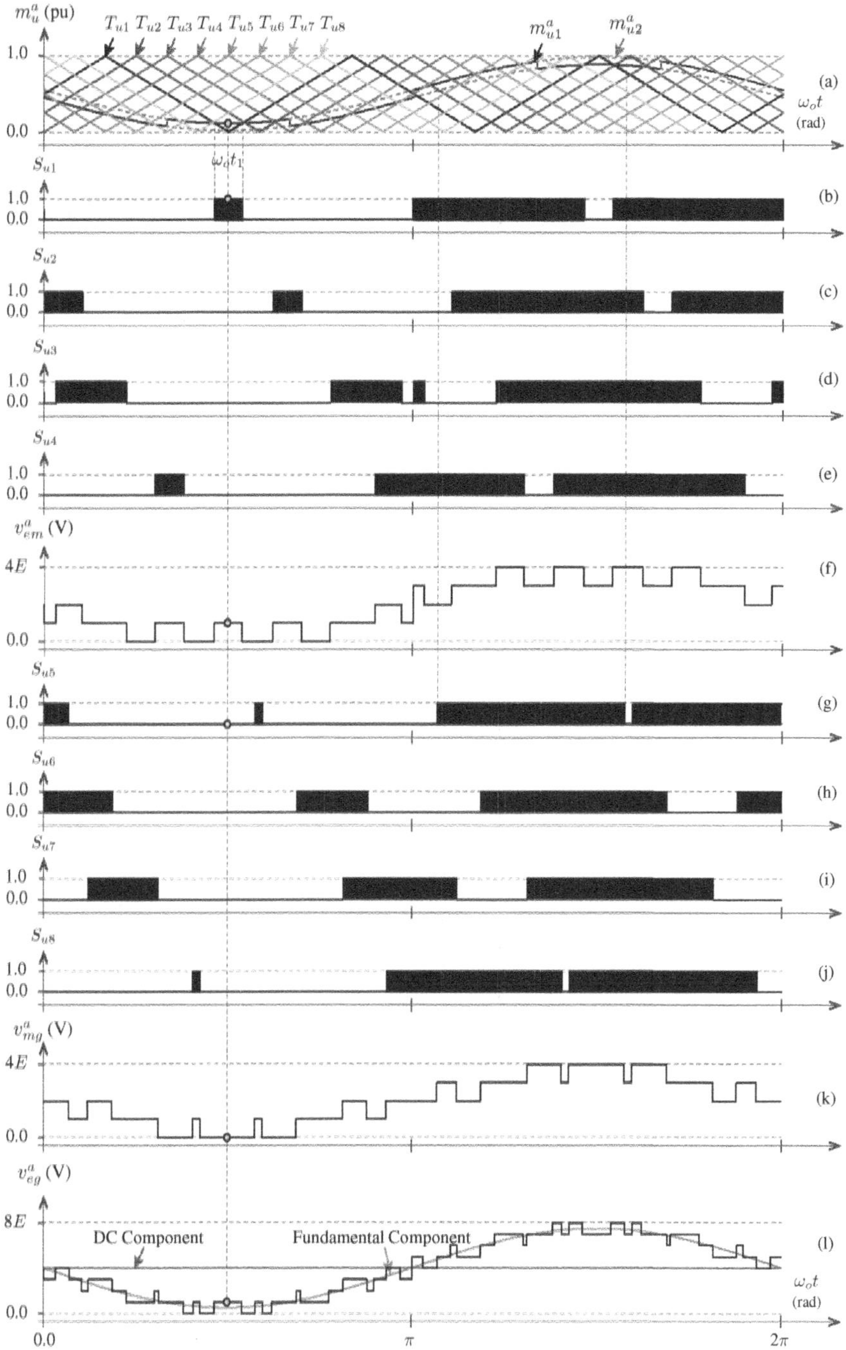

Figure 7.3 Implementation of PSC-PWM for main arms in star-channel MMC.

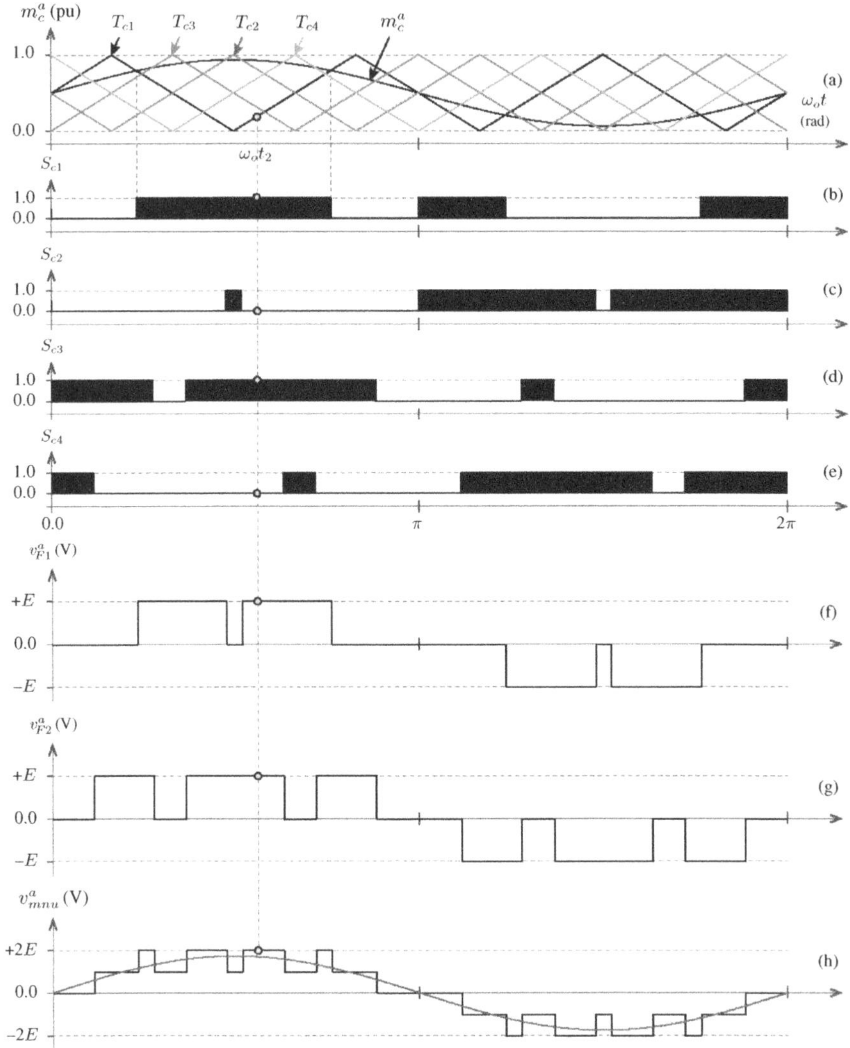

Figure 7.4 Implementation of PSC-PWM for star-channel arm.

Similarly, the modulation signals of phase-a, sub-arms III and IV are generated as per the equations given in (7.13). The nine-level star-channel MMC require eight triangular signals, which are generated with a carrier frequency of $f_c = 165$ Hz as shown in Figure 7.3(a). The carrier signals are divided into two groups and arranged with an initial phase angle shift of 45°. The first group of carrier signals T_{u1}, T_{u3}, T_{u5}, and T_{u7} are assigned to sub-arm I and generated with an initial phase angle of 0, 90°, 180°, and 270°, respectively. The second group of carrier signals T_{u2}, T_{u4}, T_{u6}, and T_{u8} are assigned to sub-arm II and generated with an initial phase angle of 45°, 135°, 225°, and 315°, respectively.

The modulation signal m_{u1}^{a} is compared with the carrier signals T_{u1}, T_{u3}, T_{u5}, and T_{u7}, and generates the gating signal of submodules in the sub-arm I as shown in Figure 7.3(b)–(e). When the modulation signal m_{u1}^{a} is greater than the carrier signals, then the submodules are turned ON and generate a voltage level of "E." Otherwise, the submodules remain turned OFF and generates a voltage level of "0." The output voltages of each submodule are added together, and form the sub-arm I voltage (v_{em}^{a}) as shown in Figure 7.3(f). The sub-arm I voltage waveform has 5 voltage levels with a voltage step of "E." The maximum output voltage is equal to "$4E$" and its minimum value is "0."

Similarly, the modulation signal m_{u2}^{a} is compared with the carrier signals T_{u2}, T_{u4}, T_{u6} and T_{u8}, and generates the gating signal of submodules in sub-arm II as shown in Figure 7.3(g)–(j). Similarly, the sub-arm II voltage (v_{mg}^{a}) is obtained from summation of all the submodule voltages as shown in Figure 7.3(k). The output voltages of sub-arms I and II are added, which result in the upper arm voltage (v_{eg}^{a}) as shown in Figure 7.3(l). The upper arm voltage has a total of nine voltage levels corresponding to the modulation index of $m_a = 0.9$. The output voltage has a maximum value of "$8E$" and a minimum value of "0." Each voltage step is equal to the rated value of submodule capacitor voltage "E." The arm voltage has a DC component along with a fundamental frequency of 55 Hz component as shown in Figure 7.3(l). The above principle can be applied to the lower arm submodules to generate the required output voltage.

In this example, the modulation signal of star-channel arm is generated with a modulation index of $m_a = 0.9$ and fundamental frequency of $f_o = 55$ Hz, and is given by

$$m_c^a = 0.9 \sin(110\,\pi\,t) \tag{7.15}$$

The star-channel arm consists of two full-bridge submodules and requires four carrier signals to modulate them. The carrier signals are divided into two groups of T_{c1} and T_{c2} as group-I and T_{c3} and T_{c4} as group-II, and applied to two full-bridge submodules. The four carrier signals are arranged with a phase angle shift of 90° and the carrier signal belonging to the same submodule are arranged with a phase angle shift of 180° as shown in Figure 7.4(a). The modulation signal m_c^a is compared with the carrier signals T_{c1} and T_{c2} to generate the gating signals for the devices S_{c1} and S_{c2} as shown in Figure 7.4(b)–(c). Similarly, the gating signal of devices S_{c3} and S_{c4} are generated by comparing the modulation signal m_c^a with the carrier signals T_{c3} and T_{c4} as shown in Figure 7.4(d)–(e). Depending on the state of the devices, the submodules I and II generate one of the three-voltage levels of 0, E, and $-E$ as shown in Figure 7.4(f)–(g). The output voltage of submodules I and II is added together, and results in a star-channel arm voltage (v_{mnu}^{a}). The star-channel arm voltage has bipolar voltage levels to control the positive and negative half-cycle of the current flowing through the star-channel arm. For example, at $\omega_o t_2$, the devices S_{c1} and S_{C3} are ON in the full-bridge submodules I and II, respectively. These submodules generate output voltages of $v_{F1}^{a} = E$ and $v_{F2}^{a} = E$. The summation of submodules output voltage gives the star-channel arm voltage, which is equal to $v_{mnu}^{a} = 2E$ as shown in Figure 7.4(h). ■

7.4 LOW-FREQUENCY OPERATION OF STAR-CHANNEL MMC

The conventional modular multilevel converter is modified to minimize the voltage ripple without increasing the motor neutral voltage or common-mode voltage during low-frequency operation [1, 3, 7–10]. The PC-MMC and AC-MMC are one of their kind, and briefly presented in Chapters 5 and 6, respectively. These topologies have cross-connected arm between the upper and lower arms in each converter leg. Other possibility is to interconnect the converter legs using full-bridge submodules in star configuration. The star-channel arm helps to balance the power fluctuations with 120° between the converter legs. A detailed analysis and capacitors voltage fluctuations in the star-channel MMC during low-frequency are presented in the following subsections.

7.4.1 Equivalent Circuit

The equivalent model of star-channel MMC is developed according to the mathematical model given in Section 7.3.1. The sub-arms of star-channel MMC are represented with a fundamental frequency (v_{xf}), high-frequency (v_{xh}), and DC voltage (v_d) components as shown in Figure 7.5. The star-channel arms are represented with a fundamental frequency voltage (v_{xf}). From equations (7.3), (7.9), (7.10), and (7.12), the magnitude of these voltage components is given by

$$\begin{aligned} v_d &= \frac{1}{4} V_{dc} \\ v_{xf} &= \frac{m_a}{4} V_{dc} \sin\left(\omega_o t + \theta_x\right) \\ v_{xh} &= \frac{1 - m_a}{4} V_{dc} S_h \end{aligned} \tag{7.16}$$

The DC voltage component in each sub-arm is controlled to maintain the power balance between the main DC-bus voltage V_{dc} and converter legs. The fundamental component in the upper and lower arms is controlled to generate a required voltage across the load (v_{xi}). The high-frequency voltage component v_{xh} is introduced in each sub-arm to minimize the submodule capacitor voltage ripple and to achieve the power balance amongst the converter legs. The high-frequency voltage component is arranged in opposite phase and cancel each other. Therefore, the high-frequency voltage does not appear in the motor neutral voltage.

The fundamental frequency voltage v_{xf} generates a current component i_{xi} in the arm. The interaction between the fundamental (low) frequency current and DC voltage component in the arm leads to a larger charging and discharging periods of submodule capacitors. The voltage fluctuation becomes worse when the DC current component (i_{xd}) interacts with the low-frequency fundamental voltage component (v_{xf}) in the submodule output voltage. The high-frequency voltage v_{xh} is introduced into each sub-arm to reduce the charging and discharging periods of submodule capacitors voltage. The high-frequency voltage induces a high-frequency current i_{xh} in each sub-arm, and flows through the star-channel arms as shown in Figure 7.5. The interaction of high-frequency current with the star-channel fundamental frequency voltage leads to a larger ripple in the star-channel submodule capacitors voltage. However, these ripples are relatively small due to the larger frequency of i_{xh}.

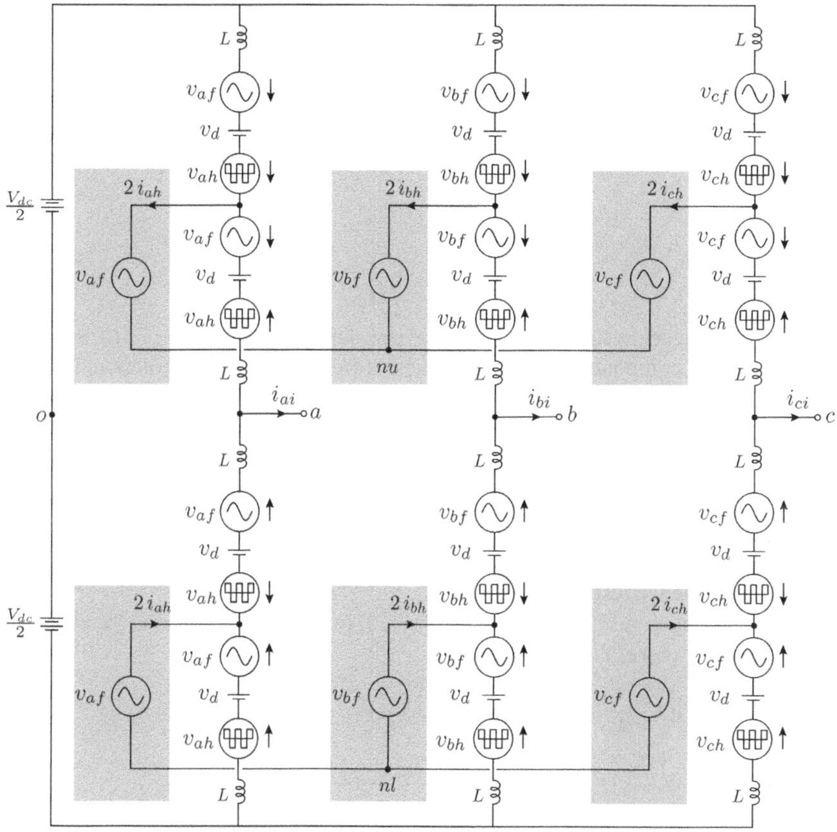

Figure 7.5 Equivalent circuit of a three-phase star-channel MMC.

7.4.2 Submodule Capacitor Voltage Ripple

The submodule capacitor voltage ripple depends on the instantaneous energy stored in the submodule capacitor. The stored energy is directly proportional to the submodule AC output power under the assumption that the power losses are neglected. For the analysis purpose, the energy stored in the submodule I capacitor is considered as

$$v^{x}_{Cy1} = \sqrt{\frac{2}{C}\left[\frac{1}{2}C\,V_{C}^{2} + \int p^{x}_{em1}\,dt\right]} \tag{7.17}$$

where v^{x}_{Cy1} represents the instantaneous value of submodule I capacitor voltage, V_C represents the average value of submodule I capacitor voltage, and p^{x}_{em1} represents the submodule I AC output power.

The submodule capacitor voltage ripple is obtained from the relationship between submodule instantaneous AC and DC-side power. During this analysis, it is assumed that each sub-arm power is equally distributed among the submodules. The submodule-I AC output power in the sub-arm I is given by

$$p_{em1}^{x} = \frac{2}{N}\,[v_{em}^{x}\,i_{xu}^{m}] \tag{7.18}$$

From equations (7.2) and (7.3), the steady-state sub-arm I voltage and current are given by

$$\begin{aligned} v_{em}^{x} &= \frac{V_{dc}}{4} - \frac{v_{xi}}{2} - v_{xh} \\ i_{xu}^{m} &= i_{xd} + \frac{1}{2}\,i_{xi} + i_{xh} \end{aligned} \tag{7.19}$$

The components in the sub-arm voltage are defined in equation (7.16), whereas the DC current component (i_{xd}) , fundamental frequency current component (i_{xi}), and high-frequency current component (i_{xh}) in the sub-arm current are defined as

$$\begin{aligned} i_{xd} &= I_{dcx} \\ i_{xh} &= I_h\,S_h \\ i_{xi} &= I\sin(\omega_o\,t + \theta_x + \varphi) \end{aligned} \tag{7.20}$$

where I_h and I are the peak values of high-frequency and fundamental frequency current components, respectively.

Substitute equations (7.16), (7.19), and (7.20) in equation (7.18), and the submodule I AC output power is given by

$$\begin{aligned} p_{em1}^{x} = \frac{2}{N}\,[(&\frac{V_{dc}\,I_{dcx}}{4} - \frac{v_{xi}\,i_{xi}}{4}) + (\frac{V_{dc}\,i_{xi}}{8} - \frac{v_{xi}\,I_{dcx}}{2} - V_h\,I_h\,S_{hw}) \\ &+ (\frac{V_{dc}\,i_{xh}}{4} - \frac{v_{xi}\,i_{xh}}{2} - \frac{v_{xh}\,i_{xi}}{2} - v_{xh}\,I_{dcx})] \end{aligned} \tag{7.21}$$

where $S_{hw} = 1$ represents the root mean square (RMS) value of high-frequency square wave voltage.

At steady state, the average submodule power becomes zero in one high-frequency cycle ($\frac{1}{f_h}$). This requirement can be met by balancing the AC- and DC-side powers of the converter using the first term on the right-hand side, and redistributing the fundamental frequency power fluctuations using the second term on the right-hand side. Using this criterion, the following steady state operating conditions are obtained from equation (7.21) as

$$\begin{aligned} \frac{V_{dc}\,I_{dcx}}{4} - \frac{v_{xi}\,i_{xi}}{4} &= 0 \\ \frac{V_{dc}\,i_{xi}}{8} - \frac{v_{xi}\,I_{dcx}}{2} - V_h\,I_h &= 0 \end{aligned} \tag{7.22}$$

From equation (7.22), the magnitude of DC and high-frequency current components at stead-state are given by

$$\begin{aligned} I_{dcx} &= \frac{v_{xi}\,i_{xi}}{V_{dc}} \\ I_h &= \frac{0.25\,V_{dc}\,i_{xi} - v_{xi}\,I_{dcx}}{2\,V_h} \end{aligned} \tag{7.23}$$

The instantaneous value of submodule power is further simplified by using the steady-state current components given in equation (7.23). After expanding the voltage and current components v_{xi}, v_{xh}, i_{xi}, and i_{xh} with reference to equations (7.10) and (7.20), the submodule I AC output power is given by

$$\begin{aligned} p_{em1}^{x} &= \frac{2}{N}\left[\frac{V_{dc}\,I_h}{4}S_h - \frac{V_f\,I_h}{2}\sin(\omega_o t + \theta_x)\,S_h \right. \\ &\left. - V_h\,I_{dcx}\,S_h - \frac{V_h\,I}{2}\sin(\omega_o t + \theta_x + \varphi)\,S_h\right] \end{aligned} \tag{7.24}$$

From equations (7.17) and (7.24), it is observed that the submodule capacitor voltage ripple is directly proportional to the high-frequency component S_h. The frequency of S_h is selected such that the charging and discharging periods of submodule capacitor voltage are greatly minimized.

CASE STUDY 7.2 Capacitor Voltage Ripple in Star-Channel MMC

Objective: In this case study, the submodule capacitor voltage ripple in star-channel MMC is analyzed mathematically and through MATLAB simulations.

Parameters: The system parameters of the star-channel MMC are as follows: rated DC-bus voltage $V_{dc} = 7$ kV, number of SMs per arm $N = 8$, number of star-channel submodules $N_c = 2$, SM capacitor voltage $v_C = 875$ V, SM capacitor $C = 4100$ μF, load current $I = 200$ A, power factor angle $\varphi = 0°$, arm inductor $L = 2$ mH, and carrier frequency $f_c = 1000$ Hz.

Analysis:
Initially, the submodule capacitor voltage ripple is analyzed using the mathematical model presented in Section 7.4.2. According to the mathematical model, the sub-arm I voltage and current are defined for a modulation index $m_a = 0.1$, fundamental frequency $f_o = 6$ Hz, peak value of fundamental current $I = 200$ A, power factor angle $\varphi = 0°$, and initial phase angle $\theta_x = 0°$ as

$$\begin{aligned} v_{em}^{x} &= 0.25 \times 7000 - 0.25 \times 0.1 \times 7000 \sin(12\pi\,t) - 0.25 \times (1 - 0.1) \times 7000\,S_h \\ i_{xu}^{m} &= \frac{(0.5 \times 0.1 \times 7000 \sin(12\pi\,t)) \times (200 \sin(12\pi\,t))}{7000} + \frac{200 \sin(12\pi\,t)}{2} \\ &+ \frac{7000^2 \times (200 \sin(12\pi\,t))}{8 \times 7000^2 \times 0.25(1 - 0.1)} S_h \\ &- \frac{4 \times (0.5 \times 0.1 \times 7000 \sin(12\pi\,t))^2 \times (200 \sin(12\pi\,t))^2}{8 \times 7000^2 \times 0.25(1 - 0.1)} S_h \end{aligned} \tag{7.25}$$

The sub-arm voltage and current waveforms are plotted using equation (7.25), and shown in Figure 7.6. It is observed that the sub-arm voltage has a 120 Hz square wave component along with the fundamental frequency of 6 Hz component as shown in Figure 7.6(a). The sub-arm current consists of square wave component along with the fundamental frequency component as shown in Figure 7.6(b). The submodule capacitor voltage ripple is calculated for an SM capacitance of 4100 μF and a capacitor voltage of 875 V as

$$v_{Cy1}^{x} = \sqrt{\frac{2}{0.0041}\left[\frac{0.0041}{2} \times 875^2 + \int p_{em1}^{x}\,dt\right]} \tag{7.26}$$

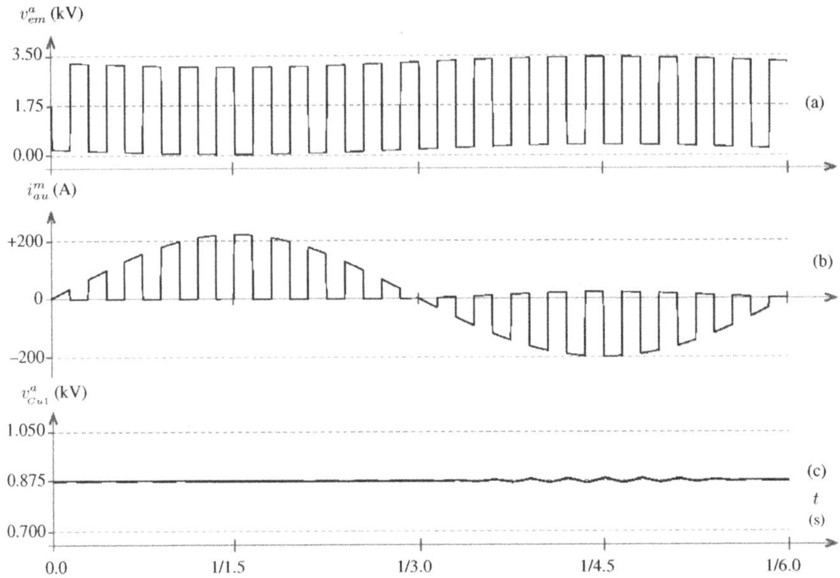

Figure 7.6 Submodule capacitors voltage in star-channel MMC obtained from mathematical model.

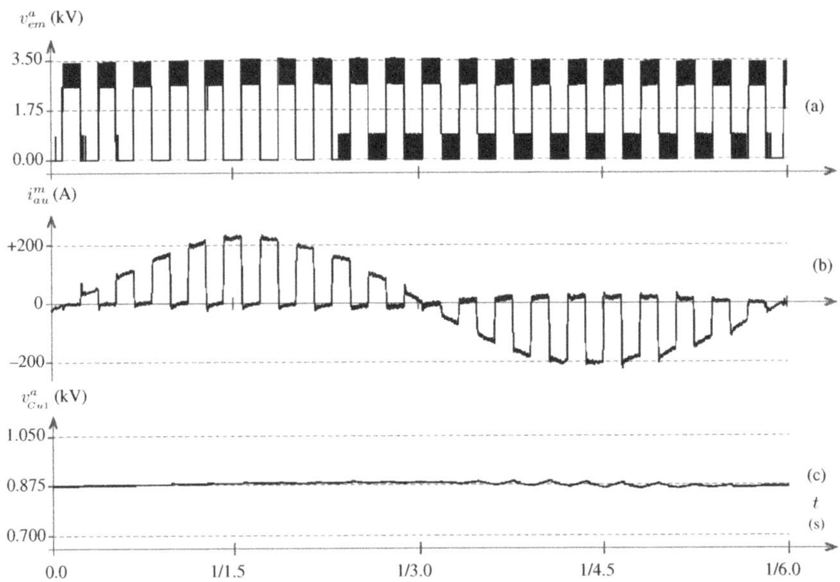

Figure 7.7 Submodule capacitor voltage in star-channel MMC obtained from simulation.

The instantaneous capacitor voltage is calculated from equation (7.26) and is shown in Figure 7.6(c). The submodule capacitor voltage has a ripple of less than 2% and fluctuates with a frequency of 120 Hz. The submodule capacitor voltage ripple is further analyzed using MATLAB simulations as shown in Figure 7.7. In this study, the DC-bus of star-channel MMC is modeled as an ideal DC voltage source of V_{dc}, and the AC terminals are connected to a passive load. The AC output currents are controlled using a closed-loop current controller, which will be discussed in Section 7.5. The magnitude and frequency of reference three-phase currents are set to 200 A and 6 Hz, respectively. Under these conditions, the star-channel MMC is operating at an amplitude modulation index (m_a) of 0.1 to supply a load current magnitude of 200 A at steady state.

Figure 7.7(a) shows the waveforms of sub-arm I voltage, which consists of fundamental and square wave voltage components along with the switching ripple. The switching ripple appears due to the pulse width modulation, which is not included in the mathematical model. Figure 7.7(b) shows the sub-arm current waveform. The submodule capacitor voltage is shown in Figure 7.7(c), which is perfectly maintained at their average value of 875 V. The results obtained from the simulations and mathematical model are quite similar, with the exception of switching ripple (which was neglected in the model). ■

7.4.3 Common-Mode Voltage

Common-mode voltage is one of the major issues in the conventional MMC-based motor drive systems. A high-frequency voltage is introduced to minimize the capacitor voltage ripple in the conventional MMC. This high-frequency voltage is added to the common-mode voltage generated by the conventional MMC, and increases the net magnitude of motor neutral voltage. This will increase the voltage stress on the motor insulation, and induce a shaft voltage which could lead to shaft and bearing failures [11]. The CMV generated by star-channel MMC is given by

$$v_{cmv} = \frac{1}{3}\left[v_{ai} + v_{bi} + v_{ci}\right] \tag{7.27}$$

where v_{ai}, v_{bi}, and v_{ci} represent the phase-to-neutral voltage of three phases. The CMV is further expressed in terms of arm voltages as

$$v_{cmi} = \frac{1}{6}\left[v_{rs}^{a} + v_{rs}^{b} + v_{rs}^{c} - v_{eg}^{a} - v_{eg}^{b} - v_{eg}^{c}\right] \tag{7.28}$$

The arm voltage consists of fundamental voltage component along with switching ripples (v_{sw}) generated by the pulse width modulation scheme. The switching ripple v_{sw} is added to the arm voltage and results in

$$\begin{aligned} v_{eg}^{x} &= \frac{V_{dc}}{2} - V_f \sin\left(\omega_o t + \theta_x\right) + v_{sw} \\ v_{rs}^{x} &= \frac{V_{dc}}{2} + V_f \sin\left(\omega_o t + \theta_x\right) + \overline{v}_{sw} \end{aligned} \tag{7.29}$$

Equation (7.29) is substituted in equation (7.28), and the CMV generated by the star-channel MMC becomes

$$v_{cmi} = \frac{1}{2}(v_{sw} + \overline{v}_{sw}) \tag{7.30}$$

The above equation shows that the common-mode voltage generated by the star-channel MMC consists of switching ripples, and there is no high-frequency voltage component. Hence, the star-channel MMC can be used in low-speed, high-torque motor drive applications.

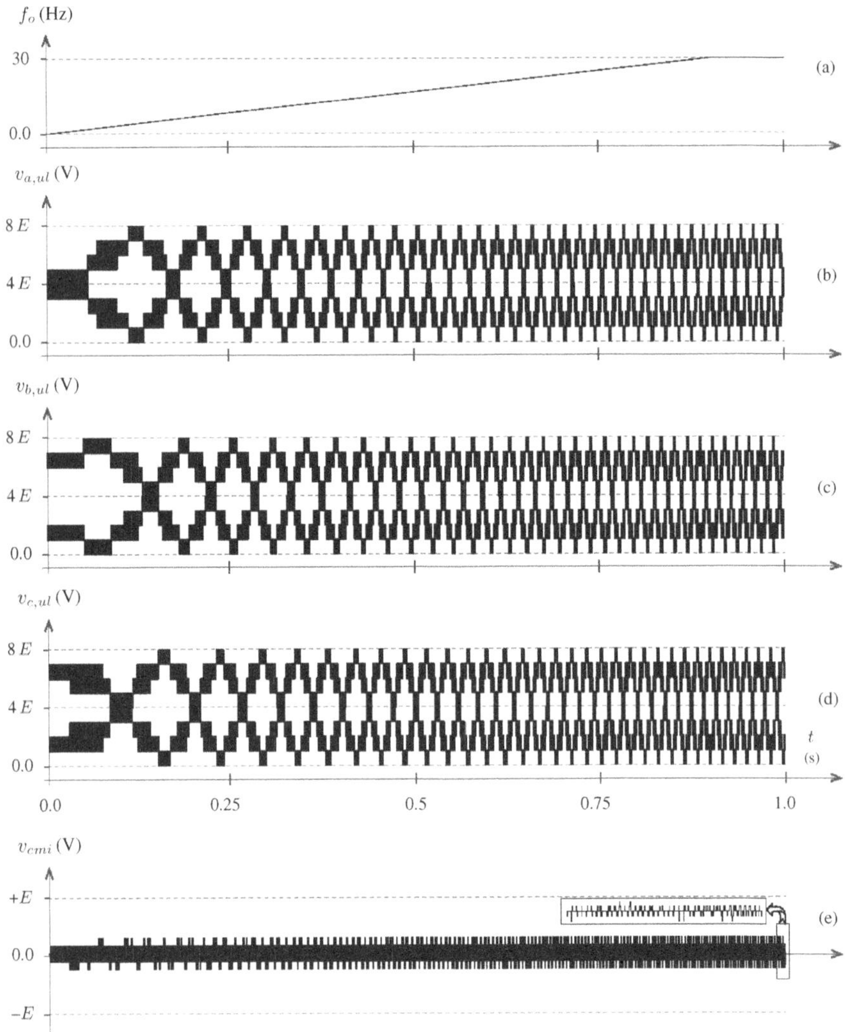

Figure 7.8 Simulated waveforms of common-mode voltage in a star-channel MMC.

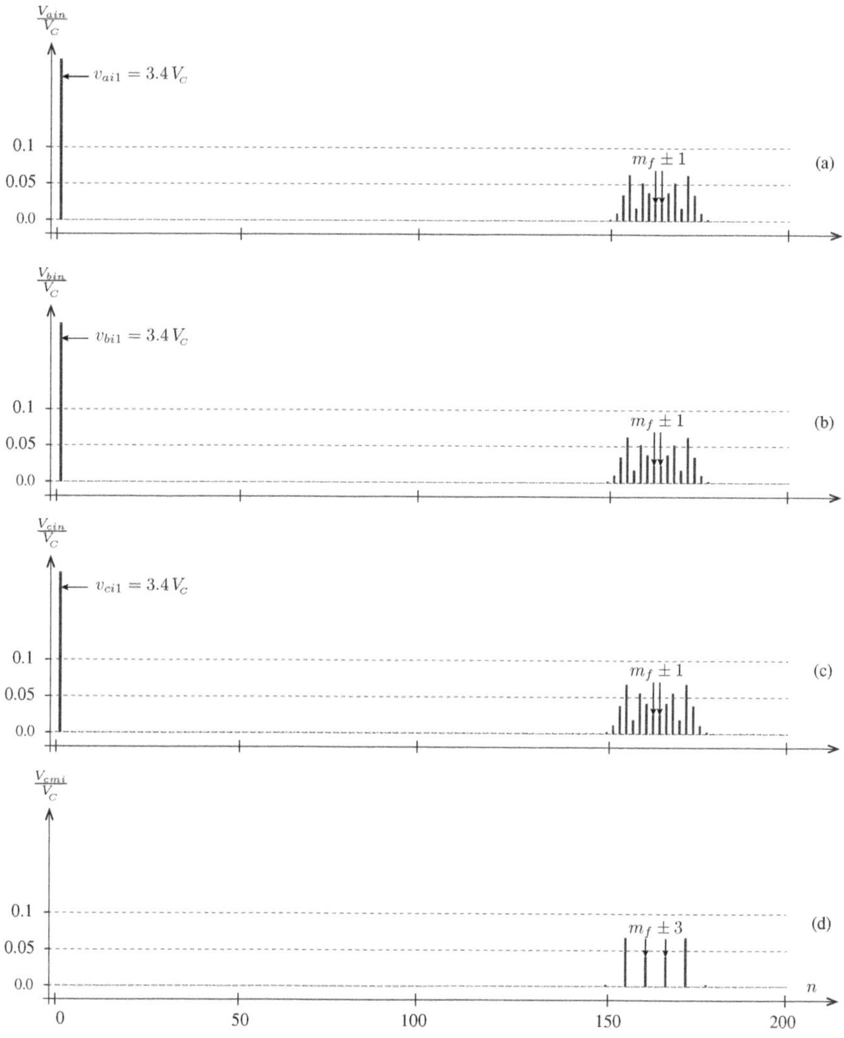

Figure 7.9 Harmonic spectrum details of three-phase output voltages.

CASE STUDY 7.3 Common-Mode Voltage of Star-Channel MMC

Objective: In this case study, the common-mode voltage generated by the star-channel modular multilevel converter is investigated under different operating frequencies using MATLAB simulation.

Parameters: The system parameters of the star-channel MMC are as follows: DC-bus voltage $V_{dc} = 7$ kV, number of SMs per arm $N = 8$, SM capacitor voltage $v_C = 875$ V, SM capacitor

$C = 4700\ \mu$F, number of SMs in star-channel arm $N_c = 2$, load inductance $L_o = 20$ mH, load resistance $r_o = 34\ \Omega$, arm inductor $L = 2$ mH, and carrier frequency $f_c = 600$ Hz.

Analysis:
In this case study, the star-channel MMC is designed with $N = 8$ submodules in main arms and $N_c = 2$ submodules in star-channel arm. Each submodule capacitor voltage is set to $E = 875$ V. The sub-arm modulation signals are generated with a modulation index $m_a = 0.85$, and a square wave voltage component with 120 Hz frequency is added to them. The modulation signals are compared with the carrier signals to generate the gating signal of submodules in the main and star-channel arms. The carrier signals are arranged according to the phase-shifted carrier pulse width modulation principle.

In this study, the fundamental frequency of modulation signals is linearly increased from 0 Hz to 30 Hz with in 0.9 s period as shown in Figure 7.8(a). From 0.9 s to 1 s, the frequency is maintained at 30 Hz itself. The phase-a, upper and lower arm voltage waveforms are shown in Figure 7.8(b), which consists of DC component ($4E$) and a fundamental frequency component along with the PWM switching ripples. The high-frequency components are arranged in opposite phase and cancel each other. Also, the phase-b and -c, upper and lower arm voltage waveforms are shown in Figure 7.8(c) and 7.8(d), respectively. These voltages have similar components as that of the phase-a voltage, but the fundamental components are phase-shifted by $120°$ and $240°$ with respect to the phase-a voltage. The common-mode voltage generated by star-channel MMC is shown in Figure 7.8(e), which consists of switching ripples with a peak value of $0.5\ E$.

The voltage components in the three-phase output voltages are analyzed at a fundamental frequency of $f_o = 30$ Hz as shown in Figure 7.9. The three-phase voltages have fundamental voltage component of $3.4\ V_C$ corresponding to a modulation index of $m_a = 0.85$ as shown in Figure 7.9(a)–(c). The dominant harmonic components appear around the frequency modulation index $m_f \pm 1$, where $m_f = \frac{f_{sw}}{f_o} = 160$ and equivalent switching frequency $f_{sw} = 4800$ Hz. On the other hand, the common-mode voltage consists of switching frequency harmonics corresponding to $m_f \pm 3$ as shown in Figure 7.9(d). ■

7.5 CLASSICAL CONTROL OF STAR-CHANNEL MMC

The star-channel MMC has half-bridge submodules in the main arm and full-bridge submodules in the star-channel arm. Hence, the star-channel MMC requires a capacitor voltage control for main and star-channel arms along with the AC output current. The generalized control block diagram is shown in Figure 7.10. The upper and lower submodule capacitors voltage controller is divided into leg voltage, arm voltage, and individual capacitor voltage controllers. The AC output current control is implemented in central controller and generates fundamental frequency modulation signal. The submodule capacitor voltage control is implemented in the local controller and generates compensation voltage signals. These compensation voltage signals are added to the fundamental frequency modulation signal and result in modulation signal of submodules. The resultant modulation signals are compared with carrier signals and generate the gating signals to the submodules. The closed-loop controller requires feedback signals of arm currents and submodule capacitor voltages.

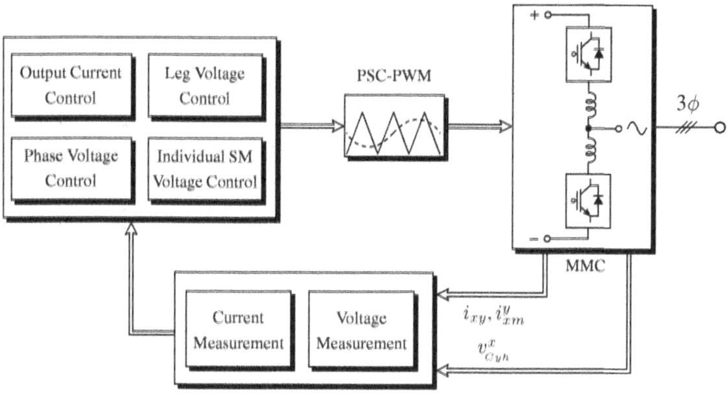

Figure 7.10 Block diagram of the star-channel control system.

7.5.1 Output Current Control

Typically, the AC output current control is implemented in the synchronous-dq reference frame rotating at the speed of output frequency [12]. The three-phase output currents are estimated from the measured arm currents. The magnitude of AC output current is equally distributed between the upper and lower arms of each leg, and flows in opposite phase. The difference of upper and lower arm currents gives the corresponding output phase current. The measured output currents are transformed into synchronous-dq frame, and become DC signals. The actual dq-axis output currents are compared with the reference dq-axis currents, and the error is minimized using PI-regulators. The d- and q-axis PI-regulators generate reference voltage commands of v_{di}^* and v_{qi}^*, respectively. The reference voltage commands are transformed to abc-frame using the dq/abc-transformation matrix given in Chapter 3. The transformation theory and control block diagram of current control are presented in Chapter 3.

7.5.2 Submodule Capacitor Voltage Control

In order to generate an output voltage with symmetrical voltage steps, a tight control of the submodule capacitor voltage is mandatory. With this control, a significant reduction of output voltage and current harmonic distortions can be obtained [13]. Also, it is possible to maintain a uniform voltage stress on semiconductor devices, which significantly enhance the converter reliability and operational capability of the converter. The star-channel MMC has multiple submodules in both the main and star-channel arms. For reliable operation, the submodule capacitor voltage control is divided into 1) overall capacitor voltage control, 2) phase capacitor voltage control, and 3) individual submodule capacitor voltage control. The implementation of each control scheme is discussed in the following subsections.

(1) Overall Capacitor Voltage Control: The arm current consists of a DC-current component, which flows between the converter leg and DC-bus. The magnitude of DC-current component depends on the voltage difference between the DC-bus and the average value of leg voltage under the assumption that the load active power demand is constant. The current magnitude increases with the increase in the voltage difference, which leads to higher device power losses and a lower efficiency of the converter. Hence, a leg voltage control is required to maintain the average DC voltage of each leg at a given reference voltage of V_C^*. The control block diagram of overall voltage control is shown in Figure 7.11, where the actual DC voltage of each converter is compared with the reference DC voltage, and their difference is minimized using a PI-regulator. The average DC voltage (v_{Cg}) of each leg is given by

$$v_{Cg} = \frac{1}{6(N+N_c)}\left[\sum_{h=1}^{2N}(v_{Cyh}^a + v_{Cyh}^b + v_{Cyh}^c) + \sum_{h=1}^{2N_c}(v_{Fyh}^a + v_{Fyh}^b + v_{Fyh}^c)\right] \tag{7.31}$$

where y represents the arm, v_{Fyh}^x represents the capacitor voltage of full-bridge submodules in the star-channel arms, v_{Cyh}^x represents the capacitor voltage of half-bridge submodules in the main arms, and v_{Cg} represents the average DC voltage.

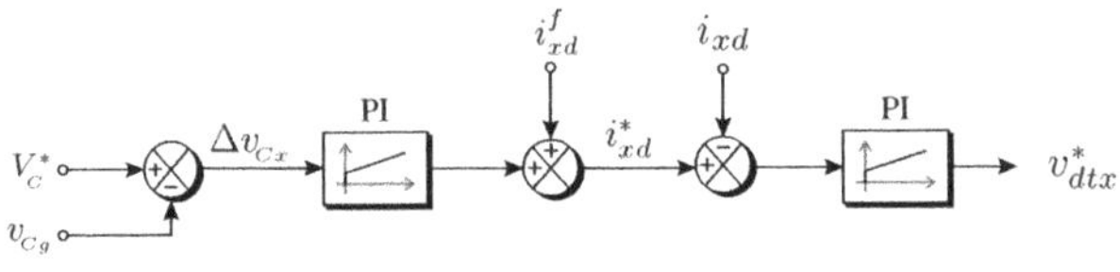

Figure 7.11 Block diagram of overall capacitor voltage control.

The output of PI-controller is added to a feed-forward current component (i_{xd}^f) corresponding to the load active power, which results in the reference DC current component. The reference and actual DC current components are compared with each other, and their error is minimized using a PI-controller. The PI-controller gives a compensation voltage command of v_{dtx}^*, and is added to the modulation signal obtained from the AC output current control.

(2) Phase Capacitor Voltage Control: The phase capacitor voltage control aims to balance the average submodule capacitors voltage among the converter legs. The voltage balancing is achieved by controlling the high-frequency current flowing among the converter legs. The block diagram of phase capacitor voltage control is shown in Figure 7.12. The outer voltage loop is formed to control the average voltage of each converter leg (v_{Cx}), which is given by

$$v_{Cx} = \frac{1}{2(N+N_c)}\left[\sum_{h=1}^{2N} v_{Cyh}^x + \sum_{h=1}^{2N_c} v_{Fyh}^x\right] \tag{7.32}$$

The average voltage of each converter leg is compared with the average of three-phase voltages, and their difference is minimized using a PI-controller. The output of

PI-controller is added to the feed-forward high-frequency current (i_{xh}^{f}) components, and the resultant output is multiplied with a unit square wave signal to obtain the reference current command (i_{xh}^{*}). The measured high-frequency current component is compared with the reference command, and their difference gives an error signal. The error signal is controlled using a PI-controller, and its output (Δv_{xh}^{*}) is added to modulation signal.

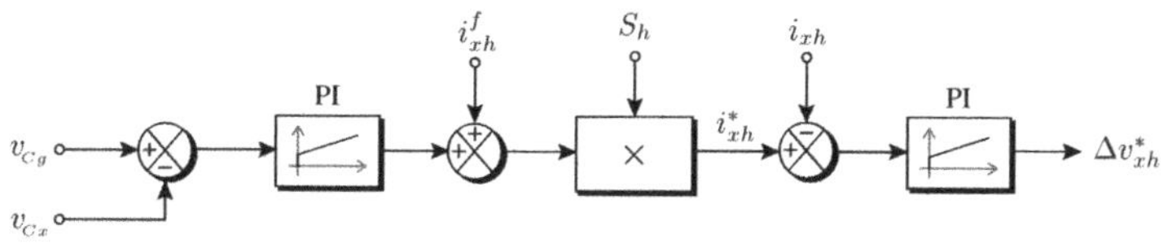

Figure 7.12 Block diagram of phase capacitor voltage control.

(3) Individual Submodule Capacitor Voltage Control: The main function of individual submodule capacitor voltage control is to regulate each submodule capacitor voltage at a given reference value of V_C^*. The control block diagram of individual submodule capacitor voltage control is shown in Figure 7.13, where the measured value of each submodule capacitor voltage is compared with the reference voltage V_C^*. The difference is minimized using a PI-controller. The PI-controller output is multiplied with the direction of the arm current to obtain the reference voltage component Δv_{xyh}^{*}, which is added to the modulation signals obtained from the AC output current control.

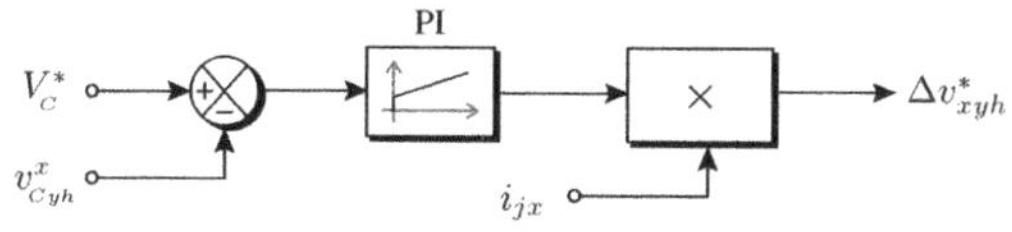

Figure 7.13 Block diagram of individual SM capacitor voltage control.

CASE STUDY 7.4 SM Capacitor Voltage Control of Star-Channel MMC

Objective: In this case study, the dynamic performance of submodule capacitor voltage control for star-channel MMC is studied through MATLAB simulation.

Parameters: The system parameters of the star-channel MMC are as follows: rated DC-bus voltage $V_{dc} = 7$ kV, number of SMs in main arm $N = 8$, number of SMs in star-channel arm $N_c = 2$, SM capacitor voltage $v_C = E = 875$ V, SM capacitor $C = 4100$ μF, load inductance $L_o = 4.7$ mH, load resistance $r_o = 1.53$ Ω, arm inductor $L = 2$ mH, fundamental frequency $f_o = 6$ Hz, high-frequency $f_h = 120$ Hz, and carrier frequency $f_c = 1000$ Hz.

Analysis:
The star-channel MMC is designed with a total DC-bus voltage of 7000 V, which is equally distributed among the submodules in each main arm. The main arms of star-channel MMC have eight submodules and each submodule capacitor is rated for 875 V. The AC terminals of star-channel MMC are connected to a resistor of (r_o) 1.53 Ω and an inductor of (L_o) 4.7 mH.

The dynamic performance of submodule capacitor voltage control for star-channel MMC is shown in Figure 7.14. Initially, the capacitor voltage control is activated, and the submodule capacitors in the upper and lower arms are perfectly maintained at their nominal value as shown Figure 7.14(a). The converter output voltage is shown in Figure 7.14(b). The output voltage forces a current of $I = 200$ A through the load. The three-phase currents are perfectly balanced and sinusoidal in shape as shown in Figure 7.14(c). At $t = 0.4$ s, the capacitor voltage control is disabled. The submodule capacitors voltage is diverging from their nominal value, which affects the output voltage and current flowing through the converter. The peak value of output current exceeds the rated current, which leads to a device failure and effects the converter operation. At $t = 0.41$ s, the voltage control is activated, and the submodule capacitor voltages are converging towards their rated value. Also, the converter generates output voltage and current waveforms with the required magnitude and frequency.

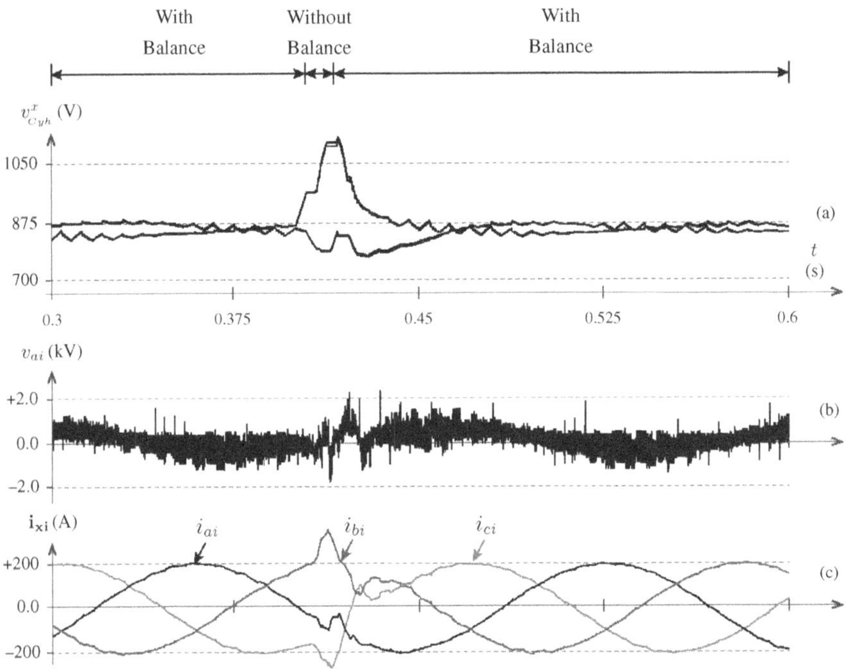

Figure 7.14 Dynamic performance of submodule capacitor voltage control.

■

7.5.3 Synthesis of Modulation Signals

Each submodule in the star-channel MMC is independently controlled using a modulation signal and a carrier signal. The modulation signal consists of several voltage components such as the DC (V_{dc}), fundamental (v^*_{xf}), and high-frequency voltage components (v^*_{xh}), and their compensation voltage components v^*_{dtx}, Δv^*_{xh}, and Δv^*_{xyh}, respectively. These voltage commands are generated using the output current control and various stages of submodule capacitor voltage control, which are discussed in Sections 7.5.1 and 7.5.2. The modulation signals for each submodule in the upper, lower sub-arms, and star-channel arms are given by

$$
\begin{aligned}
m^*_{xuh} &= \frac{1}{V_C}\left[\frac{V_{dc}}{2N} - \frac{v^*_{xf}}{N} \mp \frac{2\,v^*_{xh}}{N} - v^*_{dtx} \mp \Delta v^*_{xh} + \Delta v^*_{xyh}\right] \\
m^*_{xlh} &= \frac{1}{V_C}\left[\frac{V_{dc}}{2N} + \frac{v^*_{xf}}{N} \mp \frac{2\,v^*_{xh}}{N} - v^*_{dtx} \mp \Delta v^*_{xh} + \Delta v^*_{xyh}\right] \\
m^*_{xch} &= \frac{1}{V_C}\left[\frac{v^*_{xf}}{2N_c} - \Delta v^*_{xh} + \Delta v^*_{xyh}\right]
\end{aligned}
\tag{7.33}
$$

where the symbols $\mp$ and $\pm$ are used to indicate the top half and bottom half of the upper and lower arms, respectively, and V_C represents the nominal voltage of half-bridge and full-bridge submodules.

The voltage components V_{dc}, v^*_{xf}, and v^*_{xh} are equally shared by the submodules in each leg, arm, and sub-arms, respectively. The negative sign is assigned to the DC compensating voltage v^*_{dtx} to compensate the voltage drop due to the DC current component in the converter leg. The Δv^*_{xh} term has negative sign to compensate the voltage drop due to the high-frequency current component flowing through the star-channel arm. The compensation voltage Δv^*_{xyh} is used to adjust the individual capacitor voltage.

CASE STUDY 7.5 Steady-State Analysis of Star-Channel MMC

Objective: In this case study, the steady-state performance of a star-channel MMC is studied through MATLAB simulation.

Parameters: The system parameters of the star-channel MMC are as follows: DC-bus voltage V_{dc} = 7 kV, number of SMs in main arm N = 8, number of SMs in star-channel arm N_c = 2, SM capacitor voltage $v_C = E$ = 875 V, SM capacitor C = 4100 μF, load inductance L_o = 4.7 mH, load resistance r_o = 1.53 Ω, arm inductor L = 2 mH, fundamental frequency f_o = 6 Hz, high-frequency f_h = 120 Hz, and carrier frequency f_c = 1000 Hz.

Analysis:
The star-channel MMC is configured as DC–AC converter, and its DC system is modeled as a stiff DC source of 7000 V. The AC terminals are connected to an inductor (L_o) of 4.7 mH and a resistor (r_o) of 1.53 Ω. The nominal voltage of each submodule capacitor is set to 875 V. The current control is implemented in the synchronous dq-frame, and the reference dq-currents are set to i^*_{di} = 180 A and i^*_{qi} = 87 A. The steady-state converter voltage and current waveforms are shown in Figures 7.15 and 7.16.

The normalized modulation signal of star-channel MMC is shown in Figure 7.15(a), which consists of square wave voltage component along with the fundamental frequency component. The fundamental frequency modulation signal is obtained from the current control and other compensation voltage commands are generated using capacitor voltage control. Similarly, the modulation signal for submodules in the star-channel arm is shown in Figure 7.15(b). The modulation signal mainly consists of high-frequency compensation signals along with the fundamental frequency component. These modulation signals are used along with the carrier signals to generate the gating signals. The converter generates an output voltage with a fundamental voltage of 1 kV corresponding to the modulation index (m_a) of 0.1 as shown in Figure 7.15(c). The d- and q-axis current components are perfectly regulated at 180 A and 87 A, respectively, as shown in Figure 7.15(d). The dq-axis current components are transformed to abc-frame, which have a peak value of 200 A at a fundamental frequency of 6 Hz, and power factor is 0.9 as shown in Figure 7.15(d).

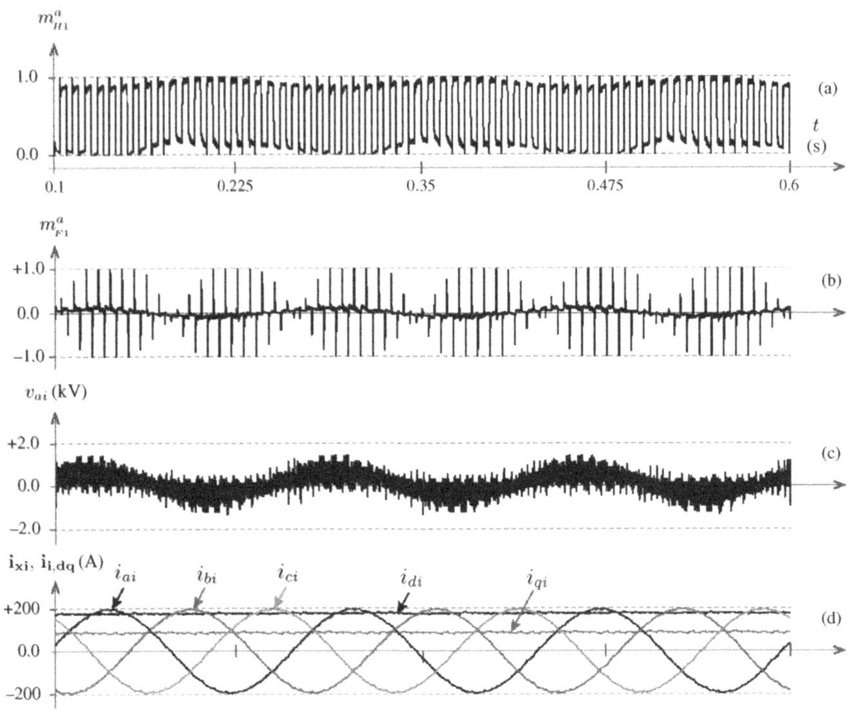

Figure 7.15 Steady-state performance of the star-channel MMC.

The DC and square wave high-frequency components in the modulation signal generates corresponding current components in the sub-arms as shown in Figure 7.16(a). The square wave high-frequency current component in the arm current is used to maintain the power balance between the upper and lower arm submodule capacitor voltages. Figure 7.16(b) shows the waveform of star-channel arm current, which consists of the bipolar high-frequency square wave current component. This current component is controlled to minimize the voltage ripple during low-frequency operation. The high-frequency voltage does not appear in the converter

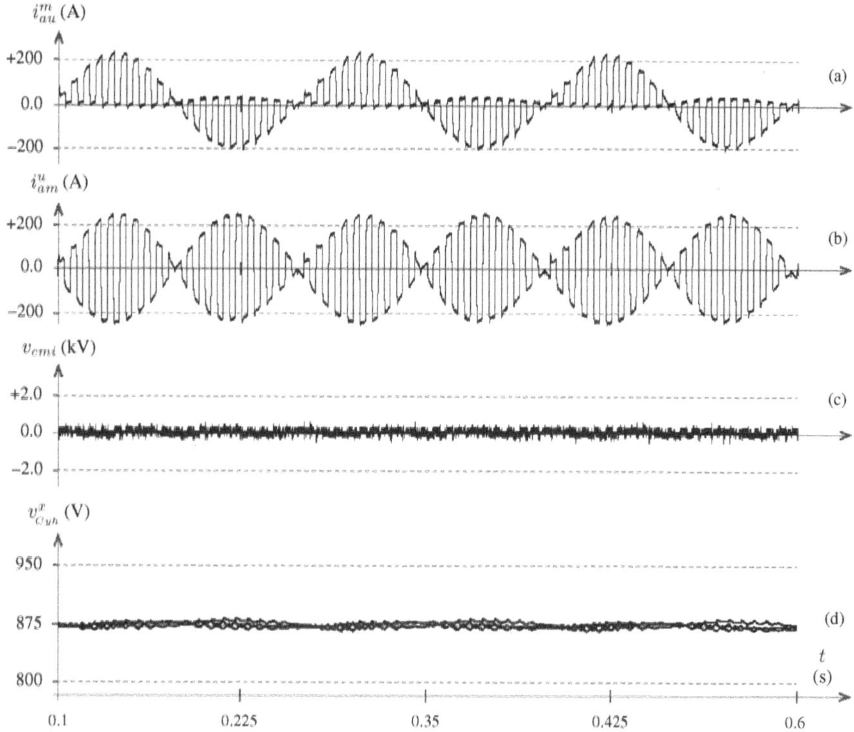

Figure 7.16 Steady-state performance of the star-channel MMC.

common-mode voltage as shown in Figure 7.16(c). It mainly consists of PWM switching ripples with a peak value of ± 423 V. The upper and lower arm submodule capacitors voltage are shown in Figure 7.16(d). The capacitor voltages are perfectly regulated at their nominal value of 875 V.

■

7.6 DELTA-CHANNEL MODULAR MULTILEVEL CONVERTER

The delta-channel modular multilevel converter is an alternative to the star-channel modular multilevel converters. The mid-point of the upper arms in the conventional modular multilevel converter is interconnected using submodules and an inductor as shown in Figure 7.17. Similarly, the mid-point of the lower arms is connected in the delta using submodules and an inductor. The resultant configuration appear as a delta-channel modular multilevel converter. Typically, the full-bridge submodules are used in the delta-channel arms to control the bidirectional current flowing through them. These arms consist of coupled inductors to limit the circulating currents flowing through them. Due to the mutual coupling, the net inductor size is greatly reduced [14]. The delta-channel arms provide an alternative path for the fluctuating energy among the three converter legs, which leads to a smaller submodule voltage ripple during low/zero-frequency operation under rated load condition. In addition, there is no increase in the motor neutral voltage.

For high efficiency, the main arms of delta-channel modular multilevel converter are designed with half-bridge submodules. Each main arm is further divided into two sub-arms, which creates the mid-point to connect the additional delta-channel arms as shown in Figure 7.17. The net arm inductor is divided into two halves and inserted in each sub-arm to minimize the circulating current flowing through the arms.

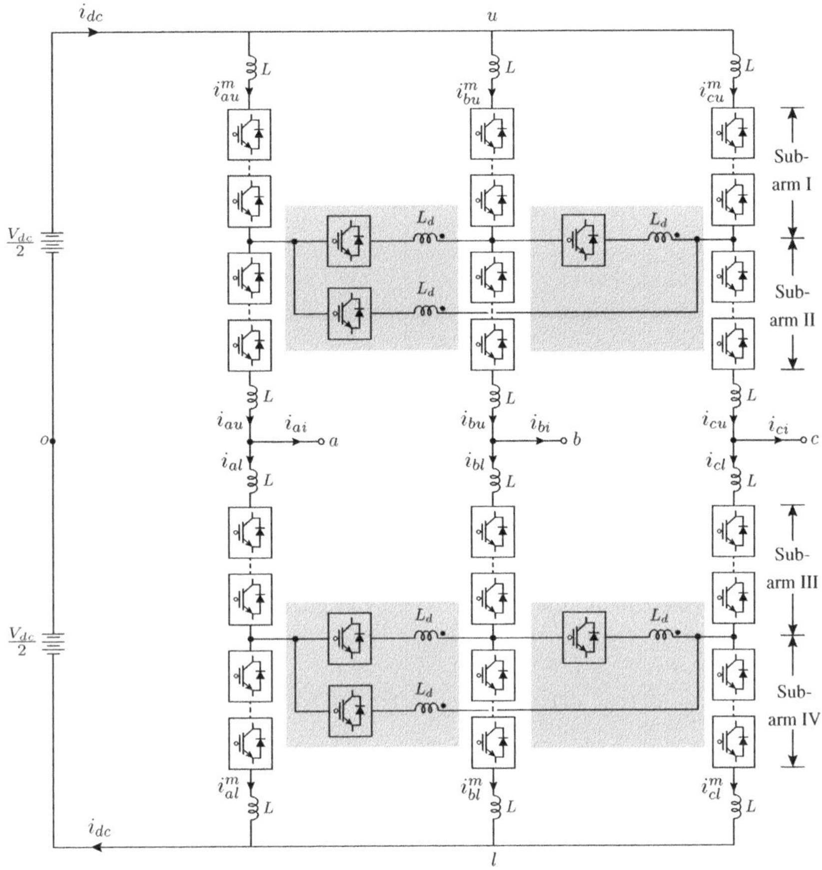

Figure 7.17 Power circuit configuration of delta-channel MMC.

The current flowing through the delta-channel arms can be shaped into square wave to reduce the current stress of the semiconductor devices. The number of submodules used in the delta-channel MMC to form the delta-channel arms is $\sqrt{3}$ times that of the star-channel arms. Also, the device current rating in star-channel MMC is around $\sqrt{3}$ times the delta-channel MMC. The delta-channels are required to handle 50% of the line-to-line voltage, which is 73.2% higher than the voltage handled by the star-channel arms.

7.7 COMPARISON OF ADVANCED MODULAR MULTILEVEL CONVERTERS

In this section, the comparison of advanced modular multilevel converters in terms of number of devices (1700 V, IGBTs), the number of capacitors (875 V/4100 μF), and the inductors used in the main and cross-connected arms, and device current stress are presented in Table 7.3. The advanced modular multilevel converters are designed with a DC-bus voltage of V_{dc}. The upper and lower arms are configured with half-bridge submodules and generate output voltage with identical number of voltage levels. Each half-bridge submodule consists of two IGBT devices and a single DC capacitor. The three-phase advanced MMC topologies require 96 IGBT devices of 1700 V rated capacity to configure main arms. In the case of cross-connected arms, the PC-MMC does not use any semiconductor devices. The AC-MMC and star-channel MMCs require 48 devices, and 96 devices are used in the delta-channel MMC. The delta-channel MMC use twice the number of devices to form the cross-connected arms compared to other advanced MMC topologies as shown in Table 7.3.

Table 7.3 Comparison of advanced modular multilevel converters

Parameter	PC-MMC		AC-MMC		Star-Channel MMC		Delta-Channel MMC	
	Arm	Branch	Arm	Branch	Arm	Branch	Arm	Branch
No. of devices	96	–	96	48	96	48	96	96
No. of arm DC capacitors	48	–	48	–	48	–	48	–
No. of branch DC capacitors	–	3	–	24	–	12	–	24
Device RMS current (A)	115	–	104	91	104	93	104	54
No. of inductors	12	–	12	–	12	–	12	2

The number of DC capacitors used in the main arms are proportional to the number of submodules. The advanced modular multilevel converters consist of 48 half-bridge submodules and uses 48 DC capacitors with a voltage rating of 875 V and a capacitance of 4100 μF. In the case of cross-connected arms, the PC-MMC requires a single DC capacitor to interconnect the upper and lower arms of each converter leg, and, in total, three DC capacitors are required for a three-phase system. The AC-MMC uses half-bridge submodules and require 24 DC capacitors as shown in Table 7.3. The star and delta-channel MMCs are equipped with full-bridge submodules, and requires 12 and 24 DC capacitors, respectively. Compared to other topologies, the star-channel MMC requires least number of DC capacitors to form the cross-connected arms.

The current stress on the semiconductor devices vary with the type of high-frequency current component such as sinusoidal and square wave components. In this comparison, the best suitable high-frequency signal is used to elevate the features of advanced modular multilevel converters. In PC-MMC, the sinusoidal signal is used as a high-frequency component, and the square wave signal is used in other advanced topologies. Hence, the device current stress is higher in a PC-MMC compared to other advanced MMCs. In the case of device current stress in the cross-connected arms, the delta-channel MMC has 50% smaller device current stress compared to the AC-MMC and star-channel MMC as shown in Table 7.3. However, if we use identical high-frequency signals then all the topologies have identical current stress in the main arms. On the other hand, the PC-MMC does not use any devices in the cross-connected arms, then it becomes the best topology compared to the other topologies.

The advanced modular multilevel converters employ arm inductors to mitigate the switching ripples and to suppress the inrush and circulating currents. All the topologies require two inductors per arm, and, in total, 12 inductors are required for a three-phase system. In addition, the delta-channel MMC requires another two three-phase coupled inductors in the delta-channel arms to mitigate the circulating currents. Overall, the PC-MMC has the least device and component count, and higher device current stress.

7.8 SUMMARY

In this chapter, a novel star-channel modular multilevel converter for medium-voltage motor drives is presented. The operation of main and star-channel arms are analyzed using a PSC-PWM scheme. During low-frequency operation, the star-channel arms provide a physical path to the exchange the fluctuating energy among the three converter legs. The operation of star-channel MMC during low-frequency is analyzed using a three-phase equivalent model. Further, the voltage ripple and common-mode voltage are investigated through mathematical modeling and MATLAB simulations. Finally, a generalized control scheme is presented to control various objectives such as output currents and submodule capacitors voltage. The steady-state and dynamic performance is validated using simulation study.

The star-channel arm is modified into delta arm, and it becomes a delta-channel MMC. The modulation and control scheme of star-channel MMC are directly applied to delta-channel arm. Hence, the delta-channel MMC is briefly introduced in this chapter. Also, a comparison of several advanced modular multilevel converters is presented. The PC-MMC has lowest device and component count. But, the device current stress is much higher due to the use of sinusoidal high-frequency voltage component. On the other hand, the star-channel MMC requires fewer number of submodule capacitors and semiconductor devices compared to the delta-channel MMC.

REFERENCES

1. S. Du, B. Wu, N. Zargari, and Z. Cheng, "A flying-capacitor modular multilevel converter (fc-mmc) for medium-voltage motor drive," *IEEE Trans. Power Electron.*, vol. 32, no. 3, pp. 2081–2089, Mar 2017.
2. S. Du, B. Wu, K. Tian, N. R. Zargari, and Z. Cheng, "An active cross-connected modular multilevel converter (ac-mmc) for a medium-voltage motor drive," *IEEE Trans. Ind. Electron.*, vol. 63, no. 8, pp. 4707–4717, Aug 2016.
3. L. He, K. Zhang, J. Xiong, S. Fan, and Y. Xue, "Low-frequency ripple suppression for medium-voltage drives using modular multilevel converter with full-bridge submodules," *IEEE Trans. Emerg. Sel. Topics Power Electron.*, vol. 4, no. 2, pp. 657–667, Jun 2016.
4. J. Kolb, F. Kammerer, M. Gommeringer, and M. Braun, "Cascaded control system of the modular multilevel converter for feeding variable-speed drives," *IEEE Trans. Power Electron.*, vol. 30, no. 1, pp. 349–357, Jan 2015.
5. K. Wang, Y. Li, Z. Zheng, and L. Xu, "Voltage balancing and fluctuation-suppression methods of floating capacitors in a new modular multilevel converter," *IEEE Trans. Ind. Electron.*, vol. 60, no. 5, pp. 1943–1954, May 2013.
6. M. Hagiwara, I. Hasegawa, and H. Akagi, "Start-up and low-speed operation of an electric motor driven by a modular multilevel cascade inverter," *IEEE Trans. Ind. Appl.*, vol. 49, no. 4, pp. 1556–1565, Jul 2013.
7. A. Korn, M. Winkelnkemper, and P. Steimer, "Low output frequency operation of the modular multi-level converter," in *Energy Conversion Congress and Exposition (ECCE), 2010 IEEE*, Sep 2010, pp. 3993–3997.
8. M. Huang, J. Zou, and X. Ma, "An improved phase-shifted carrier modulation for modular multilevel converter to suppress the influence of fluctuation of capacitor voltage," *IEEE Trans. Power Electron.*, vol. 31, no. 10, pp. 7404–7416, Oct 2016.
9. J.-J. Jung, H.-J. Lee, and S.-K. Sul, "Control strategy for improved dynamic performance of variable-speed drives with modular multilevel converter," *IEEE Trans. Emerg. Sel. Topics Power Electron.*, vol. 3, no. 2, pp. 371–380, Jun 2015.
10. B. Li, Y. Zhang, G. Wang, W. Sun, D. Xu, and W. Wang, "A modified modular multilevel converter with reduced capacitor voltage fluctuation," *IEEE Trans. Ind. Electron.*, vol. 62, no. 10, pp. 6108–6119, Oct 2015.
11. F. Wang, "Motor shaft voltages and bearing currents and their reduction in multilevel medium-voltage pwm voltage-source-inverter drive applications," *IEEE Trans. Ind. Appl.*, vol. 36, no. 5, pp. 1336–1341, Sep 2000.
12. A. Antonopoulos, L. Angquist, S. Norrga, K. Ilves, L. Harnefors, and H.-P. Nee, "Modular multilevel converter ac motor drives with constant torque from zero to nominal speed," *IEEE Trans. Ind. Appl.*, vol. 50, no. 3, pp. 1982–1993, May 2014.
13. A. Dekka, B. Wu, R. Lizana, and N. R. Zargari, "A comparison of voltage balancing versus energy balancing approach for modular multilevel converters," in *IECON 2016 – 42nd Annual Conference of the IEEE Industrial Electronics Society*, Oct 2016, pp. 3117–3122.
14. M. Hagiwara, K. Nishimura, and H. Akagi, "A medium-voltage motor drive with a modular multilevel pwm inverter," *IEEE Trans. Power Electron.*, vol. 25, no. 7, pp. 1786–1799, Jul 2010.

PART III

APPLICATIONS OF MODULAR MULTILEVEL CONVERTERS

CHAPTER 8

MODULAR MULTILEVEL CONVERTER BASED MEDIUM-VOLTAGE MOTOR DRIVES

8.1 INTRODUCTION

Several multilevel converters are commercially available for medium-voltage motor drive applications [1, 2]. Among them, the integrated multilevel converters can handle a voltage of 2.3–4.16 kV without using the series connection of semiconductor devices. These topologies require significant modifications to increase the operating voltage and output voltage levels, which is not cost-effective [3–5]. Alternatively, the cascaded multilevel converters, also referred to as multi-cell converters, are constructed with a cascade connection of identical low-power submodules. These topologies can handle a voltage of 6.6–13.8 kV and are capable of fault-tolerant operation [6–8]. Among them, the HNPC and CHB topologies require a phase-shifting transformer with multiple secondary windings to generate the isolated DC sources. The phase-shifting transformer minimizes the line (grid) current harmonic distortion and blocks the common-mode current. However, it increases the size and cost of the motor drive system.

Recently, the modular multilevel converters (MMCs) have become an attractive solution for motor drive applications due to their modular construction, voltage scalability, and a single DC-bus system. The MMC-based motor drive systems are com-

Modular Multilevel Converters: Analysis, Control, and Applications, First Edition.
Sixing Du, Apparao Dekka, Bin Wu, and Navid Zargari.

mercially available with an operating voltage range of 2.3–7.2 kV [9, 10]. These systems are available with a passive (or) an active front-end converter (AC–DC converter) on the grid side, and an MMC on the motor side. The typical system configurations of the MMC-based motor drive system, and their control schemes are presented in Section 8.2. The voltage-oriented control (VOC) scheme is used to control the grid-side MMC. The principle and block diagram of VOC are discussed in Section 8.3.

On the other hand, the motor-side MMC is controlled using a field-oriented control (FOC) scheme to meet the motor requirements. Depending on the estimation of flux vector, the FOC scheme is divided into direct and indirect FOC approaches. The principle and implementation of indirect FOC (IFOC) are presented in Section 8.4. The performance of motor-side converter during zero/low-frequency operation is analyzed. Particularly, the submodule voltage ripple in the motor-side converter, and its minimization techniques are discussed in Section 8.5. Also, the common-mode voltage generated by the motor-side converter, and its mitigation techniques are discussed in Section 8.6. Finally, a transformer-less MMC-based motor drive system is presented in Section 8.7. The summary is given in Section 8.8.

8.2 FUNDAMENTALS OF MMC-BASED MOTOR DRIVE

In this section, the possible system configurations of MMC-based motor drive systems and their control schemes are discussed.

8.2.1 System Configurations

In motor drive applications, the power conversion takes place in two stages. In the first stage, the fixed voltage and fixed frequency of AC supply (grid supply) are converted into DC supply using an AC–DC converter (rectifier system). In the second stage, the DC supply is converted into variable-voltage, variable-frequency AC supply (motor terminals) using a DC–AC converter (inverter system). In MMC-based motor drives, the MMC is used as a DC–AC conversion stage, and there are two possibilities for AC–DC conversion stage. The first possibility is to use a diode bridge converter in the rectifier system as shown in Figure 8.1. Typically, 12, 18, (or) 24-pulse diode bridge rectifiers are employed to eliminate the lower-order harmonics in the line current and improve power quality on the AC grid side. Also, the transformer on the AC grid side blocks the common-mode voltage (CMV) generated by the converter and protects the motor from the CMV. However, it is not possible to control the AC grid power factor.

Another possibility is to use a modular multilevel converter in both AC–DC and DC–AC conversion stages as shown in Figure 8.2. This kind of system is suitable for regenerative applications, and it is possible to control the grid power factor as well. The transmission line (L_g) and transformer leakage inductances act as a grid filter, which minimizes the switching frequency harmonics and improves the quality of the grid currents. During low-speed operation, the submodule voltage ripples are quite high in motor-side MMC. To minimize the voltage ripple, a high-frequency voltage is externally injected into the system. The high-frequency voltage component appears in the form of a common-mode voltage (CMV) across motor neutral terminals and increases the peak value of motor common-mode voltage. To block the common-mode voltage, a simple transformer is added in the AC–DC conversion stage as shown in Figure 8.2. The transformer-less operation is one of the major challenges in the MMC-based motor drive system, which is discussed in Section 8.7.

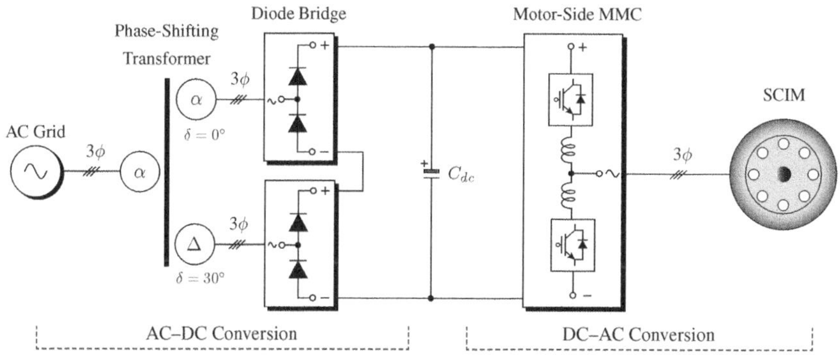

Figure 8.1 MMC-based motor drive with 12-pulse passive rectifier.

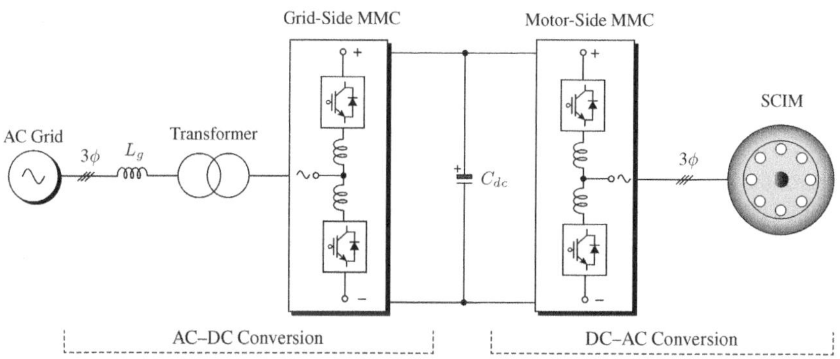

Figure 8.2 MMC-based motor drive with active rectifier.

8.2.2 Control Schemes

The back-to-back configuration of the MMC-based motor drive system is shown in Figure 8.3, in which, the grid-side MMC is used to regulate the DC-bus voltage (V_{dc}) and to improve the grid power factor (reactive power control). These objectives can be achieved using a VOC approach [11, 12]. On the other hand, the motor-side MMC is used to meet the drive requirements such as speed/torque and flux. To achieve these objectives, several control methods such as direct field orientation control (DFOC), indirect field orientation control (IFOC), and direct torque control (DTC) methods are discussed in the literature [13–15]. These control methods are well established and used in the commercial motor drive systems. In the following sections, the VOC for grid-side MMC, and IFOC schemes for motor-side MMC are discussed.

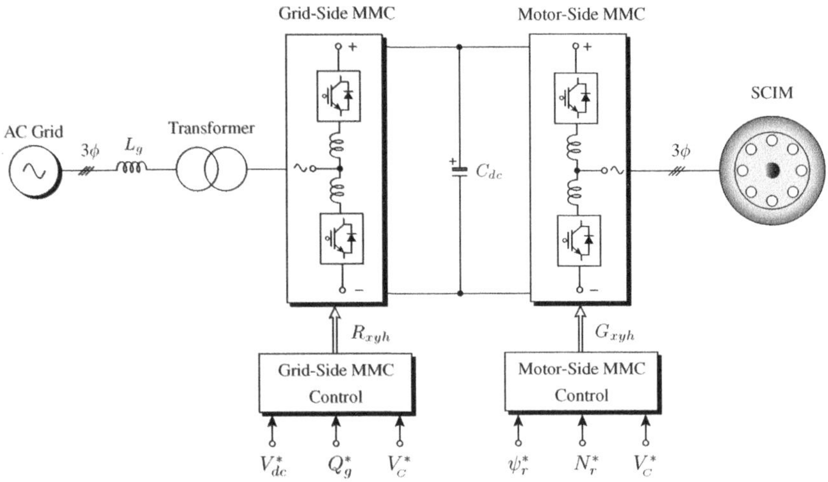

Figure 8.3 Generalized control block diagram of MMC-based motor drive.

8.3 VOLTAGE-ORIENTED CONTROL OF GRID-SIDE MMC

One of the popular control schemes for grid-tied MMC is known as voltage-oriented control (VOC) approach. This scheme is mainly implemented based on the transformation between the stationary-abc frame and the synchronous-dq frame as discussed in Chapter 3. The control block diagram of grid-tied MMC is shown in Figure 8.4. To realize the VOC, the grid voltage is measured, and its angle θ_g is obtained using a phase-locked loop (PLL). The grid angle (θ_g) is used to transform the three-phase voltages and currents from abc-frame to dq-frame, where these variables become DC signals and easy to control using a simple PI-regulators. These signals are used in the implementation of VOC. The VOC is designed to control the DC-bus voltage (V_{dc}) and grid reactive power (Q_g), and generates three-phase modulation signals ($\mathbf{v}_{\mathbf{r,abc}}^*$). In addition, the MMC requires a submodule voltage control to maintain each submodule capacitor voltage at their nominal value (v_C^*). The operating principle and implementation of submodule voltage control are presented in Chapter 3. The submodule voltage control generates compensation modulation signals (Δv_{xyh}^*), which are added to the main modulation signals ($\mathbf{v}_{\mathbf{r,abc}}^*$) obtained from the VOC. The resultant modulation signals are used along with the pulse width modulation scheme discussed in Chapter 2, to generate the submodule gating signals (R_{xyh}).

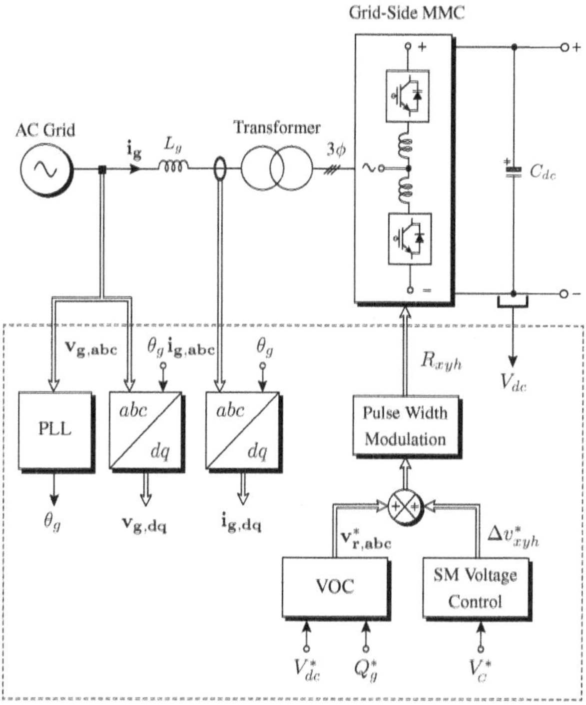

Figure 8.4 Control block diagram of grid-side MMC.

8.3.1 Principle of voltage orientation

The VOC approach performs the decoupled current control in the synchronous-dq frame in which, the measured grid currents (i_{ag}, i_{bg}, and i_{cg}), and grid voltages (v_{ag}, v_{bg}, and v_{cg}) are transformed into the dq-frame, where the time-varying signals become DC signals. The grid active (P_g) and reactive (Q_g) powers in dq-frame are given by

$$\begin{aligned} P_g &= 1.5\,(v_{dg}\, i_{dg} + v_{qg}\, i_{qg}) \\ Q_g &= 1.5\,(v_{qg}\, i_{dg} - v_{dg}\, i_{qg}) \end{aligned} \tag{8.1}$$

where, v_{dg} and v_{qg} are d- and q-axis grid voltage components and i_{dg} and i_{qg} are d- and q-axis grid current components, respectively.

From equation (8.1), the P_g and Q_g depend on both i_{dg} and i_{qg} components, which increases the complexity of the controller and affects the dynamic performance of the converter. The design of controller can be simplified by aligning the grid voltage vector "$\mathbf{v_g}$" along with the d-axis of the synchronous-dq reference frame. This process is referred to as "voltage orientation" as shown in Figure 8.5. By doing so, the v_{qg} becomes zero, and v_{dg} becomes a constant value equal to $|\mathbf{v_g}|$.

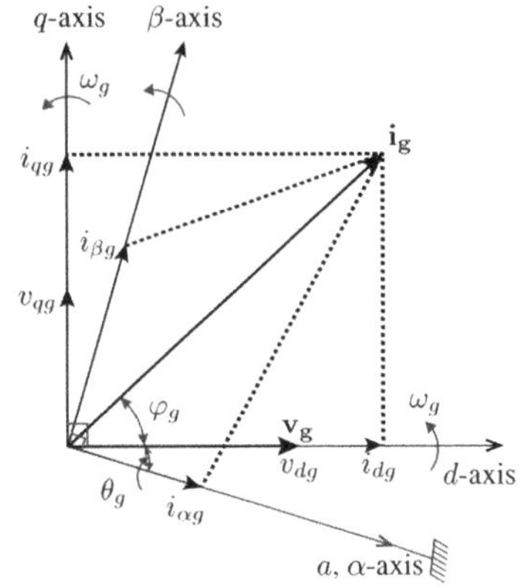

Figure 8.5 Space phasor diagram of voltage orientation.

With $v_{qg} = 0$, the P_g and Q_g are given by

$$\begin{aligned} P_g &= +1.5\, v_{dg}\, i_{dg} \\ Q_g &= -1.5\, v_{dg}\, i_{qg} \end{aligned} \tag{8.2}$$

Equation (8.2) shows the linear relationship between P_g and i_{dg}, and Q_g and i_{qg} for a constant v_{dg}. The voltage orientation approach allows the decoupled control of P_g and Q_g by controlling i_{dg} and i_{qg}, respectively. The voltage orientation approach mainly requires the grid voltage angle θ_g to achieve the perfect alignment of voltage vector $\mathbf{v_g}$ with the d-axis of the synchronous-dq reference frame. The grid voltage angle also plays a key role in the grid synchronization and transformation of variables from abc/dq-frame. The grid voltage angle θ_g is obtained from the zero-crossing detectors during non-distorted grid conditions or phase-locked loops (PLL) during distorted grid conditions.

8.3.2 Implementation of PLL

The phase-locked loop (PLL) is an important part of the grid-tied converter control system. The block diagram of three-phase PLL system in dq-frame is shown in Figure 8.6. In this approach, the measured three-phase grid voltages are transformed to the dq-frame using the feedback of grid voltage angle θ_g. A low-pass filter (LPF) is used to filter the voltage error. A PI-regulator is used to force the q-axis grid voltage component (v_{qg}) to zero. A feed-forward angular frequency term ω_f is added to the output of the PI-regulator to improve the dynamic response. The resultant output is equal to the grid angular frequency ω_g, and is converted into θ_g using an integrator. The value of θ_g varies in the range of $+\pi$ to $-\pi$, which is changed into 0 to 2π range using a modulus function. This approach is highly suitable for both ideal and distorted grid voltage conditions.

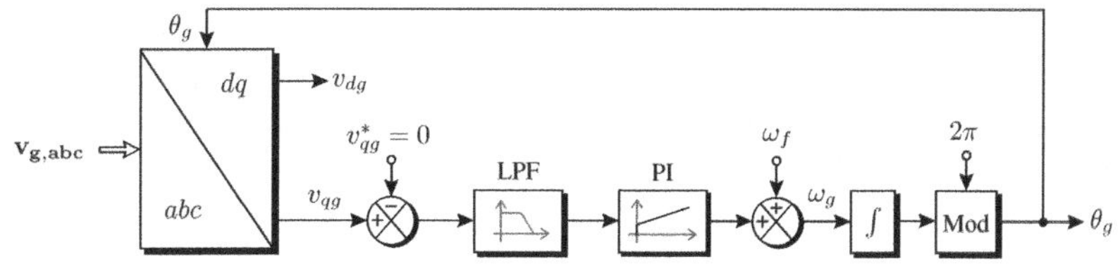

Figure 8.6 Implementation of PLL.

8.3.3 Block diagram of VOC

The control block diagram of VOC approach is shown in Figure 8.7. It consists of two inner current control loops for the accurate control of d- and q-axis current components of i_{dg} and i_{qg}, respectively, and a single outer DC-bus voltage loop for the control of DC capacitor (C_{dc}) voltage. The measured three-phase grid currents i_{ag}, i_{bg}, and i_{cg} in abc-frame are transformed into two-phase current components i_{dg} and i_{qg} in dq-frame using the grid voltage angle θ_g. The i_{dg} and i_{qg} represent the active and reactive current components in a three-phase grid current. The independent control of these two current components implies the independent control of system active and reactive powers.

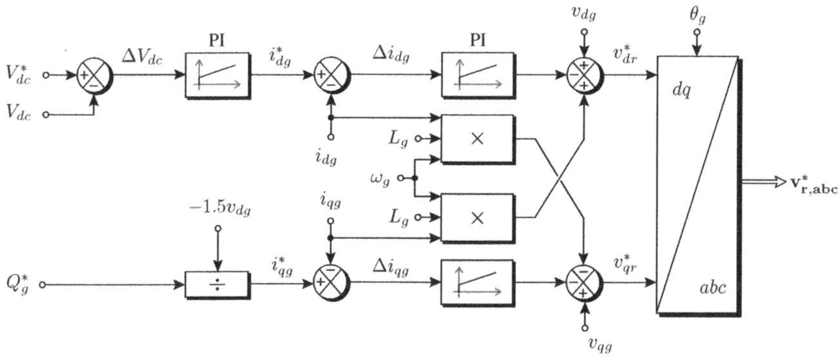

Figure 8.7 Block diagram of VOC approach.

Equation (8.2) shows that i_{dg} and i_{qg} are linearly related with P_g and Q_g, respectively. From equation (8.2), the q-axis reference current component (i^*_{qg}) is given by

$$i^*_{qg} = \frac{Q^*_g}{-1.5\, v_{dg}} \tag{8.3}$$

where Q^*_g is the reference grid reactive power, which is set to zero for unity power factor operation, a negative value for leading power factor, or a positive value for lagging power factor operation. The d-axis current reference (i^*_{dg}) represents the active power of the system, which is generated from the DC-bus voltage PI-regulator.

At steady-state operation, the DC-bus voltage V_{dc} is perfectly maintained at its reference value V_{dc}^*. The PI-regulator generates the reference current i_{dg}^* according to the operating conditions. Neglecting the power losses of the converter, the relationship between the AC- and DC-side powers is given by

$$P_g = 1.5\, v_{dg}\, i_{dg} = V_{dc}\, i_{dc} \tag{8.4}$$

The power flow in the power converter system is bidirectional. When the active power is delivered from the grid to the DC circuit, the converter operates in a rectifying mode ($P_g > 0$), whereas, when the active power is delivered from the DC circuit to the grid, the converter is operating in an inverting mode ($P_g < 0$). The controller will automatically switch between two modes of operation.

To decouple the d- and q-axis current controllers, the induced speed voltages in L_g due to the transformation from abc to dq-frame are added to the control loops. The induced speed voltages are obtained from the state-space model of the grid-side circuit. The state-space model of the grid-side circuit in abc-frame is given by

$$\frac{d\mathbf{i}_{\mathbf{g,abc}}}{dt} = (\mathbf{v}_{\mathbf{g,abc}} - \mathbf{v}_{\mathbf{r,abc}}) L_g \tag{8.5}$$

Equation (8.5) is transformed into the dq-frame as

$$\begin{aligned} \frac{d\, i_{dg}}{dt} &= (v_{dg} - v_{dr} + \omega_g\, L_g\, i_{qg})/L_g \\ \frac{d\, i_{qg}}{dt} &= (v_{qg} - v_{qr} - \omega_g\, L_g\, i_{dg})/L_g \end{aligned} \tag{8.6}$$

where "$\omega_g\, L_g\, i_{qg}$" and "$\omega_g\, L_g\, i_{dg}$" are the induced speed voltages due to the d- and q-axis current components. The current controller minimizes the error between the reference and the actual current components and gives the reference voltages. The output of decoupled current controller is given by

$$\begin{aligned} v_{dr}^* &= -(k_{pd} + k_{id}/s)\,(i_{dg}^* - i_{dg}) + \omega_g\, L_g\, i_{qg} + v_{dg} \\ v_{qr}^* &= -(k_{pq} + k_{iq}/s)\,(i_{qg}^* - i_{qg}) - \omega_g\, L_g\, i_{dg} + v_{qg} \end{aligned} \tag{8.7}$$

where "$k_{pd} + k_{id}/s$" and "$k_{pq} + k_{iq}/s$" are the transfer functions of d- and q-axis PI-regulators. The reference voltages in dq-frame are transformed to the abc-frame using the transformation matrix given in Chapter 3. The reference voltages v_{ar}^*, v_{br}^*, and v_{cr}^* are used with the pulse width modulator to generate the gating signals for the modular multilevel converter. From equations (8.6) and (8.7), the mathematical models of d- and q-axis current controller are given by

$$\begin{aligned} \frac{d\, i_{dg}}{dt} &= (k_{pd} + k_{id}/s)\,(i_{dg}^* - i_{dg})/L_g \\ \frac{d\, i_{qg}}{dt} &= (k_{pq} + k_{iq}/s)\,(i_{qg}^* - i_{qg})/L_g \end{aligned} \tag{8.8}$$

Equation (8.8) indicates that the d- and q-axis current control loops are completely decoupled, which allows the independent control of system active and reactive powers. It also simplifies the design of PI-regulators.

CASE STUDY 8.1 Voltage-Oriented Control of Grid-Tied MMC

Objective: In this case study, the performance of a grid-tied modular multilevel converter is analyzed with voltage-oriented control approach using MATLAB simulation.

Parameters: The system parameters of the grid-tied MMC are as follows: rated grid active power $P_g = 1$ MW, grid line-to-line voltage $V_g = 4160$ V, grid frequency $f_g = 60$ Hz, rated DC-bus voltage $V_{dc} = 7$ kV, number of SMs per arm $N = 20$, SM capacitor $C = 8000$ μF, SM capacitor voltage $v_C = 350$ V, arm inductor $L = 5$ mH, and carrier frequency $f_c = 500$ Hz.

Analysis:

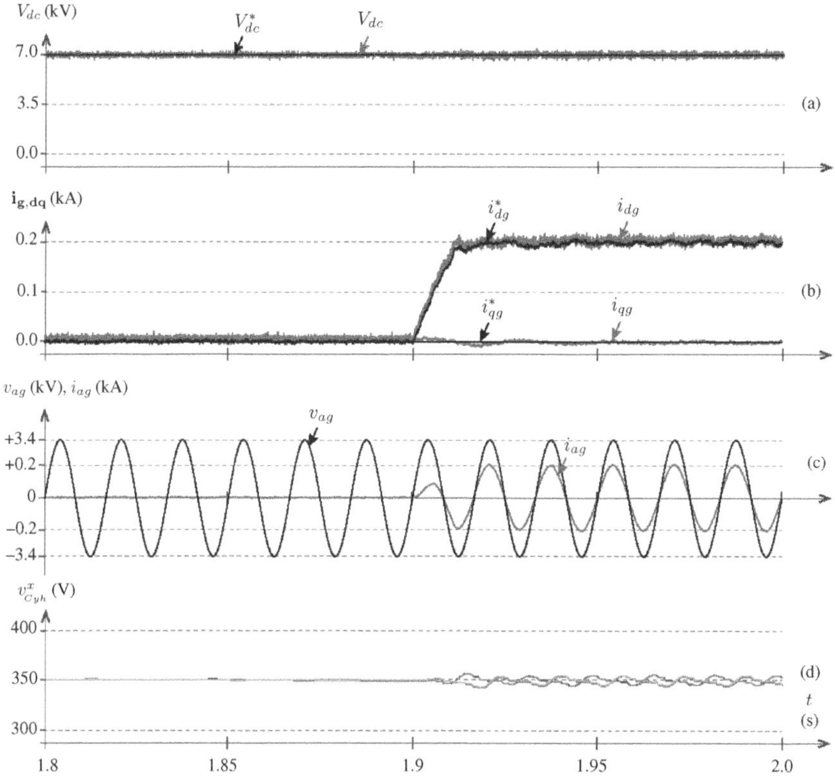

Figure 8.8 Performance of the grid-tied MMC with VOC approach.

In this case study, the motor-side converter is controlled using an indirect field-oriented control scheme to develop a rated rotor flux of 8.35 Wb (peak) in the motor. The motor rotational speed is controlled at 1189 RPM. The motor load torque is varied from 0 to 7490 N-m at $t = 1.9$ s. The grid-side converter is controlled using a VOC scheme. The reference DC-bus voltage V_{dc}^* is set to 7 kV, and the reactive power Q_g^* is set to zero for unity power factor operation. The DC-bus voltage is equally distributed between the submodules in each arm. Each submodule rated voltage is equal to 350 V. The grid angle is obtained using a PLL and used in the abc-frame to dq-axis transformation of three-phase voltage and current signals.

The performance of the grid-side MMC is shown in Figure 8.8. The DC-bus voltage controller regulates the DC-bus voltage at 7 kV as shown in Figure 8.8(a). The voltage controller generates a reference d-axis grid current component i_{dg}^*. The magnitude of i_{dg}^* varies with the load torque demand on the motor drive. From $t = 1.8$ s to $t = 1.9$ s, the load torque is zero, and i_{dg}^* is also equal to zero. From $t = 1.9$ s to $t = 2.0$ s, the load torque demand is maintained at 7490 N-m, correspondingly the i_{dg}^* is generated and equal to 200 A as shown in Figure 8.8(b). The reference and actual d-axis grid current components are perfectly following each other. On the other hand, the reference q-axis grid current is set to zero, corresponding to the zero reactive power as shown in Figure 8.8(b). The actual q-axis current component is tracking the reference q-axis grid current component. The dq-axis grid current components are transformed into abc-frame, and shows that the peak value of i_{ag} is equal to i_{dg} as shown in Figure 8.8(c). The grid phase voltage and grid currents are exactly in-phase and operate at unity power factor. Also, the submodule capacitor voltage control maintains the submodule capacitors voltage at their nominal value of 350 V as shown in Figure 8.8(d). There submodule capacitor voltage is a pure DC signal from $t = 1.8$ s to $t = 1.9$ s, due to the zero current flowing through the grid (i.e., $T_L = 0$). At $t = 1.9$ s onwards, the current flowing through the converter is around 200 A, which leads to a ripple in the submodule voltage as shown in Figure 8.8(d). ■

8.4 INDIRECT FIELD-ORIENTED CONTROL OF MOTOR-SIDE MMC

The DC-bus voltage of motor-side MMC is maintained at the rated value of V_{dc} using the grid-side MMC. The total DC-bus voltage is equally distributed between the submodules in each arm of motor-side MMC. These submodules are controlled to generate AC voltage with the required magnitude and frequency at the motor terminals, corresponding to the speed/torque and flux. The general control block diagram of motor-side MMC is shown in Figure 8.9. This control scheme is implemented in rotor flux reference frame, where the measured currents are transformed into rotor flux reference frame rotating at a speed of ω_s. The motor speed and rotor flux are controlled using an indirect field-oriented control (IFOC). The actual speed/torque and the magnitude of flux and its position are obtained from the rotor/torque calculator. The IFOC controller generates reference modulation signals ($\mathbf{v}_{\mathbf{s,abc}}^*$). In addition, the submodule voltage control used to maintain each submodule capacitor voltage at their rated value. The output of voltage control gives a compensation voltage command (Δv_{xyh}^*), which is added to the output of IFOC. The resultant modulation signals are used with the pulse width modulation scheme to generate the submodule gating signals (G_{xyh}).

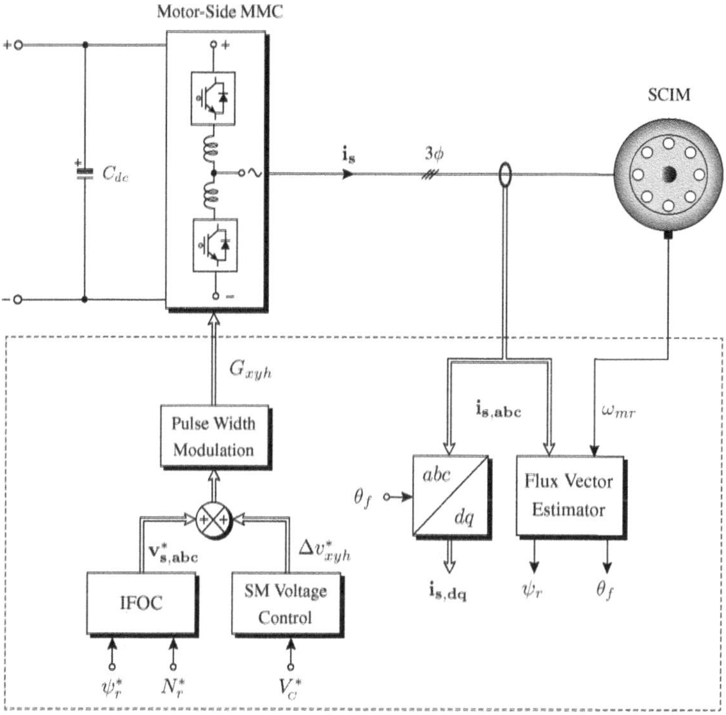

Figure 8.9 General control block diagram of motor-side MMC.

8.4.1 Principle of Field Orientation

The field orientation is similar to the voltage orientation in the control of grid-connected systems. Typically, the field orientation can be classified into stator flux, air-gap flux, and rotor flux orientations. Among them, the rotor flux orientation is simple and easy to implement for motor drives [13–15]. The electromagnetic torque developed by the motor drive system is given by

$$T_e = \frac{3\,P\,L_m}{2\,L_r}\,(\psi_{dr}\,i_{qs} - \psi_{qr}\,i_{ds}) \tag{8.9}$$

which are a function of d- and q-axis rotor flux components ψ_{dr} and ψ_{qr}, respectively. The decoupled control of rotor flux ψ_r and electromagnetic torque T_e can be achieved by aligning the rotor flux vector $\boldsymbol{\psi}_\mathbf{r}$ along with the d-axis of the synchronous-dq frame rotating at a speed of ω_s given by

$$\omega_s = 2\pi\,f_s \tag{8.10}$$

where f_s is the stator frequency of the motor drive. The process of rotor flux vector alignment with d-axis is referred to as field orientation. As a result, the q-axis component of rotor flux ψ_{qr} becomes zero. The d-axis component of rotor flux ψ_{dr} is equal to ψ_r. Finally, the simplified equation of electromagnetic torque is given by

$$T_e = K_T\, \psi_r\, i_{qs} \tag{8.11}$$

where "$K_T = 3PL_m/2L_r$" is the torque constant, ψ_r is the magnitude of rotor flux vector $\boldsymbol{\psi}_\mathbf{r}$, and i_{qs} is the q-axis stator current component. With the field orientation, the electromagnetic torque has a linear relationship with the q-axis stator current component (i_{qs}).

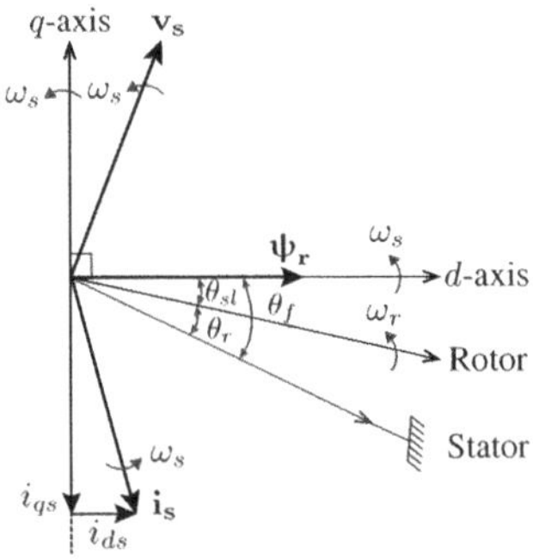

Figure 8.10 Space phasor diagram of field orientation.

The stator current vector $\mathbf{i_s}$ can be resolved into two orthogonal current components (d- and q-axis current components) as shown in Figure 8.10. The d-axis current component (i_{ds}) is the flux-producing component, while the q-axis current component (i_{qs}) is the torque-producing component. In field orientation, the rotor flux ψ_r is maintained constant at its rated value. Therefore, the i_{ds} becomes constant and equal to its rated value. The i_{qs} is independently controlled to meet the required torque demand. With the decoupled control, it is easy to stabilize the system.

In the field orientation, the determination of rotor flux vector position θ_f is one of the key issues. Different methods can be used to determine the position of rotor flux vector. If the θ_f is determined through the measured stator voltages and stator currents, then the method is referred to as a direct field-oriented control (DFOC). On the other hand, the rotor flux vector position is obtained by

$$\theta_f = \theta_r + \theta_{sl} \tag{8.12}$$

then the method is referred to as an indirect field-oriented control (IFOC), where θ_r and θ_{sl} are the measured rotor position angle and calculated slip angle, respectively. In this chapter, the implementation of indirect field orientation control approach is presented for the MMC-based motor drive system.

8.4.2 Rotor Flux Vector Estimator

The indirect field-oriented control requires the magnitude of rotor flux vector ψ_r and its position θ_f, which are determined from the measured stator currents, rotor angular speed (ω_r), and calculated angular slip speed (ω_{sl}). To calculate the ψ_r and θ_f, the induction motor model in the arbitrary reference frame is utilized.

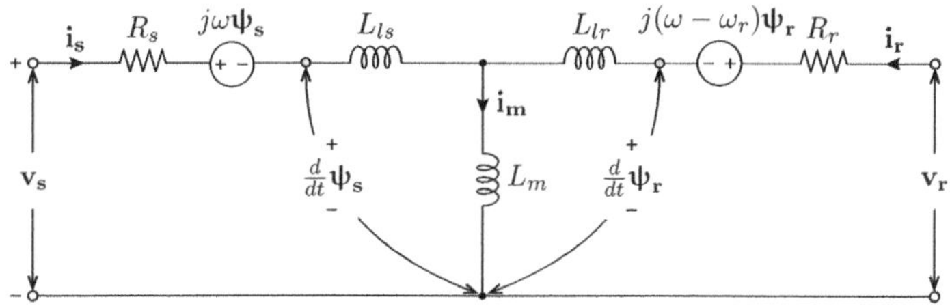

Figure 8.11 Space vector model of induction motor in the arbitrary reference frame.

From the equivalent circuit shown in Figure 8.11, the stator and rotor voltage vectors are given by

$$\begin{aligned}\mathbf{v_s} &= R_s\,\mathbf{i_s} + \frac{d\boldsymbol{\psi}_\mathbf{s}}{dt} + j\omega\,\boldsymbol{\psi}_\mathbf{s}\\ \mathbf{v_r} &= R_r\,\mathbf{i_r} + \frac{d\boldsymbol{\psi}_\mathbf{r}}{dt} + j(\omega-\omega_r)\,\boldsymbol{\psi}_\mathbf{r}\end{aligned} \tag{8.13}$$

where, $\mathbf{v_s}$ and $\mathbf{v_r}$ are the stator and rotor voltage vectors, $\mathbf{i_s}$ and $\mathbf{i_r}$ are the stator and rotor current vectors, $\boldsymbol{\psi}_\mathbf{s}$ and $\boldsymbol{\psi}_\mathbf{r}$ are the stator and rotor flux vectors, R_s and R_r are the stator and rotor winding resistances, ω is the rotating speed of arbitrary reference, and ω_r is the rotor angular speed.

The stator and rotor flux vectors are given by

$$\begin{aligned}\boldsymbol{\psi}_\mathbf{s} &= L_s\,\mathbf{i_s} + L_m\,\mathbf{i_r}\\ \boldsymbol{\psi}_\mathbf{r} &= L_r\,\mathbf{i_r} + L_m\,\mathbf{i_s}\end{aligned} \tag{8.14}$$

where $L_s = L_{ls} + L_m$ is the stator self-inductance, $L_r = L_{lr} + L_m$ is the rotor self-inductance, L_{ls} and L_{lr} are the stator and rotor leakage inductances, and L_m is the magnetizing inductance.

In this analysis, the squirrel induction motor in the synchronous-dq reference frame is considered by replacing the ω with the ω_s, and $\mathbf{v_r} = 0$ in equation (8.13); then the resultant equation becomes

$$\begin{aligned}\mathbf{v_s} &= R_s\,\mathbf{i_s} + \frac{d\boldsymbol{\psi}_\mathbf{s}}{dt} + j\omega_s\,\boldsymbol{\psi}_\mathbf{s}\\ 0 &= R_r\,\mathbf{i_r} + \frac{d\boldsymbol{\psi}_\mathbf{r}}{dt} + j\omega_{sl}\,\boldsymbol{\psi}_\mathbf{r}\end{aligned} \tag{8.15}$$

where ω_s is the angular speed of a synchronous-dq frame and ω_{sl} is the angular slip speed ($\omega_{sl} = \omega_s - \omega_r$).

The angular slip speed (ω_{sl}) is calculated by using the rotor voltage equation given in equation (8.15), from which

$$\frac{d\boldsymbol{\psi}_\mathbf{r}}{dt} = -R_r\,\mathbf{i_r} - j\omega_{sl}\,\boldsymbol{\psi}_\mathbf{r} \tag{8.16}$$

From equation (8.14), the rotor current is given by

$$\mathbf{i_r} = \frac{1}{L_r}(\boldsymbol{\psi}_\mathbf{r} - L_m\,\mathbf{i_s}) \tag{8.17}$$

which is substituted in equation (8.16) and results in

$$\frac{d\boldsymbol{\psi}_\mathbf{r}}{dt} = -\frac{R_r}{L_r}\,(\boldsymbol{\psi}_\mathbf{r} - L_m\,\mathbf{i_s} - j\omega_{sl}\,\boldsymbol{\psi}_\mathbf{r}) \tag{8.18}$$

from which

$$\boldsymbol{\psi}_\mathbf{r}\,(1 + \tau_r\,(p + j\,\omega_{sl})) = L_m\,\mathbf{i_s} \tag{8.19}$$

where p is the derivative operator ($p = \frac{d}{dt}$) and τ_r is rotor time constant ($\tau_r = \frac{L_r}{R_r}$).
Considering the field orientation ($\psi_{qr} = 0$ and $\psi_{dr} = \psi_r$), equation (8.19) is decomposed into two orthogonal components in the synchronous-dq frame as

$$\begin{aligned} \psi_r\,(1 + p\,\tau_r) &= L_m\,i_{ds} \\ \omega_{sl}\,\tau_r\,\psi_r &= L_m\,i_{qs} \end{aligned} \tag{8.20}$$

From equation (8.20), the magnitude of rotor flux vector and the angular slip speed are given by

$$\begin{aligned} \psi_r &= \frac{L_m}{1 + p\,\tau_r}\,i_{ds} \\ \omega_{sl} &= \frac{L_m}{\tau_r\,\psi_r}\,i_{qs} \end{aligned} \tag{8.21}$$

From the measured angular rotor speed (ω_r) and calculated angular slip speed (ω_{sl}), the position of rotor flux vector (θ_f) is given by

$$\theta_f = \int (\omega_r + \omega_{sl})\,dt \tag{8.22}$$

The digital implementation of rotor flux magnitude and its position estimator is shown in Figure 8.12. The three-phase stator currents in abc-frame are transformed into the synchronous dq-frame using the transformation matrix given in Chapter 3. Rest of the blocks are simple constant blocks and derived from equations (8.21) and (8.22). The output of the rotor flux estimator is the rotor flux magnitude "ψ_r" and angular slip speed "ω_{sl}." The measured rotor mechanical speed is multiplied with the pole pairs (P), which gives the rotor electrical angular speed (ω_r). The angular slip speed is added to the rotor electrical angular speed, and results in the angular speed of rotor flux vector in the synchronous dq-frame. The angular position of rotor flux vector (θ_f) is obtained by integrating its angular speed as shown in Figure 8.12.

8.4.3 Block diagram of IFOC approach

The control block diagram of the IFOC approach for induction motor drive is shown in Figure 8.13. The IFOC uses a cascaded control structure, which consists of outer

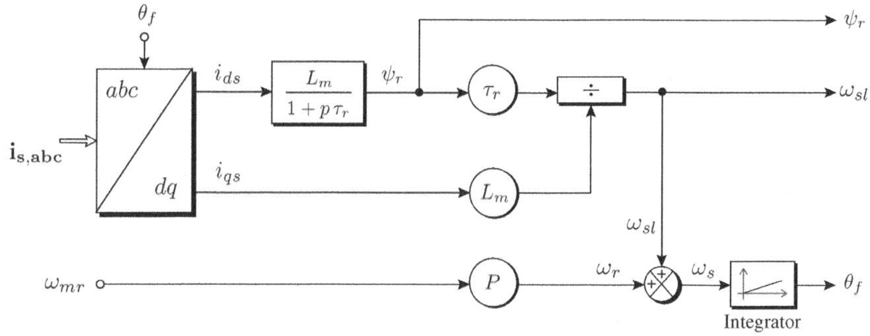

Figure 8.12 Estimation of rotor flux magnitude and position (ψ_r and θ_f).

rotor flux and speed control loops, and inner current control loops. In rotor flux control, the reference rotor flux ψ_r^* is generated with respect to the reference speed N_r^*. For the below rated speed operation, the reference rotor flux is maintained at their rated value. During the above rated speed operation, the rotor flux reduced such that the motor is operating in flux weakening region. The reference rotor flux ψ_r^* is compared with the estimated rotor flux ψ_r, and their difference is minimized using PI-regulator and given a reference d-axis stator current component i_{ds}^*. On the other hand, the speed control loop minimizes the error between the reference and actual speed. The speed controller output is a reference electromagnetic torque T_e^*, which is multiplied with the torque constant "K_T" to obtain the reference q-axis stator current component i_{qs}^* as shown in Figure 8.13.

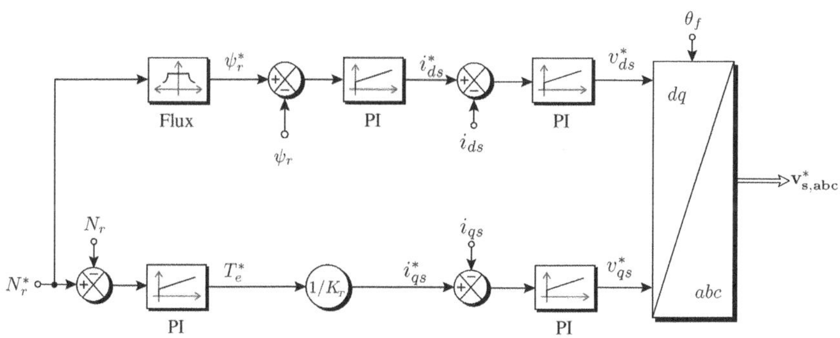

Figure 8.13 Control block diagram of IFOC.

The feedback dq stator current components are obtained by transforming the measured three-phase stator currents i_{as}, i_{bs}, and i_{cs} in abc-frame to the rotor flux synchronous dq-frame. The feedback i_{ds} and i_{qs} are compared with their references, and corresponding errors are given to the PI-regulators. The PI-regulators generate the reference stator voltages v_{ds}^* and v_{qs}^* in dq-frame. The dq voltages are transformed back to the *abc*-frame using the transformation matrix given in Chapter 3. The reference stator voltages v_{as}^*, v_{bs}^*, and v_{cs}^* are used with the pulse width modulator to generate the gating signals for the modular multilevel converter.

CASE STUDY 8.2 Indirect Field-Oriented Control of Motor-Side MMC

Objective: In this case study, the performance of a motor-side modular multilevel converter is analyzed with the indirect field-oriented control (IFOC) approach using MATLAB simulation.

Parameters: The system parameters of the motor-side MMC are as follows: rated active power $P_s = 1$ MW, line-to-line voltage $V_s = 4160$ V, stator frequency $f_s = 0$–60 Hz, rated DC-bus voltage $V_{dc} = 7$ kV, number of SMs per arm $N = 20$, SM capacitor $C = 8000$ μF, SM capacitor voltage $v_C = 350$ V, arm inductor $L = 1$ mH, and carrier frequency $f_c = 500$ Hz. The parameters of the squirrel cage induction motor are given in Table 8.1.

Table 8.1 Three-phase Squirrel Cage Induction Motor Rating and Parameters

Motor Rating	**SI Unit**	**Motor Parameter**	**SI Unit**
Output Power (P_s)	1250 hp	**Stator Resistance** (R_s)	0.21 Ω
Line-to-Line Voltage (V_s)	4160 V	**Rotor Resistance** (R_r)	0.146 Ω
Stator Current (I_s)	150 A (RMS)	**Stator Leakage Inductance** (L_{ls})	5.2 mH
Speed (N_r)	1189 RPM	**Rotor Leakage Inductance** (L_{lr})	5.2 mH
Torque (T_e)	7490 N-m	**Magnetizing Inductance** (L_m)	155 mH
Stator Flux Linkage (ψ_s)	9.0 Wb (peak)	**Moment of Inertia** (J)	10 kg-m^2
Rotor Flux Linkage (ψ_r)	8.35 Wb (peak)	**Number of Pole Pairs** (P)	3

Analysis:
The motor-side modular multilevel converter is designed with 20 SMs per arm. Each submodule capacitor is rated for 350 V. The DC terminals are connected to a rectifier system, which maintains the DC-bus voltage at their nominal value of 7 kV. The AC output terminals are connected to a squirrel cage induction motor. The motor speed/torque and flux are controlled using an indirect FOC scheme. In this case study, the IFOC is designed with a reference rotor flux of 8.35 Wb and a rotor speed of 1189 RPM. From $t = 1.8$ s to $t = 1.9$ s, the motor load torque is maintained at zero.

The performance of the MMC-based motor drive with the IFOC scheme is shown in Figure 8.14. The flux controller maintains the rotor flux at their rated value of 8.35 Wb as shown in Figure 8.14(a). The speed controller perfectly regulates the motor rotational speed at the given reference value of 1189 RPM. The controller generates reference torque command, corresponding to the load torque demand ($T_L = 0$) as shown in Figure 8.14(c). The reference *d*-axis current of 66 A is obtained from the flux controller to maintain the rotor flux at 8.35 Wb. The reference *q*-axis current is maintained at zero to generate the zero electromagnetic torque as shown in Figure 8.14(d). The *dq*-axis current components are transformed into *abc*-frame. The three-phase currents are generated with a frequency of 60 Hz, and a peak value of 66 A as shown. The *dq*-axis current controller generates the reference *dq*-axis voltage components of $v_{ds} = 0$ V and $v_{qs} = 3396$ V as shown in Figure 8.14(e). The reference phase-*a*

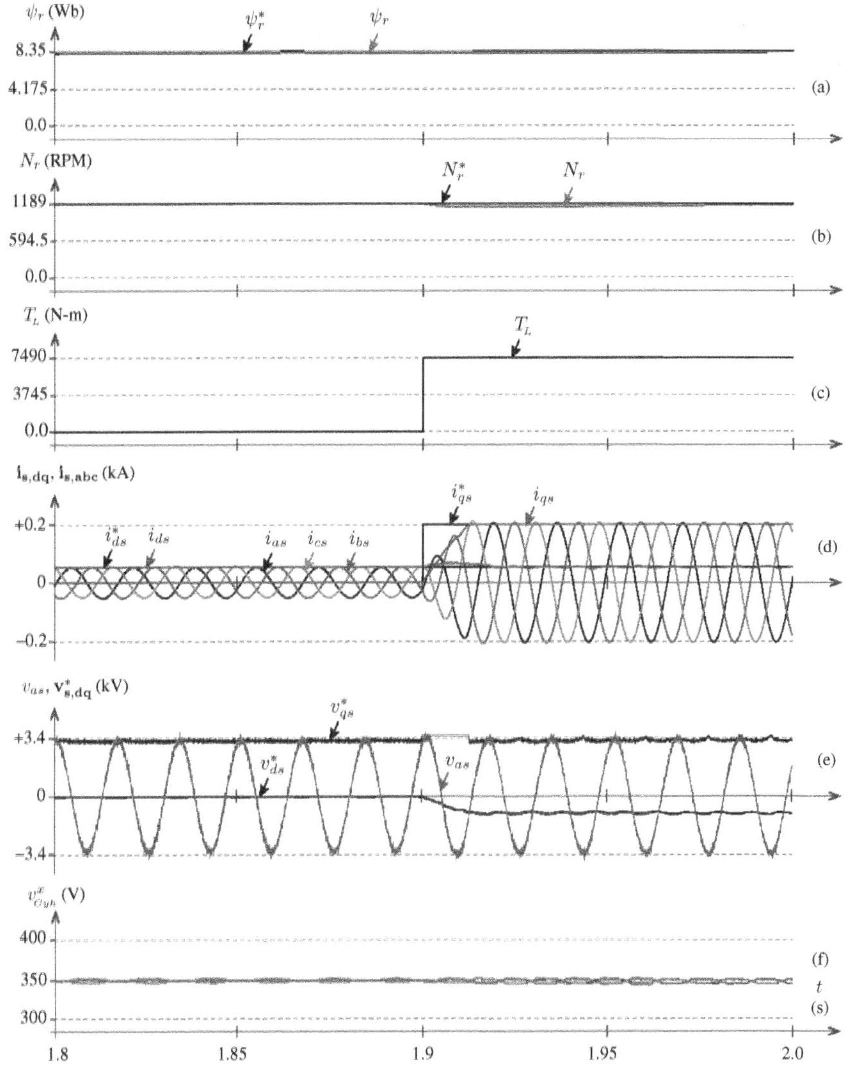

Figure 8.14 Performance of the motor-side MMC with FOC approach.

voltage has a peak value of 3396 V and frequency of 60 Hz as shown in Figure 8.14(e). The SM voltage control is perfectly maintaining each submodule capacitor voltage at their rated value of 350 V as shown in Figure 8.14(f).

At $t = 1.9$ s, the motor load torque demand is increased from 0 to 7490 N-m. The controller is maintaining the motor rotor flux and the rotational speed at their rated values. During the step-change in the load torque, the current controller output saturates to its maximum value for a short time, and its response becomes sluggish. To meet the required torque demand, the magnitude of q-axis current component is increased from 66 A to 212 A, and the d-axis current component is maintained at zero as shown in Figure 8.14(d). The three-phase stator currents have a peak value of 212 A. The reference dq-axis, and abc-frame voltage components are maintained at their rated value, corresponding to the operating conditions as shown in Figure 8.14(e). During the step change, the q-axis voltage is limited to a peak value of 3500 V for a duration 1.9–1.912 s. Later, the controller response becomes normal and is able to generate the rated voltage as shown in Figure 8.14(e). In the complete operation range, the submodule capacitor voltages are regulated at 350 V. But, there is an increase in the voltage ripple due to the increase in the magnitude of stator current as shown in Figure 8.14(f). ■

8.5 LOW-SPEED OPERATION OF MMC-BASED MOTOR DRIVE

In MMC-based motor drive, the submodule capacitor voltage ripple varies inversely with the motor rotational speed (fundamental frequency) under constant load torque. At low-speed operation, the voltage ripples are large enough to threaten the reliability of MMC [16, 17]. In addition, there is a significant effect on the output power quality and converter efficiency. Traditionally, the submodules are designed with larger size capacitors to limit the voltage ripple within the prescribed limits [18, 19]. This approach is costly and increases the size of the converter.

Alternatively, a high-frequency voltage component is added to the main modulation signals in the control stage. Theoretically, the high-frequency voltage modifies the voltage ripple frequency, resulting in a smaller voltage ripple [20, 21]. The high-frequency voltage appears in the form of a common-mode voltage (CMV) at the motor neutral terminal. To block the CMV, the conventional MMC-based motor drive system requires a transformer, which is incorporated between the AC grid and rectifier system as shown in Figure 8.2. The voltage ripple and its minimization techniques are discussed in the following subsections.

8.5.1 Analysis of Submodule Capacitor Voltage Ripple

The submodule capacitor voltage ripple is analyzed using a per-phase configuration of MMC as shown in Figure 8.15(a). The submodules in each arm are controlled to generate an arm voltage (v_{xu} and v_{xl}). The arm voltage consists of DC voltage ($v_d = \frac{V_{dc}}{2}$) and motor stator voltage (v_{xs}) components as shown in Figure 8.15(b). The upper and lower arm voltages are given by

$$\begin{aligned} v_{xu} &= \frac{V_{dc}}{2} - v_{xs} - L\frac{d\,i_{xu}}{dt} \\ v_{xl} &= \frac{V_{dc}}{2} + v_{xs} - L\frac{d\,i_{xl}}{dt} \end{aligned} \tag{8.23}$$

The derivative component in the above equation represents the voltage drop across the arm inductor. This voltage component is very small compared to DC and AC

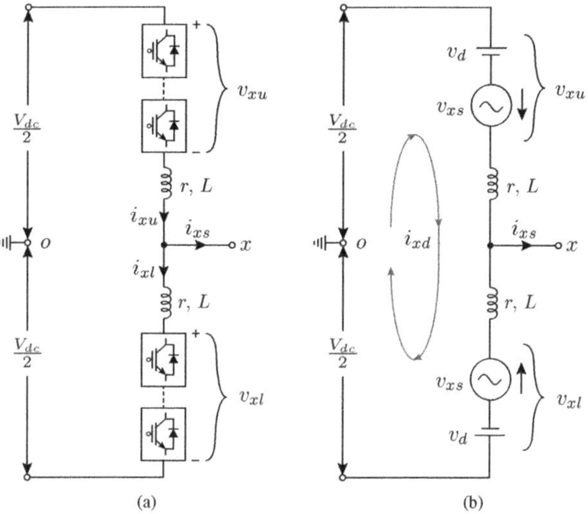

Figure 8.15 Per-phase MMC system: (a) converter configuration and (b) equivalent model

voltage components and is neglected. The simplified arm voltages are given by

$$\begin{aligned} v_{xu} &= \frac{V_{dc}}{2} - v_{xs} \\ v_{xl} &= \frac{V_{dc}}{2} + v_{xs} \end{aligned} \tag{8.24}$$

The arm voltage components force DC (i_{xd}) and AC current (i_{xs}) components, respectively. The DC current component flows among the converter legs and the main DC-bus. The AC current component (i_{xs}) flows through the motor and is equally distributed between the upper and lower arms. The upper and lower arm currents are given by

$$\begin{aligned} i_{xu} &= i_{xd} + i_{xs} \\ i_{xl} &= i_{xd} - i_{xs} \end{aligned} \tag{8.25}$$

The AC voltage and current components are defined as

$$\begin{aligned} v_{xs} &= \frac{m_a V_{dc}}{2} \sin(\omega_s t) \\ i_{xs} &= I \sin(\omega_s t - \varphi) \end{aligned} \tag{8.26}$$

The magnitude of DC current component is obtained from the power balance between AC and DC sides of MMC. According to this criteria, the DC current component is given by

$$i_{xd} = \frac{v_{xs} i_{xs}}{V_{dc}} \tag{8.27}$$

Equations (8.24) and (8.25) are plotted for one fundamental cycle ($\frac{1}{f_s}$) using a simulation tool, and the corresponding waveforms are shown in Figure 8.16. As described in the model, the arm voltage and current waveforms consist of AC components, which are offset by a DC value as shown in Figure 8.16(a)–(b). The total arm power is equal to the product of arm voltage and corresponding arm current. The total arm power is equally distributed among N submodules. The output power of submodule-h in the arm is given by

$$p_{xyh} = \frac{1}{N}\,(v_{xy}\,i_{xy}) \tag{8.28}$$

The instantaneous value of submodule-h power is shown in Figure 8.16(c). The shaded area of the waveform represents the energy stored in the submodule-h, which is mathematically given by

$$E_{xyh}(t) = \int p_{xyh}\,dt + E_{xyh}(0) \tag{8.29}$$

where $E_{xyh}(0)$ represents the initial value of energy stored in the submodule-h.

The positive area of the waveform leads to charging of the submodule-h capacitor for a duration of T_1, while the negative area leads to discharging of the submodule-h capacitor for a duration of T_2 as shown in Figure 8.16(d). The charging and discharging of the submodule-h capacitor in the upper and lower arms are opposite in-phase. The submodule capacitor voltage ripple ($\Delta v^x_{_{Cyh}}$) depends on the energy fluctuations (ΔE_{xyh}) as shown in Figure 8.16(c). The mathematical relationship between the capacitor voltage and energy fluctuations are given by

$$\Delta E_{xyh} = \frac{1}{2}\left(V_{_C} + \Delta v^x_{_{Cyh}}\right)^2 - \frac{1}{2}\left(V_{_C} - \Delta v^x_{_{Cyh}}\right)^2 \tag{8.30}$$

where ΔE_{xyh} represents the submodule-h energy fluctuation, $\Delta v^x_{_{Cyh}}$ represents the submodule-h capacitor voltage ripple, and $V_{_C}$ represents the average value of submodule capacitor voltage.

In motor drive applications, the modulation signals are generated proportional to operating frequency to maintain the "Volt/Hertz" ratio constant. According to this criteria, the modulation index is given by

$$m_a = \frac{f_s}{f_{sr}} \tag{8.31}$$

where m_a represents the amplitude modulation index, f_s represents the stator operating frequency, and f_{sr} represents the stator rated frequency.

The capacitor voltage ripple is analyzed for one submodule in the arm. The submodule voltage ripple is obtained by substituting equations (8.24)–(8.28) and (8.31) in equation (8.29) as

$$E_{xyh}(t) = \frac{1}{4N f_s}\,V_{dc}\,I\,e + E_{xyh}(0) \tag{8.32}$$

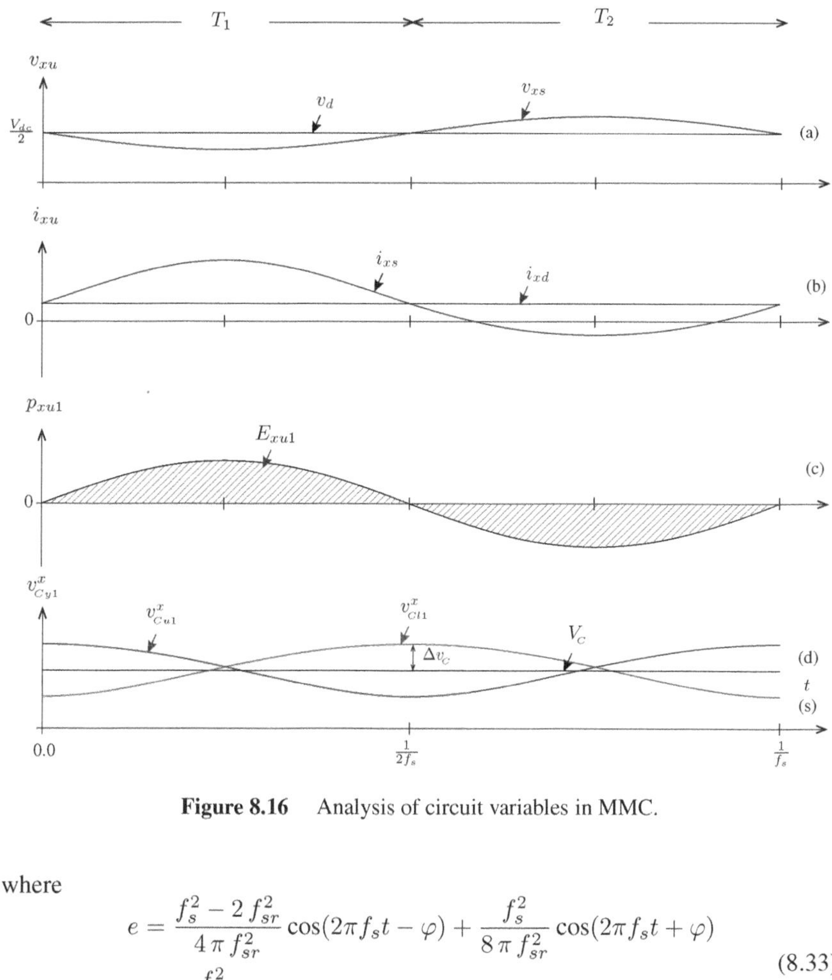

Figure 8.16 Analysis of circuit variables in MMC.

where

$$
\begin{aligned}
e = & \frac{f_s^2 - 2\,f_{sr}^2}{4\,\pi\,f_{sr}^2}\cos(2\pi f_s t - \varphi) + \frac{f_s^2}{8\,\pi\,f_{sr}^2}\cos(2\pi f_s t + \varphi) \\
& + \frac{f_s^2}{24\,\pi\,f_{sr}^2}\cos(6\pi f_s t - \varphi)
\end{aligned}
\tag{8.33}
$$

The fluctuations in the energy (ΔE_{xyh}) is obtained by observing the maximum and minimum values of instantaneous energy (E_{xyh}) as

$$
\Delta E_{xyh} = \frac{1}{4N f_s} V_{dc}\, I\, (e_{max} - e_{min}) \tag{8.34}
$$

From equations (8.30) and (8.34), the submodule capacitor voltage ripple is given by

$$
2\,C\,\Delta v^x_{Cyh} V_C = \frac{1}{4N f_s} V_{dc}\, I\, (e_{max} - e_{min}) \tag{8.35}
$$

Equation (8.35) is simplified by considering the relationship between the total DC-bus voltage and submodule capacitor voltage as $V_{dc} = N\,V_C$, and resultant submodule voltage ripple is given by

$$\Delta v^{x}_{Cyh} = \frac{I}{8\,C\,f_s}\,(e_{max} - e_{min}) \qquad (8.36)$$

From the above equation, it is observed that the submodule capacitor voltage ripple is inversely proportional to the motor operating frequency (f_s) and proportional to the motor current (I). Hence, the voltage ripple is most severe in the motor drive during low-speed operation.

CASE STUDY 8.3 Capacitor Voltage Ripple in Motor-Side MMC

Objective: In this case study, the submodule capacitor voltage ripple is analyzed in the motor-side modular multilevel converter using MATLAB simulation.

Parameters: The system parameters of the motor-side MMC are as follows: rated power $P_s = 1$ MW, line-to-line voltage $V_s = 4160$ V, power factor angle $\varphi = 14.5°$, operating frequency $f_s = 25$ Hz, rated motor frequency $f_{sr} = 60$ Hz, DC-bus voltage $V_{dc} = 7$ kV, number of SMs per arm $N = 20$, SM capacitor $C = 8000$ μF, SM capacitor voltage $v_C = 350$ V, arm inductor $L = 1$ mH, and carrier frequency $f_c = 500$ Hz. The parameters of the squirrel cage induction motor are given in Table 8.1.

Analysis:
The motor-side modular multilevel converter is designed with a total DC-bus voltage of 7 kV, which is equally distributed between 20 SMs in each arm. The rated voltage of each submodule capacitor is 350 V. The AC side of MMC is connected to a motor drive of 1 MW rated capacity. The motor drive is controlled using an FOC scheme. In this study, the FOC is designed with reference rotor flux of 8.35 Wb and rotor speed of 495 RPM (corresponding to 25 Hz). The motor is controlled to develop a rated load torque of 7490 N-m. Hence, the motor draws a rated current of

$$I = \frac{\sqrt{2} \times P_o}{\sqrt{3} \times v_o \times \cos\varphi} = \frac{\sqrt{2} \times 1 \times 10^6}{\sqrt{3} \times 4160 \times \cos(14.5)} \approx 202.73\,\text{A (peak)} \qquad (8.37)$$

Under the above operating conditions, the maximum value of energy fluctuation is given by

$$\begin{aligned} e_{max} &= \max\Big(\frac{25^2 - 2 \times 60^2}{4\,\pi\,60^2}\cos(2\pi \times 25t - 14.5) \\ &+ \frac{25^2}{8\,\pi\,60^2}\cos(2\pi \times 25t + 14.5) \\ &+ \frac{25^2}{24\,\pi\,60^2}\cos(6\pi \times 25t - 14.5)\Big) \\ &= 0.1407 \end{aligned} \qquad (8.38)$$

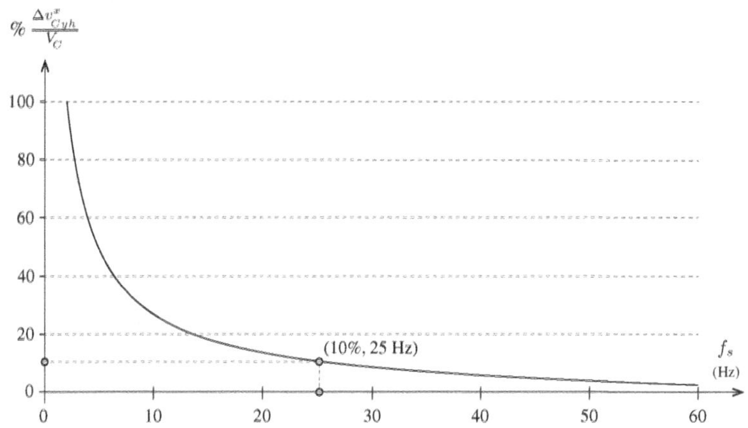

Figure 8.17 Variation of voltage ripple with frequency.

Similarly, the minimum value of energy fluctuation is given by

$$\begin{aligned} e_{min} &= \min(\frac{25^2 - 2 \times 60^2}{4\,\pi\,60^2} \cos(2\pi \times 25t - 14.5) \\ &+ \frac{25^2}{8\,\pi\,60^2} \cos(2\pi \times 25t + 14.5) \\ &+ \frac{25^2}{24\,\pi\,60^2} \cos(6\pi \times 25t - 14.5)) \\ &= -0.1407 \end{aligned} \tag{8.39}$$

The submodule capacitor voltage ripple is given by

$$\Delta v^x_{Cyh} = \frac{202.73}{8 \times 0.008 \times 25} (0.1407 - (-0.1407)) = 35.65 \text{ V} \tag{8.40}$$

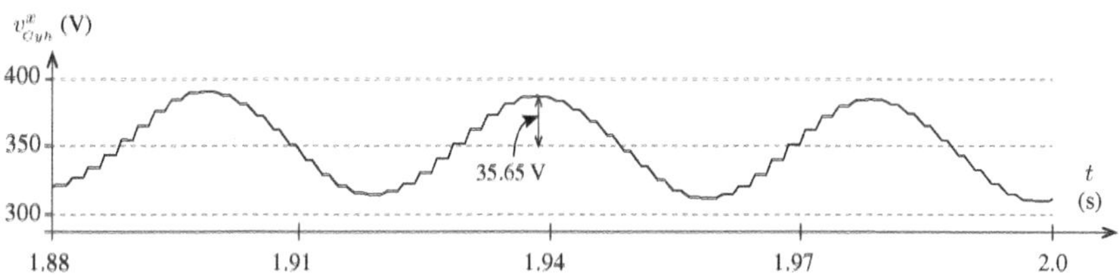

Figure 8.18 Submodule capacitor voltage at $f_s = 25$ Hz.

The mathematical analysis shows the submodule capacitor voltage has a ripple of 35.65 V, which is around 10% of the nominal voltage (350 V). The analysis is further extended to other operating frequencies, and corresponding voltage ripple in the percentage is shown in Figure 8.17. The voltage ripples are less than 10% for an operating frequency greater than 25 Hz. When the operating frequency less than 25 Hz, the voltage ripple exceeds the design limit. Hence, the MMC requires either a larger submodule capacitance or a control scheme to handle the ripples during low-frequency region. The mathematical analysis is further verified with MATLAB simulations for an operating frequency of 25 Hz. The submodule capacitor voltage has a ripple of 35.65 V as shown in Figure 8.18. The simulation results are very close to the mathematical analysis and can be used in the design of submodule capacitor. ■

8.5.2 Analysis of MMC with High-Frequency Voltage and Current Injection

The injection of high-frequency voltage and current signals into the system is one of the possible solutions to minimize the submodule capacitor voltage ripple during the low-speed region. It is possible to use sinusoidal, square, and third-harmonic waveforms as high-frequency signals to minimize the voltage ripple [22]. In this section, the MMC is analyzed with the sinusoidal high-frequency signal. The three-phase equivalent model of MMC is shown in Figure 8.19, where the high-frequency signal is arranged with opposite phase in the upper and lower arms of the converter leg. Generally, the high-frequency current signal is injected into the system along with the high-frequency voltage to mitigate the voltage ripple.

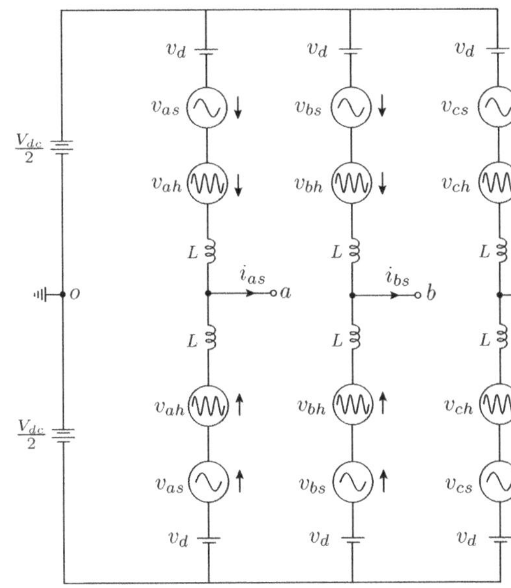

Figure 8.19 Equivalent model of a three-phase MMC in low-frequency operation.

From Figure 8.19, the simplified upper and lower arm voltages are given by

$$\begin{aligned} v_{xu} &= \frac{V_{dc}}{2} - v_{xs} - v_{xh} \\ v_{xl} &= \frac{V_{dc}}{2} + v_{xs} + v_{xh} \end{aligned} \tag{8.41}$$

where v_{xh} represents the high-frequency voltage, which has an identical magnitude with zero phase-shift in three converter legs.

Similarly, the upper and lower arm currents are given by

$$\begin{aligned} i_{xu} &= i_{xd} + i_{xs} + i_{xh} \\ i_{xl} &= i_{xd} - i_{xs} - i_{xh} \end{aligned} \tag{8.42}$$

where, i_{xh} represents the high-frequency current signal.

The sinusoidal high-frequency voltage and current signals are defined as

$$\begin{aligned} v_{xh} &= V_h \sin(2\pi f_h\, t) \\ i_{xh} &= I_h \sin(2\pi f_h\, t) \end{aligned} \tag{8.43}$$

where V_h and I_h represent the peak values of high-frequency voltage and current signals, respectively, and f_h represents the frequency of high-frequency signals.

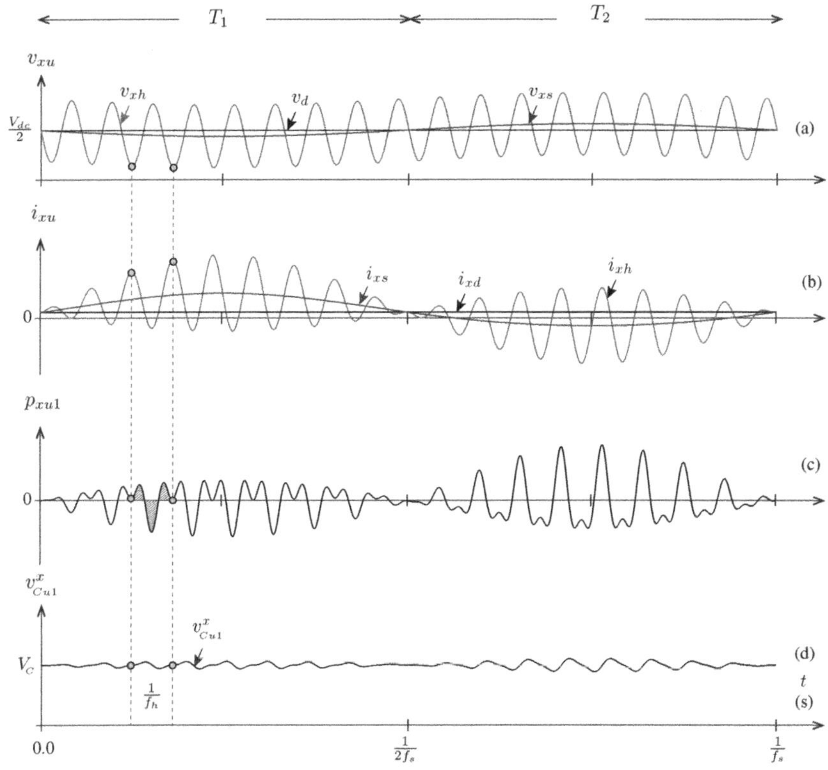

Figure 8.20 Analysis of circuit variables in MMC with high-frequency signals.

The voltage and current components of the upper arm voltage and current for one fundamental cycle are shown in Figure 8.20(a)–(b). The arm voltage and current waveforms have DC and fundamental frequency (f_s) components along with the high-frequency (f_h) components as described in equations (8.41) and (8.42). Typically, the high-frequency f_h is selected to be three times fundamental frequency f_s. The total arm power (p_{xy}) is obtained by multiplying the arm voltage (v_{xy}) with the arm current (i_{xy}), which is equally handled by each submodule. The submodule-h power is given by

$$p_{xyh} = \frac{1}{N}\,(v_{xy}\,i_{xy}) \tag{8.44}$$

The submodule-h energy (E_{xyh}) is obtained by integrating the submodule-h power (p_{xyh}) with the time as shown in Figure 8.20(c). The energy in the positive and negative shaded areas cancel each other in each high-frequency cycle, which will reduce the fluctuating energy and voltage ripple in the submodule capacitor as shown in Figure 8.20(d).

8.5.3 Estimation of High-Frequency Voltage and Current Magnitude

The magnitude of the high-frequency voltage and current signals are selected such that the converter operates in the linear modulation range without exceeding their rated limits. The high-frequency signals are designed based on the submodule power analysis. The upper arm submodule-h power given in equation (8.44) is expanded by using equations (8.41) and (8.42) as

$$\begin{aligned} p_{xuh} = &\frac{1}{N}\underbrace{(0.5\,V_{dc}\,i_{xd} - 0.5\,v_{xs}\,i_{xs})}_{\text{Term-1}} \\ &+ \frac{1}{N}\underbrace{(0.25\,V_{dc}\,i_{xs} - v_{xs}\,i_{xd} - v_{xh}\,i_{xh})}_{\text{Term-2}} \\ &+ \frac{1}{N}\underbrace{(0.5\,V_{dc}\,i_{xh} - v_{xs}\,i_{xh} - v_{xh}\,i_{xd} - 0.5\,v_{xh}\,i_{xs})}_{\text{Term-3}} \end{aligned} \tag{8.45}$$

According to the principle of power balance between the input and output, the "term-1" in equation (8.45) become zero at steady state. The "term-3" mainly consists of high-frequency voltage and current components, and are naturally balanced in each high-frequency cycle. The "term-2" consists of both fundamental and high-frequency components and leads to a ripple in the submodule capacitor. Hence, the "term-2" should become zero at steady state, and is given by

$$0.25\,V_{dc}\,i_{xs} - v_{xs}\,i_{xd} - v_{xh}\,i_{xh} = 0 \tag{8.46}$$

The high-frequency voltage and current signals given in equation (8.43) are substituted in equation (8.46), and results in

$$0.5\,V_h\,I_h - \underbrace{0.5\,V_h\,I_h\cos(4\pi\,f_h t)}_{\text{Double high-frequency term}} = 0.25\,V_{dc}\,i_{xs} - v_{xs}\,i_{xd} \tag{8.47}$$

The double high-frequency term has a negligible impact on the capacitor voltage ripple, and it is neglected. The simplified expression is given by

$$V_h\,I_h \approx 0.5\,V_{dc}\,i_{xs} - 2\,v_{xs}\,i_{xd} \tag{8.48}$$

According to equation (8.48), the high-frequency power is constant for a given operating condition. The higher amplitude of high-frequency voltage (V_h) leads to a smaller magnitude of the high-frequency current (I_h), which results in the lower current stress on semiconductor devices. The magnitude of the high-frequency voltage (V_h) is selected, such that the maximum amplitude modulation index is limited to unity. The peak value of the high-frequency voltage is given by

$$V_h = 0.5\,(1 - m_a)\,V_{dc} \tag{8.49}$$

From equations (8.48) and (8.49), the peak value of the high-frequency current is given by

$$I_h = \frac{0.5\,V_{dc}\,i_{xs} - 2\,v_{xs}\,i_{xd}}{V_h} \tag{8.50}$$

According to equations (8.31) and (8.49), the amplitude modulation index is close to unity at high-speed operation (i.e., close to rated fundamental frequency). Hence, the magnitude of the high-frequency voltage (V_h) becomes very small, which results in the infinite magnitude of the high-frequency current (I_h). However, the conventional MMC can handle voltage ripple during high-speed operation and does not require such a huge magnitude of high-frequency current. To deal with this conflict, a current limiting factor (k_c) is introduced as

$$k_c = \frac{0.25\,(1 - m_a)^2}{0.65\,m_a\,(2m_a - 0.8) + 0.25} \tag{8.51}$$

The resultant high-frequency current with a limiting factor (k_c) is given by

$$I_h = \frac{k_c\,(0.5\,V_{dc}\,i_{xs} - 2\,v_{xs}\,i_{xd})}{V_h} \tag{8.52}$$

The variation of current limiting factor (k_c) with the motor stator frequency is shown in Figure 8.21. At high-speed (high-frequency) operation, the current limiting factor k_c is zero. According to equation (8.52), the magnitude of the high-frequency current becomes zero. On the other hand, k_c is of unity magnitude at low-speed (low-frequency) operation, and results in the original value of high-frequency current obtained from equation (8.50).

8.5.4 Minimization of Submodule Capacitor Voltage Ripple

The control block diagram of motor-side MMC for low- and high-speed operation is shown in Figure 8.22. The FOC controls the motor speed/torque and rotor flux and generates the modulation signals with the required magnitude and frequency (f_s). The SM voltage control maintains each SM capacitor voltage at their nominal value. The output of SM voltage control (Δv^*_{xyh}) is added to the modulation signals as shown in Figure 8.22. In addition, the conventional MMC-based motor drive requires high-frequency voltage and current controllers to handle the voltage ripple during low/zero-speed operation.

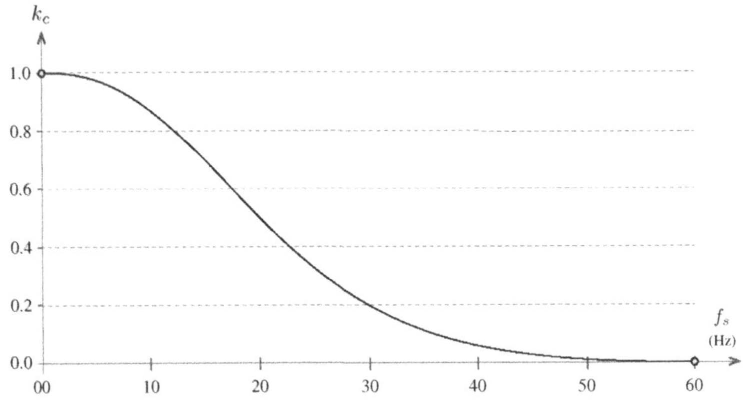

Figure 8.21 Variation of current limiting factor k_c with frequency.

The high-frequency voltage, also referred to as a common-mode voltage (CMV), and its synthesis is shown in Figure 8.23(a). The magnitude of a high-frequency voltage depends on the amplitude modulation index. The reference dq voltage components of FOC are used in the calculation of amplitude modulation index as

$$m_a = \frac{\sqrt(v_{ds}^{*2} + v_{qs}^{*2})}{0.5\, V_{dc}} \tag{8.53}$$

The peak value of reference high-frequency voltage is given by

$$V_h = 0.25\,(1 - m_a)\, V_{dc} \tag{8.54}$$

The magnitude of reference high-frequency voltage is multiplied with a unity sinusoidal signal of frequency f_h, and the resultant output is added to the fundamental modulation signals as shown in Figure 8.22. The reference high-frequency voltage is given by

$$v_{xh}^{*} = 0.25\,(1 - m_a)\, V_{dc} \sin(2\pi\, f_h t) \tag{8.55}$$

The control block diagram of high-frequency current control is shown in Figure 8.23(b), which consists of an outer voltage loop and an inner current control loop. The outer voltage loop minimizes the voltage difference between the upper and lower arms. The average voltage of upper and lower arms are given by

$$\begin{aligned} v_{Cu}^{x} &= \frac{v_{Cu1}^{x} + \cdots + v_{CuN}^{x}}{N} \\ v_{Cl}^{x} &= \frac{v_{Cl1}^{x} + \cdots + v_{ClN}^{x}}{N} \end{aligned} \tag{8.56}$$

The output of voltage control loop is multiplied with a unity sinusoidal high-frequency signal as shown in Figure 8.23(b). The resultant output is added to feed-forward current component (i_{xh}^{f}), and becomes a reference high-frequency current

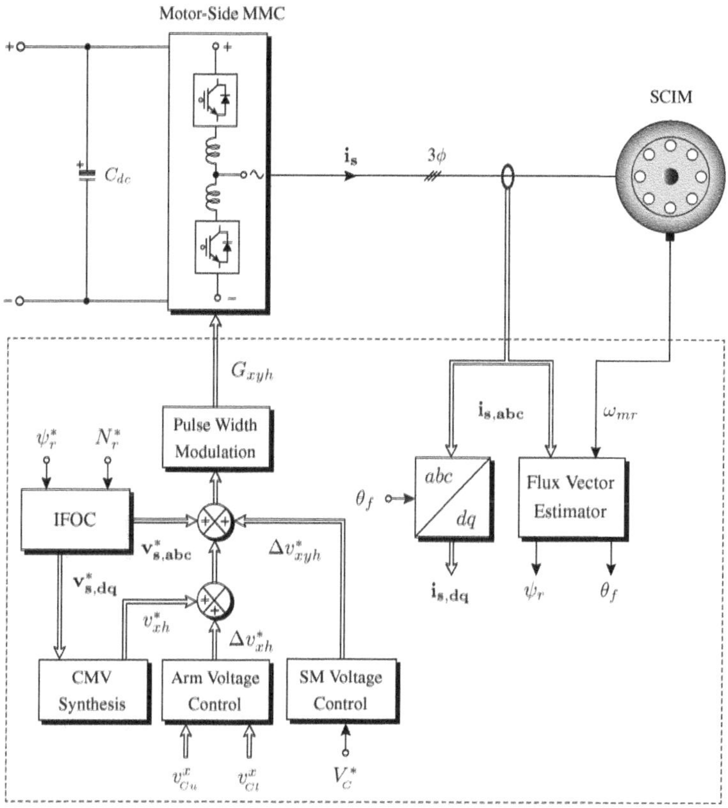

Figure 8.22 Control block diagram of motor-side MMC during low-frequency operation.

(i^*_{xh}). The feed-forward current component is given by

$$i^f_{xh} = \frac{v^*_{xs}\, i_{xs}}{V_{dc}} + \frac{k_c\,(0.5\, V_{dc}\, i_{xs} - 2\, v^*_{xs}\, i_{xd})\sin(2\pi\, f_h t)}{V_h} \tag{8.57}$$

where v^*_{xs} represents the reference stator voltage of phase-x.

The reference high-frequency current component is compared with the actual high-frequency current component. The actual high-frequency current component is estimated from the measured upper and lower arm currents. This high-frequency current is also equal to the circulating current flowing among the converter legs and is given by

$$i_{xh} = \frac{i_{xu} + i_{xl}}{2} \tag{8.58}$$

The difference between the reference and actual high-frequency current components are minimized using a PI-regulator, and its output becomes a reference voltage command Δv^*_{xh}. This voltage command is added to the modulation signals. The resultant upper and lower arm modulation signals are given by

$$\begin{aligned} v_{xu}^* &= \frac{V_{dc}}{2} - v_{xs}^* - v_{xh}^* - \Delta v_{xh}^* \\ v_{xl}^* &= \frac{V_{dc}}{2} + v_{xs}^* + v_{xh}^* - \Delta v_{xh}^* \end{aligned} \tag{8.59}$$

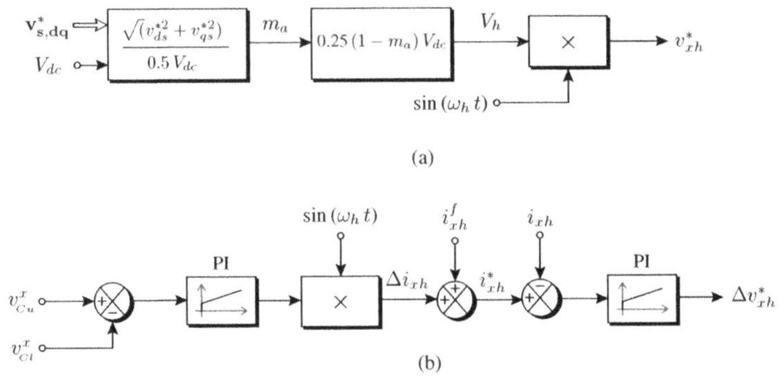

Figure 8.23 Control block diagrams: (a) CMV synthesis and (b) arm voltage control.

CASE STUDY 8.4 Performance Analysis of Motor-Side MMC

Objective: In this case study, the performance of a motor-side MMC during low- and high-speed operations is analyzed using MATLAB simulation.

Parameters: The system parameters of the motor-side MMC are as follows: rated active power $P_s = 1$ MW, line-to-line voltage $V_s = 4160$ V, fundamental frequency $f_s = 0$–60 Hz, rated DC-bus voltage $V_{dc} = 7$ kV, number of SMs per arm $N = 20$, SM capacitor $C = 8000$ μF, SM capacitor voltage $v_C = 350$ V, arm inductor $L = 1$ mH, and carrier frequency $f_c = 500$ Hz. The parameters of the squirrel cage induction motor are given in Table 8.1.

Analysis:
The motor-side MMC consists of 20 SMs per arm. Each submodule is designed with a capacitance of 8000 μF to maintain the 10% voltage ripple. The AC terminals of MMC are connected to a squirrel cage induction motor of 1250 HP and 4160 V rated capacity. The DC-bus voltage of MMC is selected to 7000 V such that the converter generates rated voltage at unity modulation index.

The motor drive is controlled with an indirect field-oriented control approach, and its performance is shown in Figure 8.24. The rotor flux is maintained at its rated value of 8.35 Wb. From $t = 1.45$ to $t = 1.5$ s, the motor speed kept at 0 RPM (standstill), then it is ramped to rated value of 1189 RPM as shown in Figure 8.24(a). The voltage components in the reference modulation signal are shown in Figure 8.24(b). During the zero-speed operation, the modulation signal mainly consists of high-frequency voltage (v_{ah}^*), and other voltage components (v_{as}^* and Δv_{ah}^*) have zero magnitudes. As the speed increases, the high-frequency voltage magnitude reduces, and it becomes zero at the rated speed of 1189 RPM. On the other hand, the

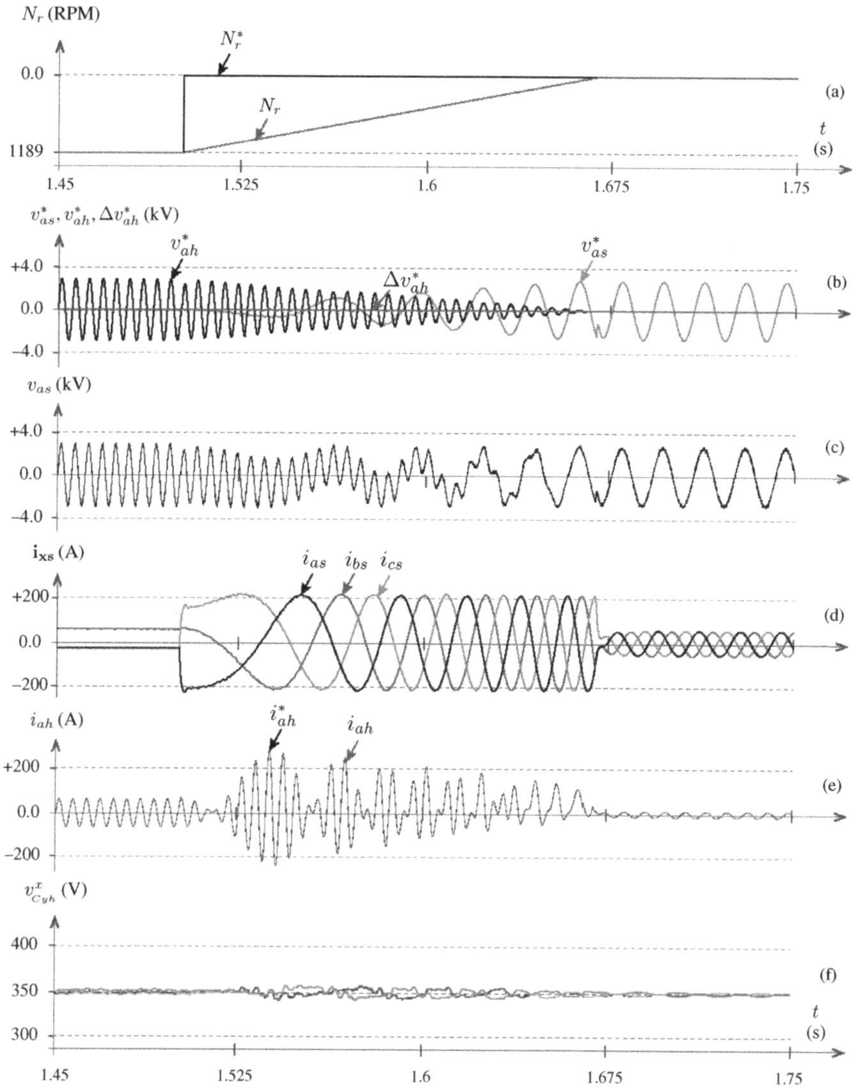

Figure 8.24 Performance of the motor-side MMC with high-frequency signal injection.

fundamental frequency component proportionally increases with the motor speed as shown in Figure 8.24(b).

The actual converter output voltage is shown in Figure 8.24(c), which is following the reference voltage command. The three-phase motor currents are shown in Figure 8.24(d). From $t = 1.45$ to $t = 1.5$ s, the currents are DC in nature corresponding to zero-speed operation. During the acceleration period (From $t = 1.5$ to $t = 1.675$ s), the motor draws a rated current of 212 A (peak). At rated speed, the motor draws a current magnitude of 66 A corresponding to the no-load condition as shown in Figure 8.24(d). The reference and actual high-frequency current waveforms are shown in Figure 8.24(e). The high-frequency current component becomes very small (approximately zero) at the rated speed operation. The submodule capacitors are maintained at their rated value of 350 V as shown in Figure 8.24(f). The high-frequency voltage and current signal help to achieve the submodule capacitor voltage balance between the upper and lower arms. In addition, the voltage ripple can be maintained within the 10% limit throughout the operation (zero to rated speed). ■

8.6 COMMON-MODE VOLTAGE ISSUES AND BLOCKING SCHEMES

In motor drive, the grid-side and motor-side MMC systems produce a common-mode voltage (CMV) due to the switching action of submodules in the converters. In addition, a high-frequency voltage is injected into the motor-side MMC to handle the voltage ripple during zero/low-speed operation. This external injected high-frequency voltage appears in the form of a CMV and increases its overall magnitude. If not mitigated, the CMV is superimposed on the motor phase voltages and causes premature failure of the motor winding insulation and bearings. Hence, the mitigation of common-mode voltage is necessary to improve the reliability of the motor drive. In this section, the conventional solutions are discussed to mitigate the CMV.

8.6.1 Definition of Common-Mode Voltage

The simplified MMC-based motor drive system configuration is shown in Figure 8.25, where v_{cmr} and v_{cmi} are the common-mode voltages of the grid-side and motor-side MMC, respectively. The inductor L_g filters out the switching harmonics produced by the grid-side MMC. On the other hand, the motor side does not require any filters due to the multilevel waveform produced by the MMC. During this analysis, it is assumed that the filters in the drive system do not have any common-mode impedance. The common-mode voltage generated by the grid-side MMC is measured between the mid-point of DC-bus (o) and the neutral point of three-phase AC grid (g), that is,

$$v_{og} = v_{cmr} \tag{8.60}$$

The common-mode voltage generated by the motor-side MMC is measured between the neutral point of three-phase motor (n) and the mid-point of DC-bus (o), that is,

$$v_{no} = v_{cmi} \tag{8.61}$$

To calculate the common-mode voltage of the motor-side MMC, the three-phase motor terminal voltages with respect to the mid-point of DC-bus (o) are considered. The three-phase voltages are given by

$$\begin{aligned} v_{ao} &= v_{an} + v_{no} \\ v_{bo} &= v_{bn} + v_{no} \\ v_{co} &= v_{cn} + v_{no} \end{aligned} \tag{8.62}$$

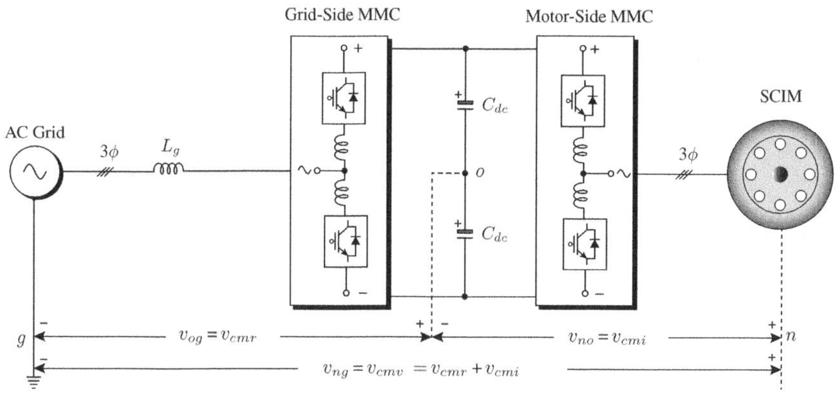

Figure 8.25 CMV issues in the MMC-based motor drive.

where v_{an}, v_{bn}, and v_{cn} are the phase voltages of the motor stator winding. Equation (8.63) is further simplified into

$$(v_{ao} + v_{bo} + v_{co}) = (v_{an} + v_{bn} + v_{cn}) + 3\, v_{no} \tag{8.63}$$

Considering the three-phase balanced system, where $v_{an} + v_{bn} + v_{cn} = 0$, the above equation can be simplified into

$$v_{no} = v_{cmi} = \frac{(v_{ao} + v_{bo} + v_{co})}{3} \tag{8.64}$$

According to the equivalent model shown in Figure 8.15(b), the converter output voltages in terms of the upper and lower arm voltages are given by

$$\begin{aligned} v_{ao} &= \frac{(v_{al} - v_{au})}{2} - 2\,L\,\frac{d\,i_{as}}{dt} \\ v_{bo} &= \frac{(v_{bl} - v_{bu})}{2} - 2\,L\,\frac{d\,i_{bs}}{dt} \\ v_{co} &= \frac{(v_{cl} - v_{cu})}{2} - 2\,L\,\frac{d\,i_{cs}}{dt} \end{aligned} \tag{8.65}$$

Considering the three-phase balanced system, where $i_{as} + i_{bs} + i_{cs} = 0$, the above equation can be simplified into

$$(v_{ao} + v_{bo} + v_{co}) = \frac{(v_{al} + v_{bl} + v_{cl}) - (v_{au} + v_{bu} + v_{cu})}{2} \tag{8.66}$$

From equations (8.64) and (8.66), the common-mode voltage of motor-side MMC is given by

$$v_{no} = v_{cmi} = \frac{(v_{al} + v_{bl} + v_{cl}) - (v_{au} + v_{bu} + v_{cu})}{6} \tag{8.67}$$

Similarly, the common-mode voltage of grid-side MMC can be calculated and is given by

$$v_{og} = v_{cmr} = \frac{(v_{Al} + v_{Bl} + v_{Cl}) - (v_{Au} + v_{Bu} + v_{Cu})}{6} \tag{8.68}$$

where A, B, and C are the three-phase terminals of grid-side MMC system; v_{Al}, v_{Bl}, and v_{Cl} are the three-phase lower-arm voltages of grid-side MMC; and v_{Au}, v_{Bu}, and v_{Cu} are the three-phase upper arm voltages of grid-side MMC.

The total common-mode voltage (v_{cmv}) generated by the motor drive is equal to the summation of common-mode voltage generated by the grid-side and motor-side MMC systems, that is

$$v_{cmv} = v_{ng} = v_{no} + v_{og} = v_{cmi} + v_{cmr} \tag{8.69}$$

8.6.2 Blocking of Common-Mode Voltage

The conventional solution to block the common-mode voltage is shown in Figure 8.26, where an isolation transformer is added between the AC grid and the grid-side MMC. The common-mode voltage of motor drive cannot be eliminated by adding the transformer; it is simply transferred from the motor to the transformer. With the grounding of motor stator winding neutral point, the common-mode voltage generated by the motor-side and grid-side MMCs are applied to the transformer neutral point as shown in Figure 8.26. Hence, the transformer windings should be designed with a higher voltage insulation to withstand the additional common-mode voltages.

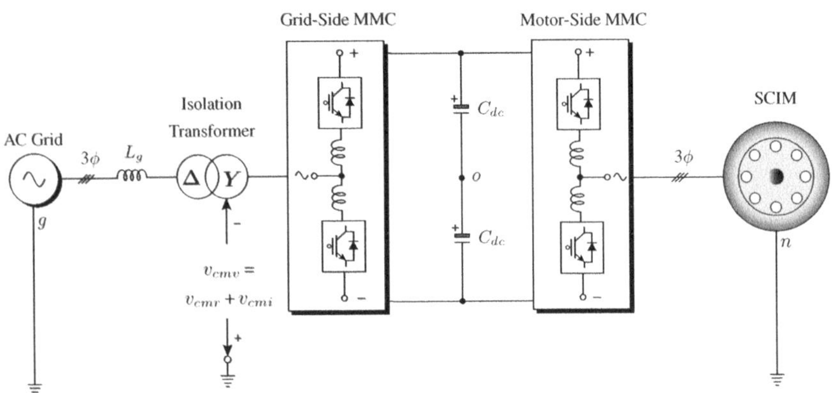

Figure 8.26 Conventional solution to block the CMV.

Normally, the neutral point of the motor stator winding is not accessible for grounding purpose. Hence, the most practical solution is to leave the neutral point of both transformer and motor ungrounded (floating). In such scenarios, the distribution of common-mode voltage between the motor and transformer depends on the parasitic capacitance of motor and transformer windings. The system configuration with winding parasitic capacitance is shown in Figure 8.27, where C_T and C_M represent the parasitic capacitance of transformer and motor, respectively. These parasitic capacitances provide a path for the current generated by the common-mode voltage (v_{cmv}).

According to the core and winding structure of the motor and transformer, the motor parasitic capacitance C_M is much larger than the transformer parasitic capacitance C_T. As a result, most of the common-mode voltage appear across the transformer C_T (i.e., between the transformer neutral to ground) as shown in Figure 8.27. The common-mode voltage across motor neutral to ground will be negligibly small [23]. For example, with a C_M of 60 nF and C_T of 0.5 nF in a motor drive [23], then the common-mode voltage across C_M is less than 1%, and rest of the common-mode voltage appear across C_T.

Other possible solutions are designing the motor winding insulation to withstand the additional common-mode voltages. This solution is suitable for applications in need of new motors, and not applicable to the retrofitting applications, where a large number of standard motors have been installed and running continuously in the field. Another possible solution is to use a large common-mode inductor as a filter to block the common-mode voltage and to limit the common-current. The common-mode filter causes LC resonant oscillations and requires a resistor to dampen these oscillations. The damping resistor leads to additional power losses in the motor drive.

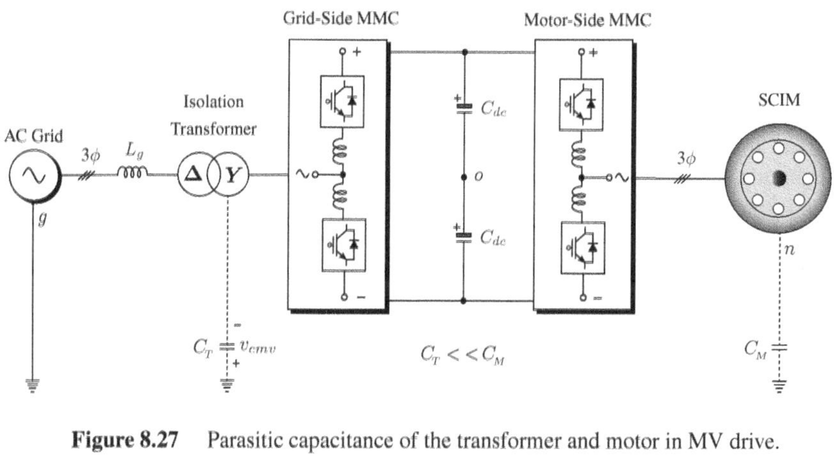

Figure 8.27 Parasitic capacitance of the transformer and motor in MV drive.

8.7 TRANSFORMER-LESS MMC-BASED MOTOR DRIVE

In the conventional solution, the common-mode voltage generated by the grid-side and motor-side MMC is blocked by the transformer and will not appear on the motor. This solution is highly effective and helps to minimize the line current harmonic distortion. However, the transformer increases the cost and physical size of the drive. The operating cost of the drive is also increased due to the transformer power losses.

Hence, the transformer should be eliminated to achieve a low-cost and reliable motor drive.

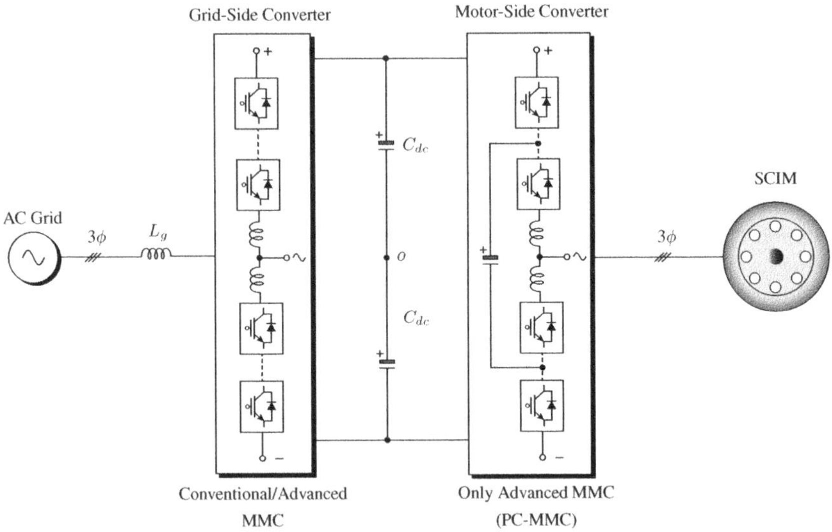

Figure 8.28 Transformer-less MMC-based motor drive.

In the case of the MMC-based motor drive, a high-frequency voltage is injected to minimize the voltage ripple at zero/low-speed operation. This high-frequency voltage appears in the converter common-mode voltage, which will increase its net magnitude. Considering the common-mode voltage and voltage ripple at zero/low-speed operation, the advanced modular multilevel converters (PC-MMC, AC-MMC, star-channel MMC, and delta-channel MMC) are presented in Chapters 5, 6, and 7. With the advanced MMC topologies, it is possible to realize the transformer-less MMC-based motor drive system as shown in Figure 8.28 in which, the grid-side converter can be realized by using conventional or advanced modular multilevel converters. On the other hand, the motor-side converter can be realized by using advanced modular multilevel converters only. With this configuration, the effect of externally injected high-frequency voltage on the motor common-mode voltage can be eliminated. Naturally, the MMCs generate common-mode voltage, which is negligibly small due to the larger number of submodules in the converter.

CASE STUDY 8.5 Performance Analysis of Transformer-Less Motor Drive

Objective: In this case study, the performance of a transformer-less motor drive system is analysed using MATLAB simulation.

Parameters: The system parameters of the motor-side MMC are as follows: rated power $P_s = 1$ MW, line-to-line voltage $V_s = 4160$ V, fundamental frequency $f_s = 0$–60 Hz, rated DC-bus voltage $V_{dc} = 7$ kV, number of SMs per arm $N = 20$, SM capacitor $C = 8000$ μF, SM capacitor voltage $v_C = 350$ V, cross-connected capacitor $C_x = 400\mu$F, high-frequency $f_h = 180$ Hz, arm inductor $L = 1$ mH, and carrier frequency $f_c = 500$ Hz. The parameters of the squirrel cage induction motor are given in Table 8.1.

Analysis:

The transformer-less motor drive consists of conventional MMC on the grid-side and the advanced MMC (PC-MMC) on the motor-side. The grid-side and motor-side MMCs are designed with 20 SMs per arm with a rated voltage of 350 V per submodule. In addition, the motor-side MMC consists of a cross-connected capacitor between the upper and lower arms to handle the voltage ripple during zero/low-speed operation. The grid-side MMC is controlled with the VOC approach, and its performance is shown in Figure 8.29.

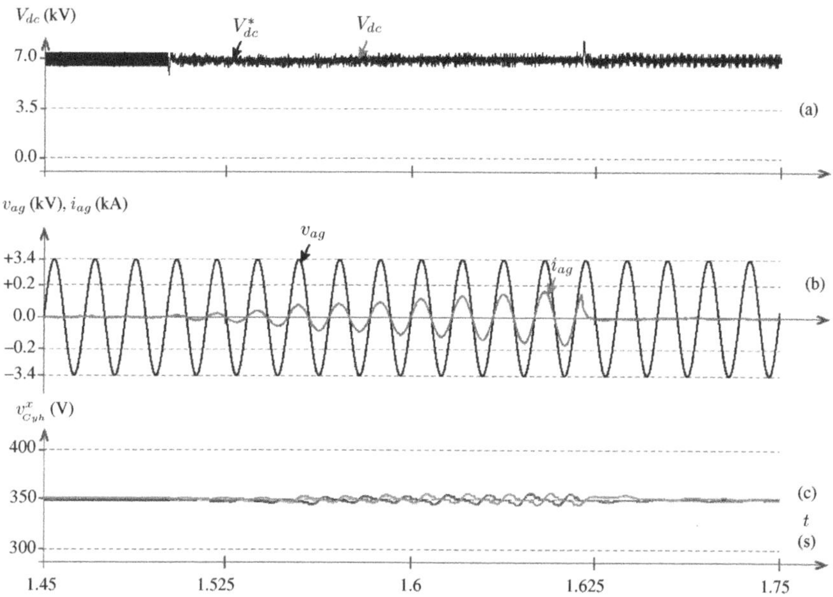

Figure 8.29 Performance of the grid-side MMC in a transformer-less MV drive.

The VOC is designed to maintain the total DC-bus voltage of 7 kV as shown in Figure 8.29(a). The grid-side converter draws zero reactive power (unity power factor) as shown in Figure 8.29(b), where the grid voltage and current signals are exactly in-phase. The total-DC-bus voltage is equally distributed between the submodules in each arm. Each submodule capacitor voltage is maintained at their nominal value of 350 V as shown in Figure 8.29(c). The grid-side MMC operates at rated frequency. Hence, the capacitor voltage ripple is directly proportional to the current magnitude. From $t = 1.45$ to $t = 1.5$ s, the grid current is zero, and the submodule voltages are perfectly DC in nature. From $t = 1.5$ to $t = 1.625$ s, the grid current magnitude is close to rated value, corresponding to the rated power on the motor side. Hence, the voltage ripple increased, which is less than the 10% design limits. From $t = 1.625$ s onwards, the grid current becomes zero, and there is no ripple in the submodule voltages as shown in Figure 8.29(c).

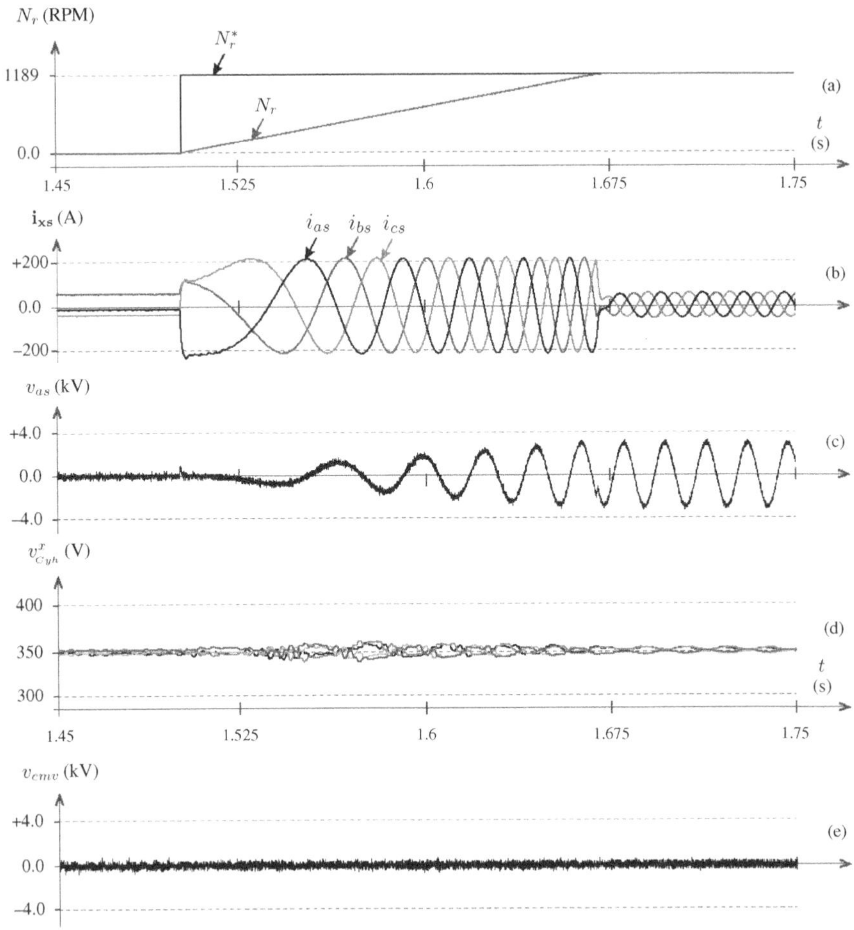

Figure 8.30 Performance of the motor-side MMC in a transformer-less drive.

The PC-MMC is used in the motor side to handle the voltage ripple during zero/low-speed operation of the medium-voltage drive. The motor drive is controlled with a FOC scheme, and its performance is shown in Figure 8.30. The rotor flux is maintained at the rated value of 8.35 Wb. The high-frequency signals are injected at a frequency of 180 Hz. From $t = 1.45$ to $t = 1.5$ s, the reference speed is set to zero RPM (zero frequency), and the controller perfectly regulates the motor speed at the given reference speed as shown in Figure 8.30(a). The three-phase stator currents are DC in nature as shown in Figure 8.30(b). The converter output voltage is shown in Figure 8.30(c), which has zero magnitude corresponding to zero-speed operation. The submodule capacitor voltages are perfectly maintained at their nominal value of 350 V as shown in Figure 8.30(d). The voltage ripples are less than the 10% design limits even under zero-speed operation. The motor common-mode voltage consists of switching ripples as shown in Figure 8.30(e).

At $t = 1.5$ s, the reference speed is changed from 0 to 1189 RPM. However, the motor accelerates slowly due to the large inertia of the motor. At $t = 1.625$ s, the motor reaches the rated speed as shown in Figure 8.30(a). During the acceleration period, the motor draws rated current of 212 A (peak) as shown in Figure 8.30(b). Once the motor reaches the rated speed, then the motor draws a current equal to the no-load condition. The magnitude of converter output voltage (both magnitude and frequency) increases linearly with the operating speed as shown in Figure 8.30(c). At rated speed, the converter generates rated voltage with a frequency of 60 Hz. The SM voltages in the upper and lower arms are balanced as shown in Figure 8.30(d). During the zero to rated speed operation, the converter generates zero common-mode voltage, and it consists of switching ripples as shown in Figure 8.30(e). Hence, the motor drive does not require a transformer to block the common-mode voltage. ■

8.8 SUMMARY

The modular multilevel converter is a promising topology for medium-voltage drives. In this chapter, the passive and active rectifier fed MMC-based motor drives are presented. With the active rectifier, the four-quadrant operation of the motor drive is possible. Also, the grid power factor (reactive power) can be adjusted. In addition, the voltage-oriented control (VOC) scheme for grid-side MMC and field-oriented control (FOC) scheme for motor-side MMC are discussed. The performance of VOC and FOC schemes are analyzed through MATLAB simulations.

Further, the submodule voltage ripple in the motor-side converter is analyzed during low/zero-speed operation. The voltage ripple varies inversely with the output frequency and exceeds the practical limit of 10% for an output frequency less than 25 Hz. To minimize the voltage ripple, a control scheme with high-frequency signal injection is presented. The variation of high-frequency signal magnitude with the output frequency is analyzed. At rated frequency, the high-frequency signal magnitude is close to zero, and vice versa. The injected high-frequency signal appears in the motor common-mode voltage and increases its net magnitude. The impact of common-mode voltage on the motor drive is discussed. To block the common-mode voltage (CMV), an isolation transformer is added between the AC grid and grid-side MMC. This solution protects the motor from the CMV but increases the cost and physical size of the drive. Alternatively, the transformer-less MMC-based motor drive is presented. In the transformer-less drive, the advanced modular multilevel converters are employed on motor-side converters, whereas the grid-side converters can be realized by using the conventional or advanced modular multilevel converters.

REFERENCES

1. H. Abu-Rub, J. Holtz, J. Rodriguez, and G. Baoming, "Medium-voltage multilevel converters: State of the art, challenges, and requirements in industrial applications," *IEEE Trans. Ind. Electron.*, vol. 57, no. 8, pp. 2581–2596, Aug 2010.
2. S. Kouro, J. Rodriguez, B. Wu, S. Bernet, and M. Perez, "Powering the future of industry: High-power adjustable speed drive topologies," *IEEE Ind. Appl. Mag.*, vol. 18, no. 4, pp. 26–39, Jul 2012.

3. J. Rodriguez, S. Bernet, B. Wu, J. Pontt, and S. Kouro, "Multilevel voltage-source-converter topologies for industrial medium-voltage drives," *IEEE Trans. Ind. Electron.*, vol. 54, no. 6, pp. 2930–2945, Dec 2007.

4. S. Kouro, M. Malinowski, K. Gopakumar, J. Pou, L. G. Franquelo, B. Wu, J. Rodriguez, M. A. Perez, and J. I. Leon, "Recent advances and industrial applications of multilevel converters," *IEEE Trans. Ind. Electron.*, vol. 57, no. 8, pp. 2553–2580, Aug 2010.

5. J. Rodriguez, S. Bernet, P. Steimer, and I. Lizama, "A survey on neutral-point-clamped inverters," *IEEE Trans. Ind. Electron.*, vol. 57, no. 7, pp. 2219–2230, Jul 2010.

6. M. Malinowski, K. Gopakumar, J. Rodriguez, and M. Pèrez, "A survey on cascaded multilevel inverters," *IEEE Trans. Ind. Electron.*, vol. 57, no. 7, pp. 2197–2206, Jul 2010.

7. J. Rodriguez, P. W. Hammond, J. Pontt, R. Musalem, P. Lezana, and M. J. Escobar, "Operation of a medium-voltage drive under faulty conditions," *IEEE Trans. Ind. Electron.*, vol. 52, no. 4, pp. 1080–1085, Aug 2005.

8. I. Etxeberria-Otadui, A. L. de Heredia, J. San-Sebastian, H. Gaztanaga, U. Viscarret, and M. Caballero, "Analysis of a h-npc topology for an ac traction front-end converter," in *2008 13th International Power Electronics and Motion Control Conference*, Sep 2008, pp. 1555–1561.

9. Benshaw, "M2l 3000 series medium voltage motor drive." [Online]. Available: http://www.benshaw.com/uploadedFiles/Literature/Benshaw_M2L_MVFD_2.3-6.6kV.pdf

10. Siemens, "Sinamics sm120." [Online]. Available: http://www.industry.siemens.com/drives/global/en/converter/mv-drives/sinamics-sm120-cm/

11. B. Wu, Y. Lang, N. Zargari, and S. Kouro, *Power Conversion and Control of Wind Energy Systems*, 1st ed., ser. IEEE Press Series on Power Engineering. Hoboken, NJ: John Wiley & Sons, Inc., Jul 2011.

12. A. Yazdani and R. Iravani, *Voltage-Sourced Converters in Power Systems*. Hoboken, NJ: John Wiley & Sons, Inc., Jan 2010.

13. B. Wu, *High-Power Converters and AC Drives*. New york: Wiley-IEEE Press, Mar 2006.

14. P. Vas, *Sensorless Vector and Direct Torque Control*. Oxford Science Publications, 1998.

15. D. Novotny and T. Lipo, *Vector Control and Dynamics of AC-Drivers*. Oxford Science Publications, 1996.

16. J. Kolb, F. Kammerer, M. Gommeringer, and M. Braun, "Cascaded control system of the modular multilevel converter for feeding variable-speed drives," *IEEE Trans. Power Electron.*, vol. 30, no. 1, pp. 349–357, Jan 2015.

17. J.-J. Jung, H.-J. Lee, and S.-K. Sul, "Control strategy for improved dynamic performance of variable-speed drives with modular multilevel converter," *IEEE Trans. Emerg. Sel. Topics Power Electron.*, vol. 3, no. 2, pp. 371–380, Jun 2015.

18. Y. Okazaki, H. Matsui, M. Hagiwara, and H. Akagi, "Design considerations on the dc capacitor of each chopper cell in a modular multilevel cascade inverters (mmci-dscc) for medium-voltage motor drives," in *Energy Conversion Congress and Exposition (ECCE), 2014 IEEE*, Sep 2014, pp. 3393–3400.

19. J. Kolb, F. Kammerer, and M. Braun, "Dimensioning and design of a modular multilevel converter for drive applications," in *Power Electronics and Motion Control Conference (EPE/PEMC), 2012 15th International*, Sep 2012, pp. LS1a–1.1–1–LS1a–1.1–8.

20. K. Wang, Y. Li, Z. Zheng, and L. Xu, "Voltage balancing and fluctuation-suppression methods of floating capacitors in a new modular multilevel converter," *IEEE Trans. Ind. Electron.*, vol. 60, no. 5, pp. 1943–1954, May 2013.

21. A. Korn, M. Winkelnkemper, and P. Steimer, "Low output frequency operation of the modular multi-level converter," in *Energy Conversion Congress and Exposition (ECCE), 2010 IEEE*, Sep 2010, pp. 3993–3997.

22. M. Hagiwara, I. Hasegawa, and H. Akagi, "Start-up and low-speed operation of an electric motor driven by a modular multilevel cascade inverter," *IEEE Trans. Ind. Appl.*, vol. 49, no. 4, pp. 1556–1565, Jul 2013.

23. B. Horvath, "How isolation transformers in mv drives protect motor insulation," *TM GE Automation Systems, Roanoke, VA*, 2004.

CHAPTER 9

ROLE OF MODULAR MULTILEVEL CONVERTERS IN THE POWER SYSTEM

9.1 INTRODUCTION

The modular multilevel converter (MMC) can be easily extended to any operating voltage level by increasing the submodules in each arm [1]. This feature is also referred to as a voltage scalability, and attractive for the high-voltage ($\geq$ 200 kV) applications as shown in Figure 9.1. One of the applications is high-voltage direct current (HVDC) transmission systems, in which the MMC is employed to transfer the power between two synchronous/asynchronous AC grids [2–4]. It is also possible to form a DC grid by interconnecting more than two MMCs at a common DC point. The resultant system is known as a multi-terminal HVDC system [5]. The power quality improvement is another important application of MMC, where the reactive power control, voltage regulation, harmonic compensation, and compensation of unbalanced grid currents are the common control objectives of a power quality conditioner [6, 7].

In this chapter, the role of modular multilevel converters in the power transmission systems and power quality improvement are discussed. For the long distance power transmission, HVDC systems are the preferred choice over a high-voltage AC (HVAC) transmission system. In Section 9.2, the two-terminal and multi-terminal

Modular Multilevel Converters: Analysis, Control, and Applications, First Edition.
Sixing Du, Apparao Dekka, Bin Wu, and Navid Zargari.

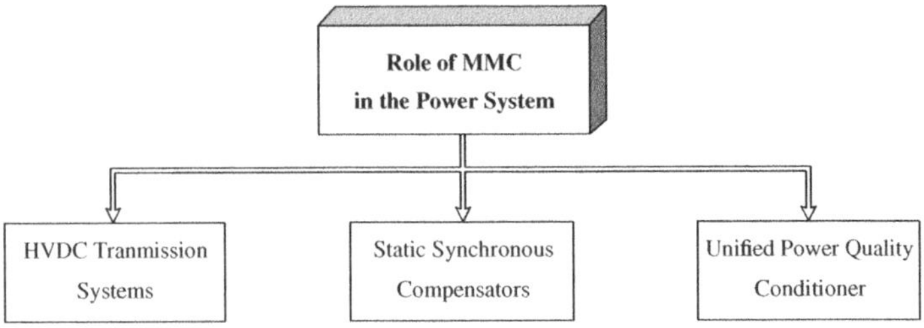

Figure 9.1 Applications of MMC in the power system.

MMC-based HVDC transmission systems are discussed. The DC-side fault is one of the major issues in the MMC-based HVDC system. The DC-side fault protection schemes including the submodule based approach and HVDC circuit breakers are presented in Section 9.2. The MMC converter stations are controlled to regulate the active and reactive power flow through the transmission system. The control schemes of two-terminal and multi-terminal systems are discussed in Sections 9.3 and 9.4, respectively.

In the case of power quality improvement, the MMCs can be used as a series and shunt compensator to compensate the reactive power and grid unbalance. The static synchronous compensator (STATCOM) and unified power quality conditioner (UPQC) are the most popular power quality improvement devices. The operation and control of STATCOM are presented in Section 9.5. The case studies are presented to analyze the performance of STATCOM under balanced and unbalanced grid conditions. The fundamentals of UPQC are presented in Section 9.6, and the summary is given in Section 9.7.

9.2 MMC-BASED HVDC TRANSMISSION SYSTEMS

Currently, the MMC (MMC) is the only solution available for the commercial VSC-HVDC systems. The MMC provides a substantial increase in the output voltage levels and output switching frequency, enabling a reduction in the filter size and power losses [1]. Even though the overall converter switching frequency is high, each submodule can be switched at a very low frequency of few hundred Hertz (50–150 Hz). Other prominent features are modularity and scalability and handling a higher DC-voltage using a low-cost and low-voltage semiconductor technology [2]. With the advent features of MMC, it is considered as a suitable candidate to form the multi-terminal HVDC systems as well. These systems are used for transmitting power from offshore wind energy systems to the onshore grids [8].

There are several MMC-HVDC configurations such as monopolar and bipolar systems in the symmetrical and asymmetrical layouts are available to realize the two-terminal and multi-terminal HVDC systems [9, 10]. Unlike monopolar configuration, the bipolar configuration has two converters at both sending and receiving ends. This configuration can achieve a higher power rating and is suitable for the fault-tolerant operation. However, the system cost is quite high. Due to the cost consideration, the symmetrical monopolar MMC-HVDC configuration is widely used in the commercial HVDC projects, including Trans Bay Cable project in the USA [11, 12], three-terminal Nanao project [13], and Zhoushan project in the China [14]. In this section, the operation and control of symmetrical two-terminal and multi-terminal HVDC systems are presented. Also, the grounding and DC-side fault issues are discussed in the following subsections.

9.2.1 Two-Terminal System

The two-terminal MMC-based HVDC system is also referred to as a symmetrical monopolar MMC-based HVDC system, and its configuration is shown in Figure 9.2. The symmetrical monopolar configuration has one converter at the sending-end and one converter at the receiving-end station (Note: the sending and receiving-end terminology is only for the reference purpose) [15]. Due to the low-cost and high-efficiency, the MMC at the sending and receiving end stations are designed with half-bridge submodules. Also, the HVDC system consists of an isolation transformer, star-point reactor with impedance ground, and DC cables. Typically, the isolation transformer is designed with a star winding connection on the AC grid while the delta winding connection is employed on the MMC AC-side. Hence, the zero sequence currents are blocked by entering into the system during the AC-side faults. Also, the transformer isolates the AC grid and the converter station, and it is used to match the AC- and DC-side voltages [15].

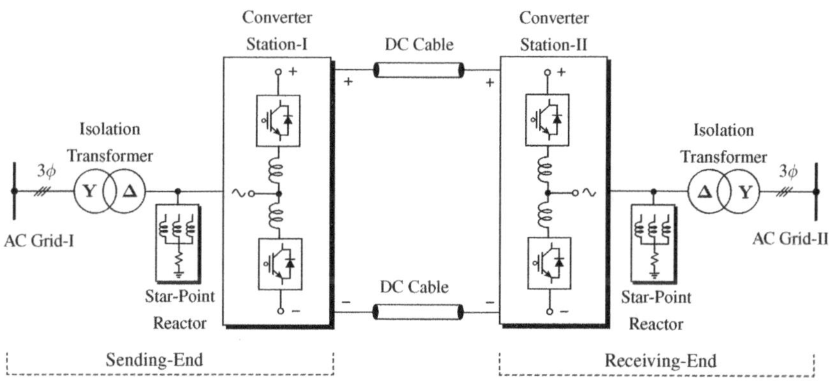

Figure 9.2 Two-terminal MMC-based HVDC system.

The monopolar system can produce a positive pole ($+0.5\,V_{dc}$) and a negative pole ($-0.5\,V_{dc}$) with respect to the ground. Typically, the grounding through a neutral point on the transformer $Y - \Delta$ winding is employed and requires less grounding devices. However, the transformer with higher technical specifications is required to handle the large fault currents and DC offset voltage during the AC-side faults [16]. On the other hand, the star-point reactor with resistive grounding is used in the MMC-HVDC systems. The star-point reactor reduce the rising speed of the fault current, and the grounding resistor limits the steady-state fault current magnitude. However, the star-point reactor absorbs a certain amount of lagging reactive power during the normal operation [16].

The DC-side of sending and receiving-end converters are connected through several hundreds of kilometer long underground cables or overhead transmission lines. In the simulation and experimental studies, the cables are modeled as an inductor with an equivalent resistor. In the offshore wind energy systems, the underground cables, also referred to as sub-sea cables, are used to interconnect the DC side of sending and receiving-end converters [17, 18]. The overhead transmission lines are used in transmitting the power on land. However, the overhead transmission lines require a special design to handle the pole-to-ground faults on the DC side [19].

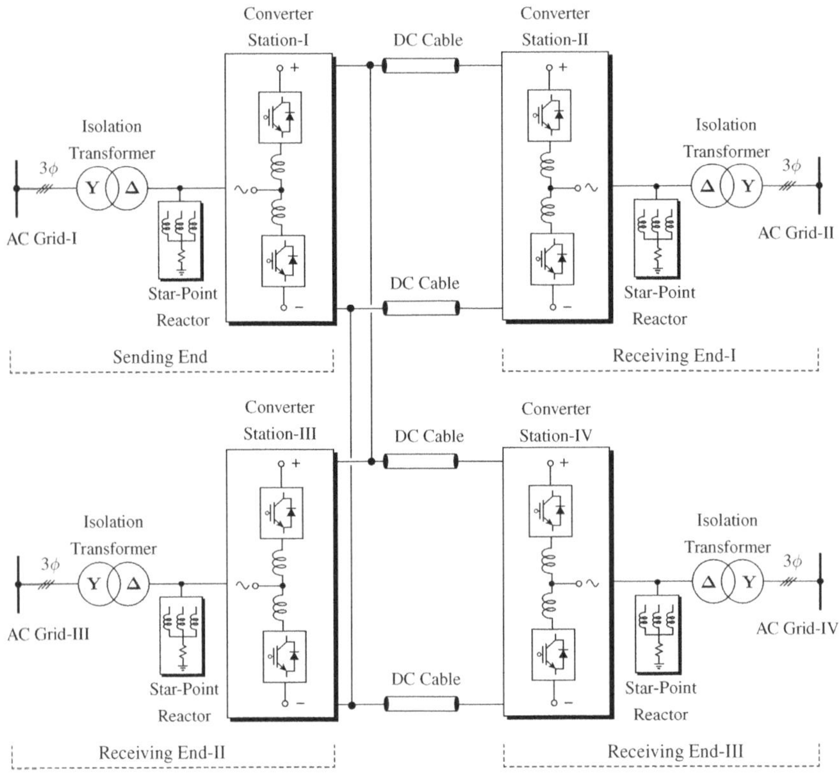

Figure 9.3 Multi-terminal MMC-based HVDC system.

9.2.2 Multi-Terminal System

The multi-terminal HVDC systems are characterized by having more than two converter stations somehow connected on the DC side of the transmission system. To form a multi-terminal network, the converter stations can be connected either in series or in parallel [20]. When connected in series, all the converters share the same DC current, while for the parallel connection, the converters share the same DC transmission voltage [21]. During permanent faults, the complete operation of series connected multi-terminal system is interrupted. Therefore, the parallel connected multi-terminal systems are a preferred choice over the series connected multi-terminal systems to form a large commercial network [22].

The multi-terminal systems can be formed by using both line-commutated current source converter (CSC) and voltage source converter (VSC) based HVDC technology. Among them, the VSC-HVDC technology becomes popular and capable of controlling both active and reactive powers. Currently, the modular multilevel converter based VSC-HVDC technology is widely used to form the multi-terminal HVDC systems. The configuration of four-terminal MMC-based HVDC system is shown in Figure 9.3. Each terminal consists of a half-bridge based MMC converter station, star-point reactor, and an isolation transformer. The function of each equipment is similar to the two-terminal system, and their details are discussed in Section 9.2.1. Some of the commercial examples of the multi-terminal system are three-terminal Nanao project [13] and Zhoushan project in China [14]. The multi-terminal systems are highly vulnerable to the DC-side short-circuit faults. The details of DC-side fault protection for MMC-HVDC systems are discussed in the following subsection.

9.2.3 DC-Side Short-Circuit Fault Protection

The DC-side short-circuit fault is one of the major issues in the half-bridge based MMC-HVDC system. The common DC-side faults are pole-to-pole (PTP) and pole-to-ground (PTG) as shown in Figure 9.4. In the event of a PTP fault, a huge fault current flows through the antiparallel diodes of HB-SM in the MMC. The fault current flows from the AC grid to the fault point on the DC side, even if the converter station is blocked immediately [16, 23]. The low impedance short-circuit path leads to a steep rise in the fault current which may cause serious damage to the converter stations, and may result in the complete shutdown of the HVDC network [16, 23]. On the other hand, the PTG fault will cause overvoltage at the AC grid and the DC side of the MMC-HVDC system. This will damage the equipment insulation and accelerate the corrosion of grounding point [23]. The impact of PTP and PTG faults on the MMC-HVDC network vary with the grounding scheme and the type of system configuration (monopolar and bipolar) [15, 16, 24].

There are two possible solutions to handle the DC-side short-circuit faults. One solution is to embed the fault-blocking capability into the converter, which is discussed in this section. Another solution is to use the HVDC circuit breakers to protect the system from the DC-side faults, and it is discussed in Section 9.2.4. In the former solution, the submodule with capable of generating the reverse (negative) voltage should be used in the MMC. Due to the cost and efficiency consideration, the half-bridge submodule (HB-SM) is widely used in various commercial MMC-based HVDC systems. The circuit configuration of HB-SM is shown in Figure 9.5(a). It

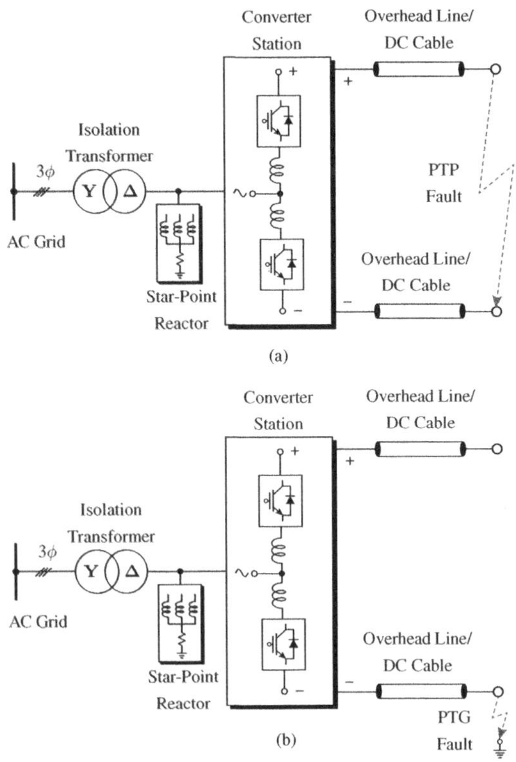

Figure 9.4 DC-side faults in the MMC-based HVDC system: (a) pole-to-pole, and (b) pole-to-ground.

is composed of two IGBT devices, two antiparallel diodes, and a capacitor. When a short-circuit fault occurs at the DC side, the fault current will flow through the antiparallel diodes, and the submodule output voltage will be zero [25].

Another possibility is to replace the HB-SM in the MMC with either full-bridge (FB-SM) or double-clamp (CD-SM) submodules [26, 27]. The circuit configuration of FB-SM and CD-SM is shown in Figure 9.5(b) and (c), respectively. When a short-circuit fault occurs at the DC side, all the IGBT devices will be blocked and the fault current will flow through the antiparallel diodes. These SMs generate a reverse voltage at the output, and the fault current will be suppressed to zero in a very short time [28–30]. This solution embeds the DC fault-blocking capability into the converter.

Under same operating conditions, the FB-SM based MMC requires twice the number of IGBT devices and the power losses increase by about 100% compared to HB-SM based MMC [31]. On the other hand, the CD-SM based MMC requires 1.25 times the amount of devices used in HB-SM based MMC and the power losses increase by about 35% [31]. Compared to the HB-SM based MMC, the CD-SM based MMC requires a higher number of devices. Hence, the CD-SM based MMC has higher power losses, which is far less than the power losses in the FB-SM based MMC. Several new submodule configurations such as asymmetrical unipolar full-bridge submodule, lattice submodule, hybrid submodules are studied for the MMC-HVDC systems [32–34]. These submodules are under research and development phase only.

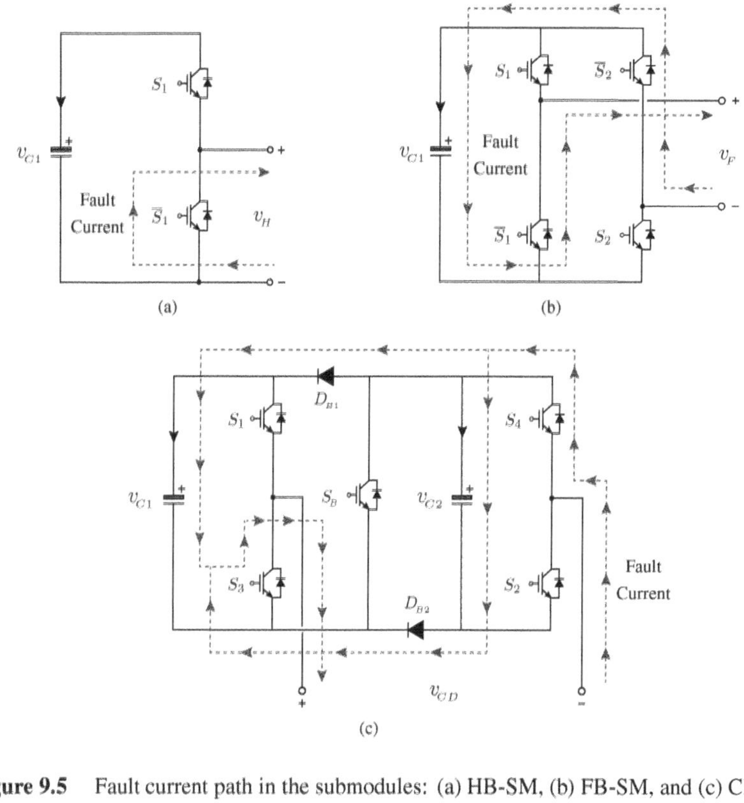

Figure 9.5 Fault current path in the submodules: (a) HB-SM, (b) FB-SM, and (c) CD-SM.

9.2.4 HVDC Circuit Breakers

Another solution is to use a HVDC circuit breaker to protect the system from the DC-side faults. Unlike the AC fault current, the DC fault current does not have the natural zero crossing, and it is hard to extinguish the arc. The conventional mechanical HVDC circuit breakers can interrupt the DC fault current but they are slow in action to fulfill the protection requirements of the HVDC grid [25, 35, 36]. In order to achieve fast operation, semiconductor switches are used in the HVDC breakers. These circuit breakers have higher transfer losses [25, 35, 36]. Currently, hybrid circuit breakers are used in the HVDC systems to interrupt the fault current with a low cost. These circuit breakers have the advantages of both conventional and semiconductor based circuit breakers.

Figure 9.6 shows the basic structure of a hybrid HVDC circuit breaker, which is developed by ABB [37]. The hybrid HVDC breaker consists of a main breaker (MB) formed by semiconductor switches and semiconductor based load commutation switch (LCS) in series with an ultra-fast disconnector (UFD). During the normal operation, the current will flow through the LCS and the UFD. When a DC-side fault is detected, then the LCS will open immediately. The fault current is transferred to the main breaker (MB), and the UFD is open within 2 ms and isolates the LCS from the faulted line. With the UFD in open position, the MB will break the fault current. Finally, the residual DC current breaker will open and isolate the faulted line from the HVDC grid [25, 35–37]. ABB has tested the hybrid circuit breaker with a breaking current of 9 kA, and a typical fault clearance time is around 5 ms [37].

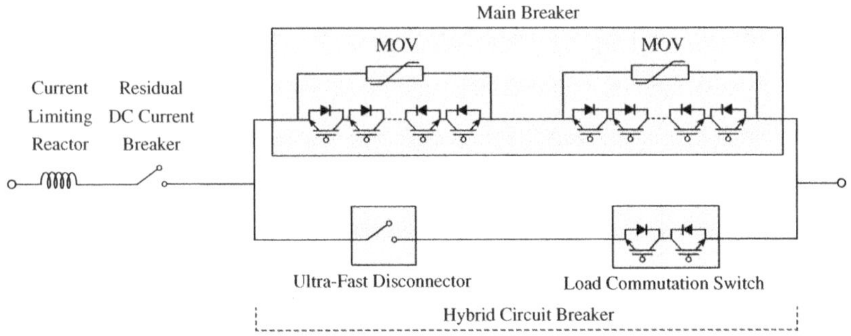

Figure 9.6 Structure of a hybrid HVDC circuit breaker.

9.3 CONTROL OF TWO-TERMINAL MMC-BASED HVDC SYSTEM

The generalized control block diagram of a two-terminal MMC-based HVDC system is shown in Figure 9.7 in which, one terminal is referred to as a sending-end, and the other terminal is referred to as a receiving-end. The sending-end converter controls the DC-bus voltage and supports its reactive power. This converter can be operated as a rectifier, and its control scheme is referred to as a "voltage mode control" [38]. On the other hand, the receiving-end converter controls the active and reactive power. This converter can be operated as an inverter, and its control scheme is referred to as a "power mode control" [38]. In addition, the sending-end and receiving-end converters require a submodule (SM) voltage control to maintain each submodule capacitor voltage at their nominal value. The details of submodule voltage control are discussed in the early chapters of the book. The output of SM voltage control is added to the voltage or power mode controller output (main modulation signals). The resultant modulation signals are used with the pulse width modulation scheme to generate the SM gating signals.

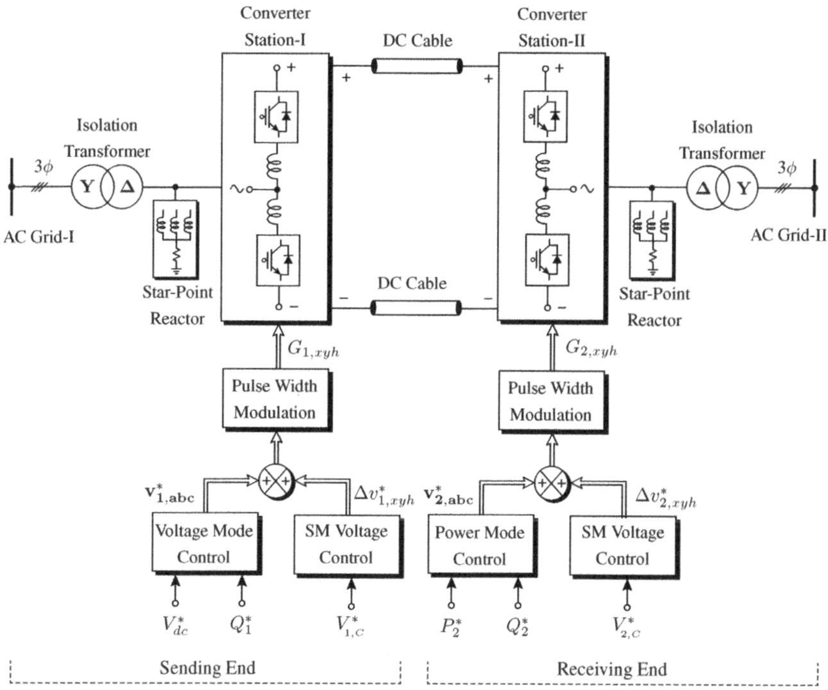

Figure 9.7 General control block diagram of the two-terminal MMC-HVDC system.

9.3.1 Sending-End Converter Control

The MMC-based HVDC system can control the active and reactive power through the transmission system in an independent way. Usually, the direct control or the vector control schemes are used to control the active and reactive powers in the HVDC system [20, 39]. In the direct control, the active and reactive powers are regulated using the AC voltage magnitude and its angle [40]. However, this control scheme is not practical since it does not have the current limit capability and its bandwidth is limited by the AC resonance frequency [40].

The vector control is also referred to as a synchronous-dq frame control approach, and is widely used in the control of MMC-based HVDC system [41]. In the synchronous-dq frame, the dq-axis is rotating at the grid frequency, and the grid voltage vector ($\mathbf{v_1}$) is aligned with the d-axis. Therefore, the q-axis voltage component (v_{q1}) becomes zero, and the d-axis voltage component is equal to the grid voltage vector ($v_{d1} = \mathbf{v_1}$) [42]. The control block diagram of the sending-end converter is shown in Figure 9.8. The control scheme consists of an outer loop to regulate the DC-bus voltage (V_{dc}) and reactive power (Q_1). The DC-bus voltage controller minimizes the error between the reference (V^*_{dc}) and actual DC-bus voltage (V_{dc}). The controller output is added to a feed-forward current component (i^f_{xd}), and results in a reference d-axis current component (i_{d1}^*). The feed-forward current improves the dynamic performance of the DC-bus current controller, and its magnitude is given based on the power balance between the AC and DC side of the converter as

$$i_{xd}^f = \frac{V_{dc}\,(i_{au} + i_{bu} + i_{cu})}{1.5\,v_{d1}} \tag{9.1}$$

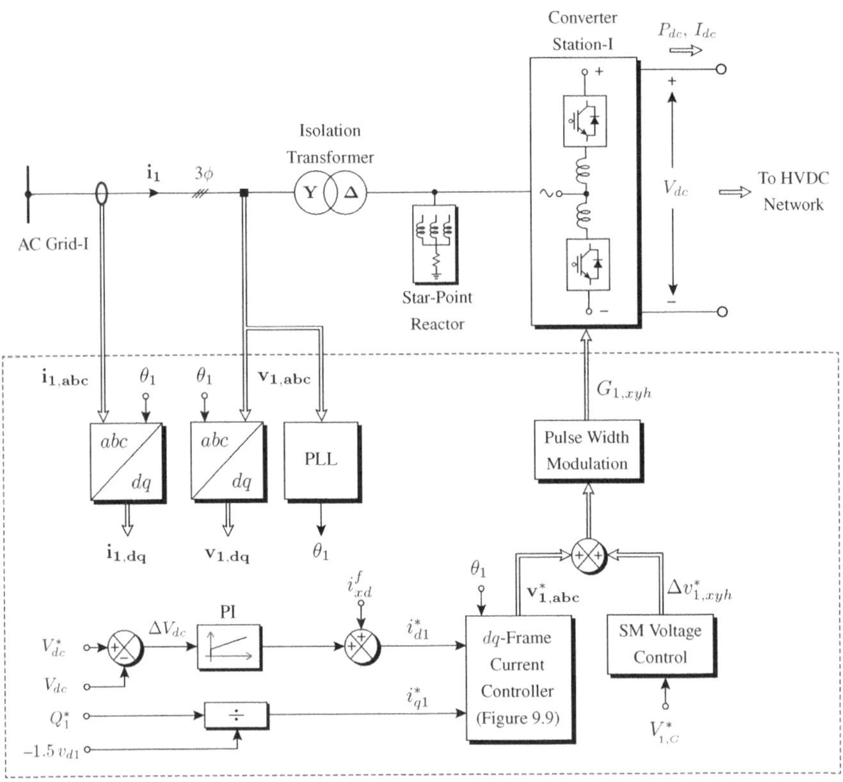

Figure 9.8 Control scheme of the sending-end converter.

In the synchronous-dq frame, the reactive power is expressed as

$$Q_1 = 1.5\,(v_{q1}\,i_{d1} - v_{d1}\,i_{q1}) \tag{9.2}$$

where v_{d1} and v_{q1} are the grid-I d- and q-axis voltage components, and i_{d1} and i_{q1} are the grid-I d- and q-axis current components.

By considering the $v_{q1} = 0$, then equation (9.2) is simplified as

$$Q_1 = -1.5\,v_{d1}\,i_{q1} \tag{9.3}$$

Equation (9.3) shows that the reactive power can be regulated by controlling the q-axis current component, under the assumption that the grid voltage is stiff (v_{d1}). The reference q-axis current component is given by

$$i_{q1}^{*} = \frac{Q_1^{*}}{-1.5\, v_{d1}} \tag{9.4}$$

In the case of weak grids, the grid voltage fluctuates and directly affects the reactive power loop. In such scenarios, the closed-loop feedback controller is an ideal option to improve the performance of reactive power control loop. The reference d- and q-axis current components are given to the dq-frame current controller, and its block diagram is shown in Figure 9.9. To implement the current controller, the three-phase grid voltages and currents are transformed to the dq-frame, which will be synchronized with the grid frequency by using a phase-locked loop (PLL). The d- and q-axis current controllers minimize the difference between their respective actual and reference currents. The controller output is added to feed-forward grid voltage components (v_{d1} and v_{q1})) and speed voltage terms ($\omega_1 L_{1t} i_{d1}$ and $\omega_1 L_{1t} i_{q1}$) as shown in Figure 9.9. Finally, the reference d- and q-axis voltage components are transformed to the stationary-abc frame using the transformation matrix given in Chapter 3. The voltage components in the abc-frame are added to the output of SM voltage control, and the resultant modulation signals are used with the pulse width modulation scheme to generate the SM gating signals ($G_{1,xyh}$).

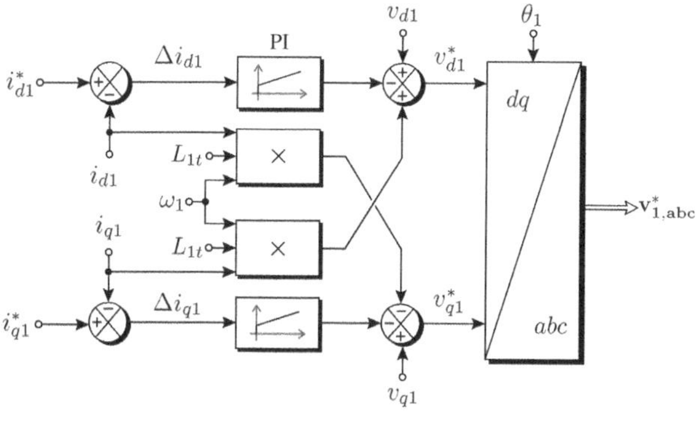

Figure 9.9 Block diagram of dq-frame current controller.

9.3.2 Receiving-End Converter Control

The receiving-end converter is operated in the power mode to regulate the active and reactive power flow through the transmission system. The control scheme of the receiving-end converter is implemented in the synchronous-dq frame as shown in Figure 9.10. The AC grid voltage and currents are transformed to the synchronous-dq frame rotating at a frequency of ω_2. In this approach, the reference active and reactive powers are generated directly [43]. From the active power, the reference d-axis current component is obtained. Similarly, the reference q-axis current component is obtained from the reactive power. The reference d- and q-axis current components are given by

$$i_{d2}^* = \frac{P_2^*}{1.5\, v_{d2}}, \quad i_{q2}^* = \frac{Q_2^*}{-1.5\, v_{d2}} \tag{9.5}$$

where P_2^* and Q_2^* are the reference active and reactive powers at the receiving-end, i_{d2}^* and i_{q2}^* are the reference d- and q-axis current components at the receiving-end, and v_{d2} is the d-axis voltage component of the receiving-end grid.

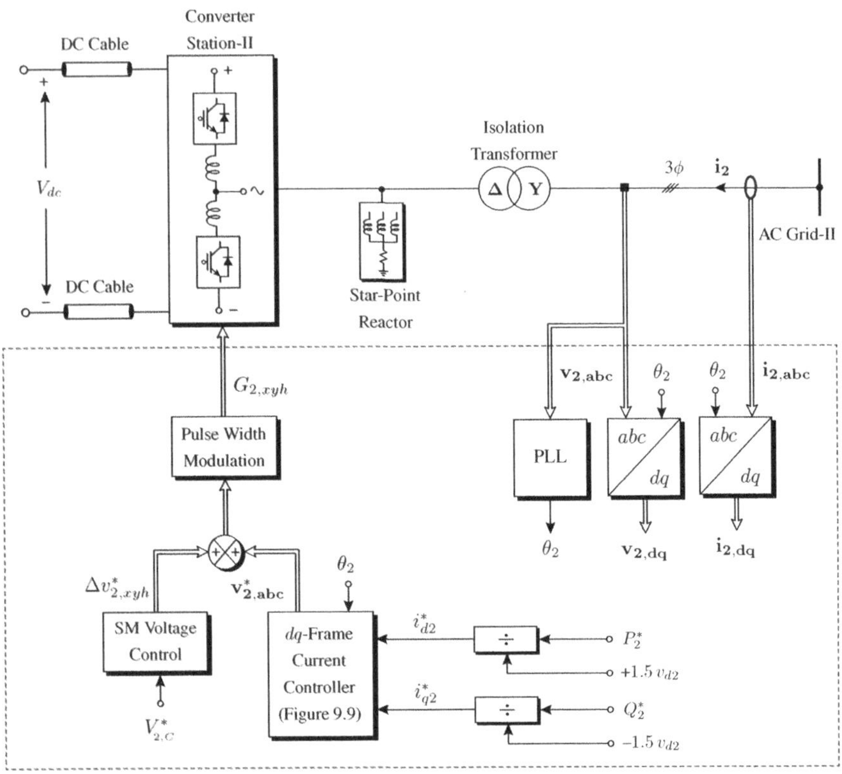

Figure 9.10 Control scheme of the receiving-end converter.

According to equation (9.5), when the grid voltage is stiff, the active and reactive powers linearly vary with the d- and q-axis current components, respectively. However, the grid voltage fluctuates in the case of weak AC systems, and this fluctuation directly affects the generation of the reference d- and q-axis current components. From the control perspective, the grid voltage fluctuation can be seen as a voltage disturbance. To handle these disturbances, the reference currents should be generated by using the closed-loop active and reactive power controllers instead of generating from the powers directly.

The reference and actual currents are applied to the dq-frame current controller. Figure 9.9 shows the structure of dq-frame current controller of sending-end converter. This structure can be used to control the receiving-end converter by replacing the sending-end grid parameters with the receiving-end grid. The d- and q-axis current controllers are realized using a simple PI-regulator. These PI-regulators minimize the d- and q-axis current error components and generate the d- and q-axis voltage commands, respectively. These voltage commands are added to the feed-forward receiving-end grid voltage components (v_{d2} and v_{q2}) and speed voltage terms ($\omega_2 L_{2t} i_{d2}$ and $\omega_2 L_{2t} i_{q2}$) as shown in Figure 9.9. The resultant voltage commands are transformed from dq-frame to abc-frame using the transformation matrix. These voltage commands are used along with the output of SM voltage control to generate the SM gating signals ($G_{2,xyh}$).

CASE STUDY 9.1 Analysis of Two-Terminal MMC-HVDC System

Objective: In this case study, the operation and performance of two-terminal MMC-based HVDC system is analyzed using MATLAB simulation.

Parameters: The system parameters are as follows: rated active power $P_1 = P_2 = 10$ MW, rated reactive power $Q_1 = Q_2 = \pm 3$ MVAr, grid line-to-line voltage $V_1 = V_2 = 10$ kV, grid frequency $f_1 = f_2 = 60$ Hz, positive pole voltage $V_{dc}^{+} = +10$ kV, negative pole voltage $V_{dc}^{-} = -10$ kV, number of SMs per arm $N = 10$, SM capacitor $C = 4000$ μF, SM capacitor voltage $v_C = 2$ kV, arm inductor $L = 5$ mH, and carrier frequency $f_c = 500$ Hz.

Analysis:

In this case study, the operation and performance of the two-terminal MMC-based HVDC system are investigated through MATLAB simulation. The rated active and reactive power rating of the sending-end and receiving-end converters are 10 MW and 3 MVAr. These converters are designed with half-bridge submodules, and each converter arm consists of 10 SMs. The sending-end converter is operated in the voltage mode to regulate the DC-side voltage and the system reactive power. The voltage mode control is designed with a reference DC-side voltage (V_{dc}^{*}) of 20 kV and reactive power (Q_1^{*}) of -3 MVAr. The SM voltage control is designed to maintain each submodule capacitor voltage at their nominal value of 2 kV.

The performance of sending-end converter is shown in Figure 9.11. The DC voltage controller maintains the positive and negative pole voltages at $+10$ kV and -10 kV as shown in Figure 9.11(a). The total DC-bus voltage (V_{dc}) is equal to 20 kV. The DC voltage controller generates the reference d-axis current component of 416 A, corresponding to the receiving-end active power demand of 5 MW from $t = 0.44$ s to $t = 0.52$ s. The reference q-axis current component is directly calculated from the reactive power, and its magnitude is given by

$$i_{q1}^{*} = \frac{-3 \times 10^6}{-1.5 \times 8165} = +244.95\text{ A} \tag{9.6}$$

At $t = 0.52$, the receiving-end active power demand is changed from 5 MW to 10 MW. Still, the DC-bus voltage controller regulates the total DC-bus voltage of 20 kV and generates a reference d-axis current component of 820 A. There is no change in the reference q-axis

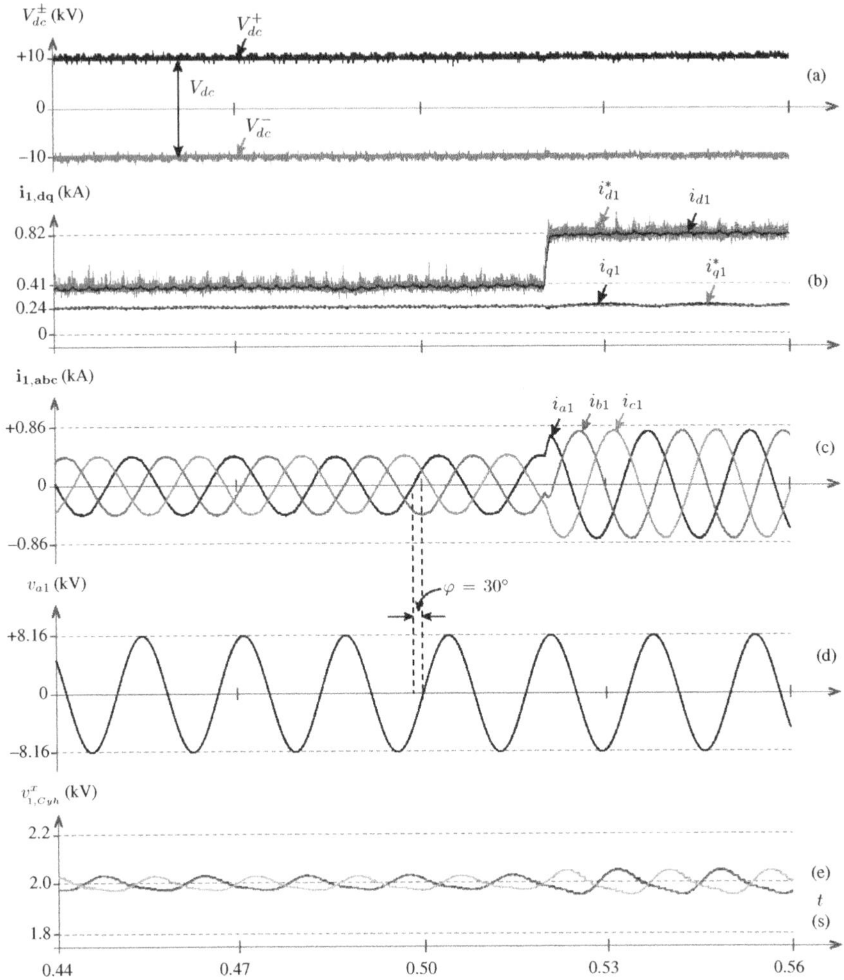

Figure 9.11 Performance of sending-end converter station.

current component. The actual d- and q-axis current components are perfectly regulated at their nominal values as shown in Figure 9.11(b). The currents in the dq-frame are transformed back to the abc-frame as shown in Figure 9.11(c). At $t = 0.55$ s, the current in the abc-frame has a peak value of 856 A as shown in Figure 9.11(c). The phase-a voltage of the AC grid-I is shown in Figure 9.11(d). It is observed that the phase-a current leads the voltage by an angle (φ) of 30° as shown in Figure 9.11(d). The upper and lower arm submodule capacitor voltages are regulated at 2000 V as shown in Figure 9.11(e).

The receiving-end converter is operated in the power mode to regulate the active and reactive power flow through the receiving-end transmission system. The power mode control scheme is designed with a reference active power (P_2^*) of −5 MW from $t = 0.44$ s to $t = 0.52$

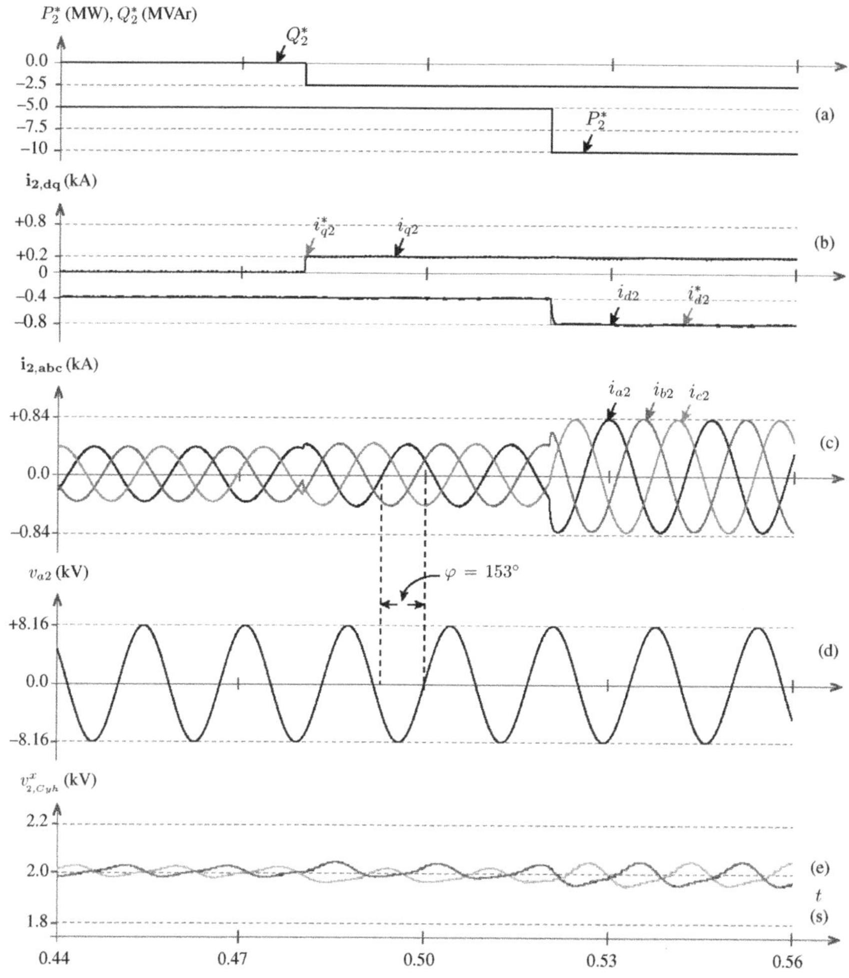

Figure 9.12 Performance of receiving-end converter station.

s. At $t = 0.52$ s, the reference active power is changed from −5 MW to −10 MW as shown in Figure 9.12(a). Similarly, the reference reactive power (Q_2^*) is set to 0 MVAr from $t = 0.44$ s to $t = 0.48$ s. At $t = 0.48$ s, the reference reactive power is changed to 2.5 MVAr as shown in Figure 9.12(a).

The reference d- and q-axis currents are directly calculated from the reference powers. At $t = 0.5$ s, the magnitude of d- and q-axis currents corresponding to $P_2^* = -5$ MW is given by

$$i_{d2}^* = \frac{-5 \times 10^6}{1.5 \times 8165} = -408.2\ \text{A} \tag{9.7}$$

Similarly, the reference q-axis current for $Q_2^* = -2.5$ MVAr is given by

$$i_{q2}^* = \frac{-2.5 \times 10^6}{-1.5 \times 8165} = +204.12\text{A} \tag{9.8}$$

The actual d- and q-axis current components are perfectly following their reference currents as shown in Figure 9.12(b). These currents are transformed to the abc-frame as shown in Figure 9.12(c). At $t = 0.5$ s, the currents in the abc-frame have a peak value of 456.39 A. The phase-a voltage of the AC grid-II is shown in Figure 9.12(d). The grid current leads the voltage by an angle (φ) of 153° as shown in Figure 9.12(d). The submodule capacitor voltages are maintained at their nominal value of 2000 V as shown in Figure 9.12(e). ■

9.4 CONTROL OF MULTI-TERMINAL MMC-BASED HVDC SYSTEM

The typical configuration of a multi-terminal MMC-based HVDC system is shown in Figure 9.13. It consists of four MMC converter stations connected to their respective AC grids. The DC systems are connected to form an HVDC grid. For the stability of the HVDC grid, the converter stations should maintain the balance between the active power going into the HVDC grid and the active power going out of the HVDC grid plus DC system losses [44–49]. To achieve this objective, the sending-end converter station should regulate the DC-bus voltage at the given reference value. This converter station is also referred to as a "slack bus," and the corresponding control scheme is named as "voltage mode control" [44–49].

In the case of a slack bus failure, the complete HVDC system will collapse, which leads to a complete "blackout." Also, when the slack bus converter reaches its maximum power capacity, the slack bus cannot regulate the DC-bus voltage, which causes an unstable operation of the HVDC grid [46–49]. The former issue can be addressed by transferring the slack bus function (i.e., DC-bus voltage control) to other converter stations (receiving end-I). This objective can be achieved by using a "voltage margin control," which is a combination of voltage mode and power mode control approaches [45–49]. When the DC-bus voltage (V_{dc}) is within the prescribed limits (upper limit (V_{dcu}) and lower limit (V_{dcl})), the voltage margin control is equivalent to the power mode control. On the other hand, when the DC-bus voltage (V_{dc}) exceeds either the upper limit (V_{dcu}) or the lower limit (V_{dcl}), then the voltage margin control is going to enter into the voltage mode control region [45–49].

The later issue can be overcome by adjusting the converter (receiving end-II) active power flow with the variation of DC-bus voltage. This objective can be achieved by using a "voltage droop control" [45–49]. Rest of the converter stations (receiving end-III) can be operated in the "power mode" to regulate the active and reactive powers injected into the AC grid [42].

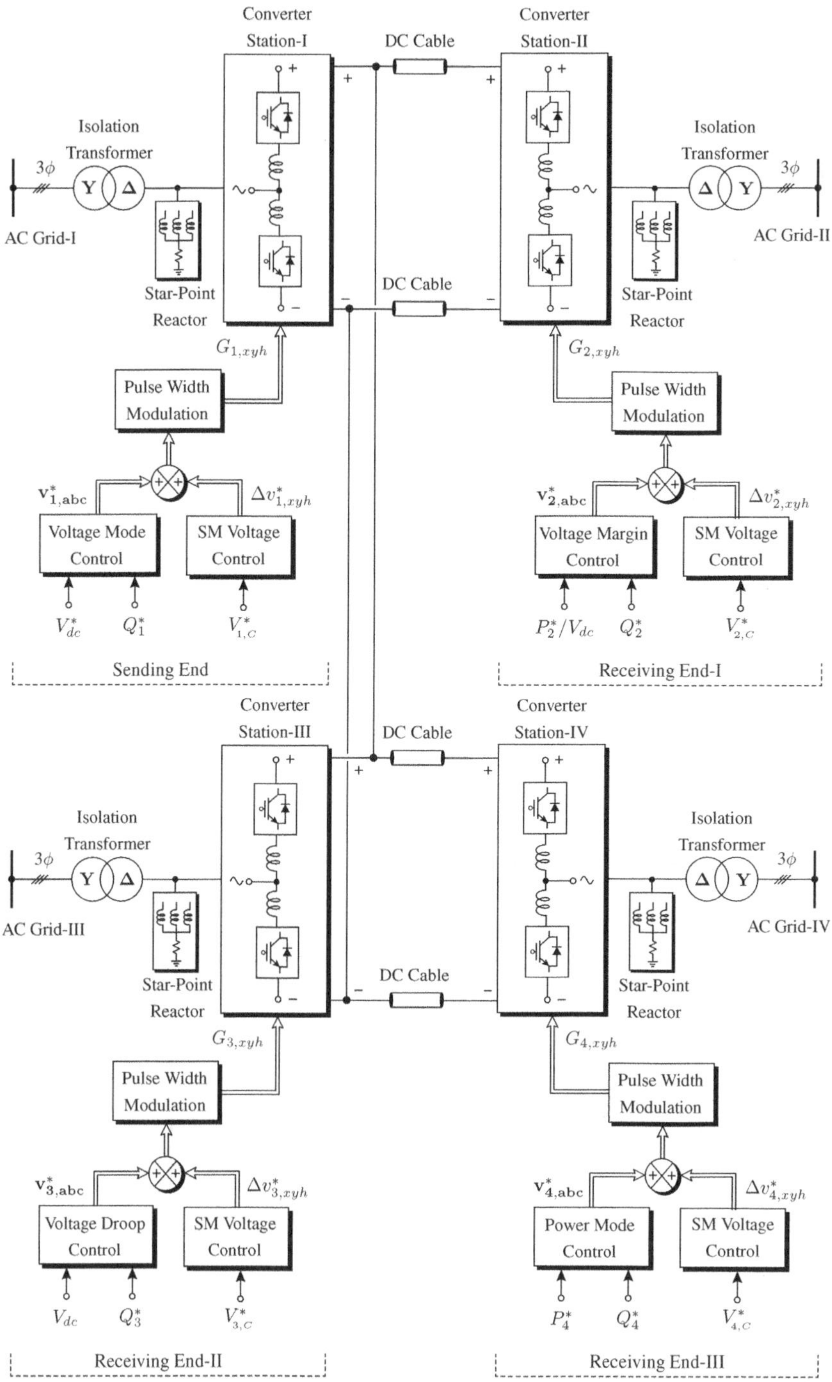

Figure 9.13 General control block diagram of multi-terminal MMC-HVDC system.

9.4.1 Voltage Margin Control Scheme

In the multi-terminal system, one of the converter stations is operated with the voltage margin control scheme and provides the back-up for the slack bus converter station in the event of a fault [8, 46–49]. The voltage margin control scheme is implemented in the synchronous-dq frame rotating at the grid frequency (ω_2) as shown in Figure 9.14. This control scheme uses either control loop-① or control loop-② to regulate the d-axis current component. The controller selects one of the control loops depending on the DC-bus voltage (V_{dc}). The $P - V_{dc}$ characteristics of voltage margin control are shown in Figure 9.15(a), which consists of a constant power region-① and constant voltage region-②.

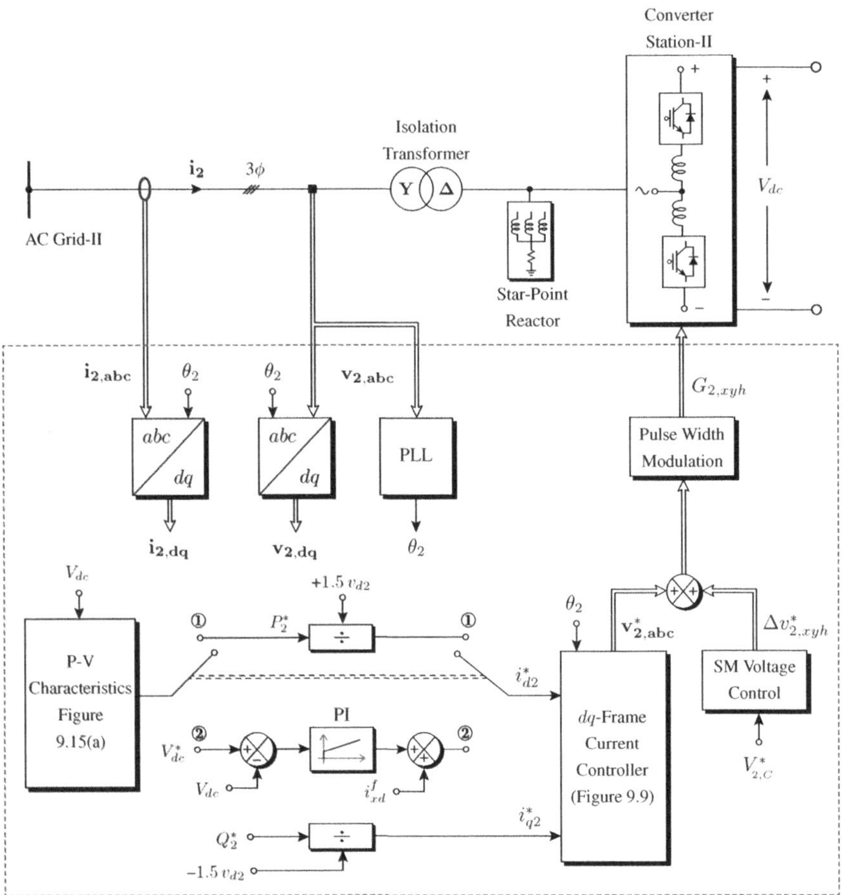

Figure 9.14 Voltage margin control of receiving-end converter.

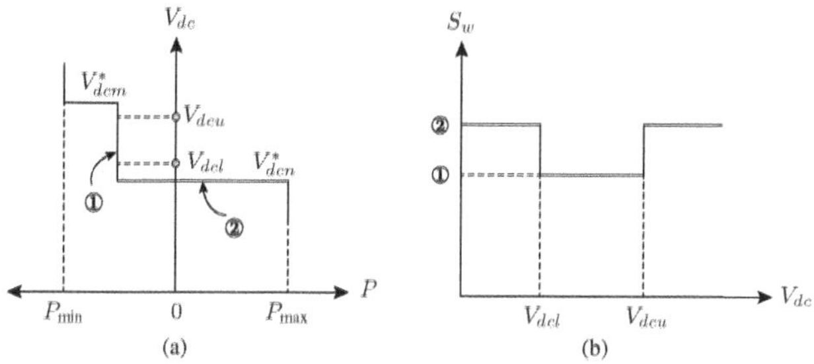

Figure 9.15 Voltage margin control: (a) $P - V_{dc}$ characteristics, and (b) selection of operating modes.

According to these characteristics, when the DC-bus voltage (V_{dc}) is within the upper (V_{dcu}) and the lower limits (V_{dcl}) then the controller selects the control loop-① as shown in Figure 9.15(b). The control loop-① is constant active power regulation loops and requires active power reference (P_2^*) as shown in 9.14. From this reference, the reference d-axis current component is calculated. When the DC-bus voltage (V_{dc}) exceeds either the upper (V_{dcu}) or the lower limits (V_{dcl}), then the controller selects the control loop-② as shown in Figure 9.15(b). The control loop-② is a DC-bus voltage regulation loop and requires a DC-bus voltage reference (V_{dc}^*). When the DC-bus voltage (V_{dc}) exceeds the upper limit (V_{dcu}), then the voltage reference jumps to its maximum value (V_{dcm}^*). Similarly, when the DC-bus voltage (V_{dc}) exceeds their lower limit (V_{dcl}), then the voltage reference jumps to its minimum value (V_{dcn}^*). The reference DC-bus voltage under different conditions are given by [8]

$$V_{dc}^* = \begin{cases} V_{dc} & \text{for } (V_{dcu} \le V_{dc} \le V_{dcl}) \\ V_{dcm}^* & \text{for } (V_{dc} \ge V_{dcu}) \\ V_{dcn}^* & \text{for } (V_{dc} \le V_{dcl}) \end{cases} \tag{9.9}$$

The reference DC-bus voltage is compared with the actual DC-bus voltage, and their difference is minimized using a PI-regulator. The output of PI-regulator is added to a feed-forward current term (i_{xd}^f), resulting in a reference d-axis current component. The reference q-axis current component is calculated from reference reactive power (Q_2^*). The reference d- and q-axis current components are applied to the dq-frame current controller as shown in Figure 9.9. The current controller consists of two independent PI-regulators to control the d- and q-axis current components. The current controller output is added to the feed-forward grid voltages and the speed-voltage terms. Finally, the reference d- and q-axis voltages are transformed to the stationary-abc frame. These reference voltages are added to the compensation terms obtained from the SM voltage control. The resultant modulation signals are compared with the carrier signals to generate the gating signals.

CASE STUDY 9.2 Analysis of Three-Terminal MMC System

Objective: In this case study, the performance of a three-terminal MMC-based HVDC system is analyzed using MATLAB simulation.

Parameters: The system parameters are as follows: rated active power $P_1 = P_2 = P_3 = \pm 10$ MW, rated reactive power $Q_1 = Q_2 = Q_3 = \pm 3$ MVAr, grid line-to-line voltage $V_1 = V_2 = V_3 =$ 10 kV, grid frequency $f_1 = f_2 = f_3 = 60$ Hz, positive pole voltage $V_{dc}^{+} = +10$ kV, negative pole voltage $V_{dc}^{-} = -10$ kV, number of SMs per arm $N = 10$, SM capacitor $C = 4000$ μF, SM capacitor voltage $v_C = 2$ kV, arm inductor $L = 5$ mH, and carrier frequency $f_c = 500$ Hz.

Analysis:
In this case study, the performance of the three-terminal MMC-based HVDC system during the normal operation is investigated through MATLAB simulation. The sending-end converter station (MMC-1) is operated in the voltage mode to regulate the DC-bus voltage. The voltage margin control is applied to the receiving-end-I converter station (MMC-2). This control scheme is designed with an upper voltage limit $V_{dcu} = 21$ kV, and the lower voltage limit $V_{dcl} = 19$ kV. The maximum and minimum value of reference DC-bus voltage in the voltage margin control is $V_{dcm}^{*} = 22$ kV and $V_{dcn}^{*} = 18$ kV, respectively. The receiving-end-II (MMC-3) converter station is operated in the power mode to regulate the active and reactive powers directly.

The performance of the three-terminal MMC-based HVDC system during the normal operation is shown in Figure 9.16. The MMC-1 perfectly regulates the positive pole voltage at $+10$ kV and the negative pole voltage at -10 kV as shown in Figure 9.16(a). The total DC-bus voltage (V_{dc}) is regulated at 20 kV, and it is kept within the upper and lower limits as shown in Figure 9.16(b). Hence, the voltage margin control becomes the power mode control and directly controls the active and reactive power of MMC-2. The active power flow through the three converter stations are shown in Figure 9.16(c). From $t = 0.44$ s to $t = 0.5$ s, the MMC-2 and MMC-3 are supplying an active power of -2 MW and -5 MW to their respective grids. During this, the MMC-1 draws an active power of 7 MW from its AC grid. So that, the DC-bus voltage is perfectly regulated at their nominal (i.e., the power balance is achieved).

The performance of the three-terminal system is verified with the step-change in the active power as shown in Figure 9.16(c). At $t = 0.5$, the active power of MMC-3 is changed from -5 MW to -8 MW, and the MMC-2 is supplying a constant active power of -2 MW. To ensure the power balance, the active power flow through the MMC-1 is increased to 10 MW as shown in Figure 9.16(c). At $t = 0.56$ s, the reference active power of MMC-2 is changed from -2 MW to 3 MW. That means, the MMC-2 drawing the active power instead of supplying the power to its grid, and the MMC-3 is supplying an active power of 8 MW to its grid. During this time, the active power flow through the MMC-1 is reduced from 10 MW to 5 MW as shown in Figure 9.16(c).

The reactive power flow in the three-terminal MMC-based HVDC system is shown in Figure 9.16(d). The MMC-1 is controlled to supply a zero reactive power throughout the operation. The MMC-2 is supplying a zero reactive power for duration of 0.3 s. At $t = 0.47$ s, the reference reactive power of MMC-2 is changed from 0 to $+3$ MVAr and the MMC-2 can generate the reactive power corresponding to its reference. Similarly, the MMC-3 is controlled to absorb a reactive power of -2 MVAr from $t = 0.47$ s to $t = 0.59$ s as shown in Figure 9.16(d). At $t = 0.59$ s, the reactive power reference of MMC-3 is changed from -2 MVAr to $+2$ MVAr. That means, the MMC-3 is supplying the reactive power instead of absorbing it. Even under the dynamic changes, the three converter stations are perfectly regulating their active and reactive powers independently.

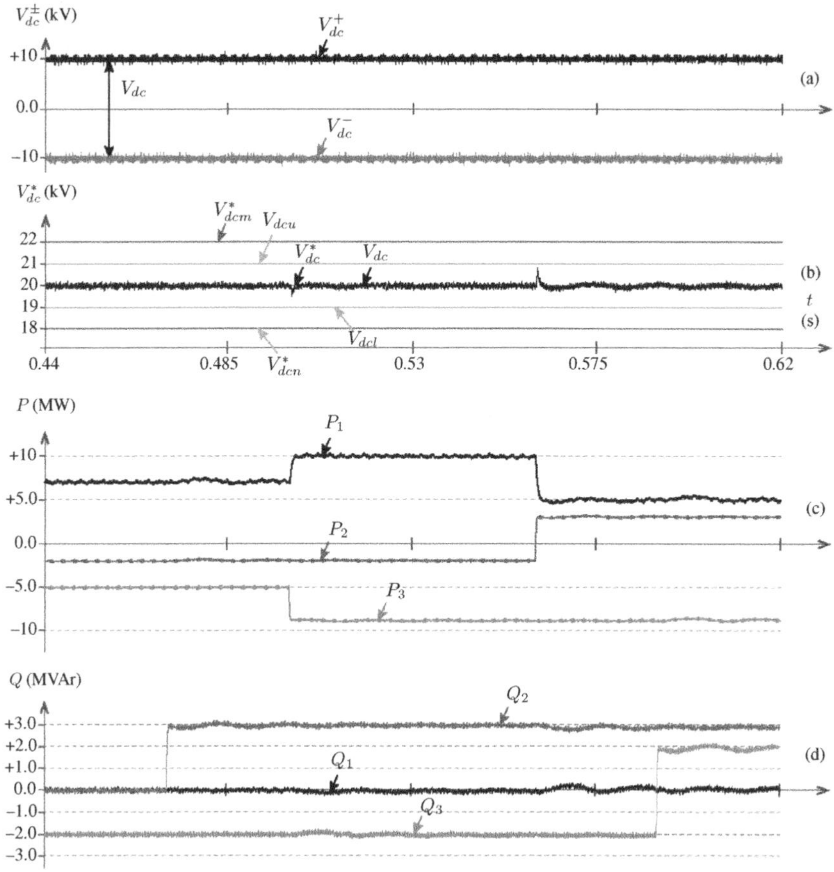

Figure 9.16 Performance of the three-terminal system during normal operation.

CASE STUDY 9.3 Three-Terminal MMC-HVDC with Slack Bus Failure

Objective: In this case study, the performance of a three-terminal MMC-based HVDC system during the slack bus failure is analyzed using MATLAB simulation.

Parameters: The system parameters are as follows: rated active power $P_1 = P_2 = P_3 = \pm$ 10 MW, rated reactive power $Q_1 = Q_2 = Q_3 = \pm$ 3 MVAr, grid line-to-line voltage $V_1 = V_2 = V_3 =$ 10 kV, grid frequency $f_1 = f_2 = f_3 = 60$ Hz, positive pole voltage $V_{dc}^{+} = +10$ kV, negative pole voltage $V_{dc}^{-} = -10$ kV, number of SMs per arm $N = 10$, SM capacitor $C = 4000$ μF, SM capacitor voltage $v_C = 2$ kV, arm inductor $L = 5$ mH, and carrier frequency $f_c = 500$ Hz.

Analysis:

In this case study, the performance of the three-terminal MMC-based HVDC system during the slack bus failure is investigated through MATLAB simulation. The sending-end converter station (MMC-1) is operated in the voltage mode to regulate the DC-bus voltage. This converter is also referred to as a slack bus. The converter stations at the receiving-end-I (MMC-2) and II (MMC-3) are controlled with the voltage margin control and power model control schemes, respectively. The voltage margin control scheme is designed with an upper voltage limit $V_{dcu} = 21$ kV, and the lower voltage limit $V_{dcl} = 19$ kV. The maximum and minimum values of reference DC-bus voltage in the voltage margin control is $V^*_{dcm} = 22$ kV and $V^*_{dcn} = 18$ kV, respectively.

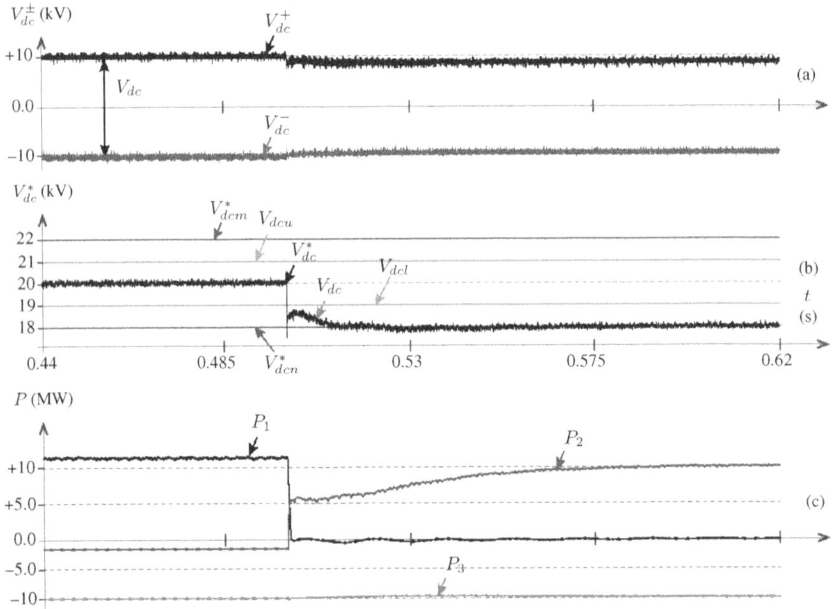

Figure 9.17 Performance of the three-terminal system during the slack bus failure.

From $t = 0.44$ s to $t = 0.5$ s, the MMC-1 establishes the DC-bus voltage, and it regulates the positive and negative pole voltages at +10 and −10 kV as shown in Figure 9.17(a). During this, the total DC-bus voltage is kept within the upper and lower limits as shown in Figure 9.17(b). The MMC-2 and MMC-3 are supplying an active power of −2 MW and −10 MW, respectively. To ensure the power balance, the MMC-1 draws an active power of −12 MW from its grid as shown in Figure 9.17(c).

At $t = 0.5$ s, the MMC-1 (slack bus) fails to regulate the DC-bus voltage due to the inner fault. When the DC-bus voltage drops below the lower limit (V_{dcl}), the MMC-2 will act as a slack bus to regulate the DC-bus voltage, and the voltage margin control becomes voltage mode control. The reference value of DC-bus voltage is equal to $V^*_{dcn} = 18$ kV as shown in Figure 9.17(b). The total DC-bus voltage is equally distributed between the positive and negative poles of +9 kV and −9 kV, respectively. Due to the fault, the active power flow of MMC-1 becomes zero as shown in Figure 9.17(c). The MMC-3 is supplying a constant active power of 10 MW to its grid. To ensure the power balance as well as to regulate the DC-bus voltage, the MMC-2 should draw an active power of 10 MW from its grid as shown in Figure 9.17(c). ■

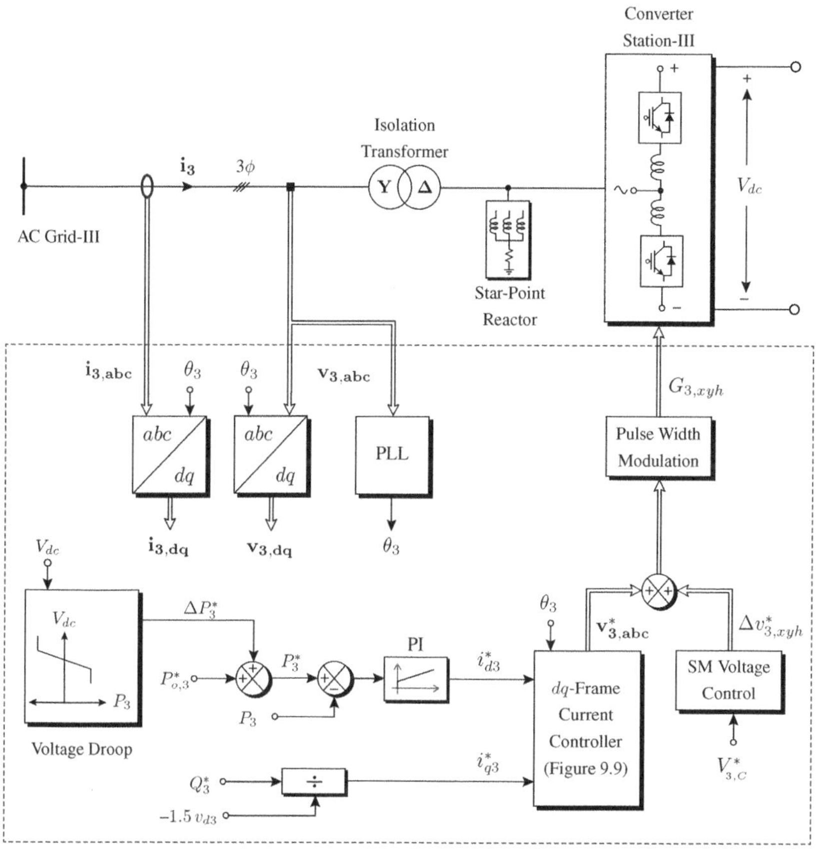

Figure 9.18 Voltage droop control of receiving-end converter.

9.4.2 Voltage Droop Control Scheme

When the sending-end converter (slack bus) reaches its maximum power capacity, the slack bus cannot regulate the DC-bus voltage. In that case, the voltage droop control schemes are applied to one of the converter stations in the multi-terminal system [8, 46–49]. These converter stations contribute the deficient active power to regulate the DC-bus voltage. The principle of voltage droop control is to adjust the active power flow of the converter station with the variation of DC-bus voltage. The block diagram of the receiving-end converter with the voltage droop control is shown in Figure 9.18. It consists of an active power loop, and its reference value is given by [8]

$$P_3^* = P_{o,3}^* + \Delta P_3^* \tag{9.10}$$

where $P_{o,3}^*$ is the initial (steady-state) value of the reference active power, and ΔP_3^* is the change in the reference active power.

Depending on the sign of ΔP_3^*, the net active power flow of the converter will increase or decrease. The value of ΔP_3^* is obtained from the droop characteristics, and is expressed as [8]

$$\Delta P_3^* = \frac{1}{K_{dc}} \left(V_{dc}^o - V_{dc} \right) \tag{9.11}$$

where V_{dc}^o is the initial (previous) value of DC-bus voltage, V_{dc} is the present value of DC-bus voltage, and K_{dc} is the slope of the voltage droop characteristics.

The reference active power (P_3^*) is compared with the actual active power (P_3), and their difference is applied to a PI-regulator. The PI-regulator minimizes their difference and generates the reference d-axis current component. Similarly, the reference q-axis current component is obtained from the reactive power as shown in Figure 9.18. The reference and actual dq currents are given to the current controller as shown in Figure 9.9. The current controller generates the voltage commands, which are added to the feed-forward grid voltages and speed voltage terms. The resultant dq voltages are transformed back to the stationary-abc frame. These voltage commands are used along with the output of the SM voltage control to generate the gating signals.

9.5 MMC-BASED STATIC SYNCHRONOUS COMPENSATOR

The modern power system networks operate under distorted and lagging power factor conditions due to the integration of wind energy systems and non-linear loads (AC traction systems). Also, voltage flicker, voltage sag/swell, harmonic compensation, and grid unbalance are common power quality problems in the power system networks [6, 7, 50–52]. In the early days, thyristor-based line-commutated converters were widely used for power quality improvement. Later, power converters based on self-commutated devices (GTO and IGBTs) were introduced in power quality technologies. Among them, the STATCOM is one of the most popular technologies. The latest STATCOM technology uses a MMC, and it is available under the trade name of `SVC-PLUS` [50]. This technology has received big acceptance due to its superior features (modularity and voltage scalability) and competitive cost [6, 7, 50–52]. The STATCOM is capable of supplying both leading and lagging reactive powers, and compensate the unbalanced grid currents. The operation and control of MMC-based STATCOM during the balanced and unbalanced grid conditions are discussed in the following subsections.

9.5.1 System Configuration

The system configuration of MMC-based STATCOM is shown in Figure 9.19. In this configuration, the AC grid is connected to a distorted and unbalanced load. These loads are inductive in nature and draw the lagging reactive power from the AC grid. To compensate the reactive power, the MMC-based STATCOM is connected to the point of common coupling (PCC) as shown in Figure 9.19. Considering the cost and efficiency, the half-bridge submodules are used to realize the MMC-based STATCOM [50]. These submodules consist of DC capacitor as a storage element. The STATCOM DC-bus voltage (V_{dc}) is distributed between the submodules in each leg and does not require a separate DC-bus capacitor (C_{dc}) as shown Figure 9.19. Also, the MMC-based STATCOM can be directly connected to the medium-voltage networks without any transformer. In the case of high-voltage networks, the step-up transformer is included in the system [50].

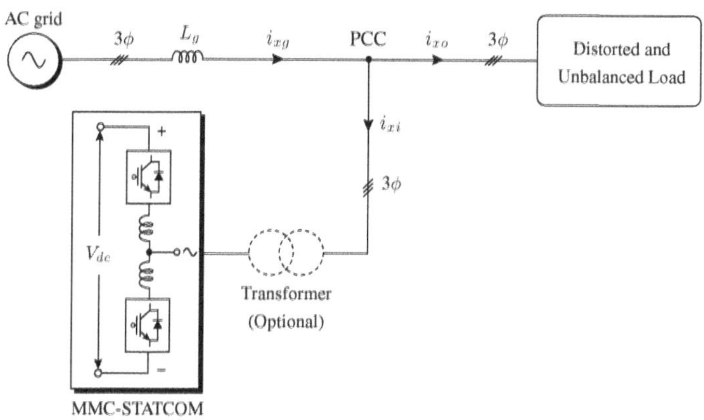

Figure 9.19 System configuration of MMC-based STATCOM.

9.5.2 Reactive Power Compensation

The reactive power compensation is one of the grid code requirements for the integration of wind energy systems. Also, the grid-connected electrical equipment and loads should be operated at the prescribed power factor by the grid operator. The grid power factor can be adjusted by controlling the reactive power supplied by the MMC-based STATCOM [6, 7, 50–52]. The control scheme of MMC-based STATCOM is implemented in the synchronous-dq frame, and its block diagram is shown in Figure 9.20. The control scheme consists of an outer loop to control the overall DC voltage of the MMC. The error between the reference and actual DC voltage (v_{C_g}) is minimized by using a PI-regulator, and its output is equal to the reference d-axis current component (i^*_{di}). This current component represents the converter power losses, and its magnitude is very small in practice. The reference q-axis current component (i^*_{qi}) is directly obtained from the reactive power.

The reference currents are applied to the dq-frame current controller as shown in Figure 9.9. The current controller requires actual d- and q-axis STATCOM current components, which are obtained by transforming the measured three-phase currents ($\mathbf{i_{i,abc}}$) from abc-frame to dq-frame as shown in Figure 9.20. The reference and actual current components are compared with each other, and their difference is minimized by using a PI-regulator. The PT-regulator output is added to the feed-forward grid voltage components and speed voltage components, and results in the fundamental modulation signals of the STATCOM ($\mathbf{v^*_{i,abc}}$).

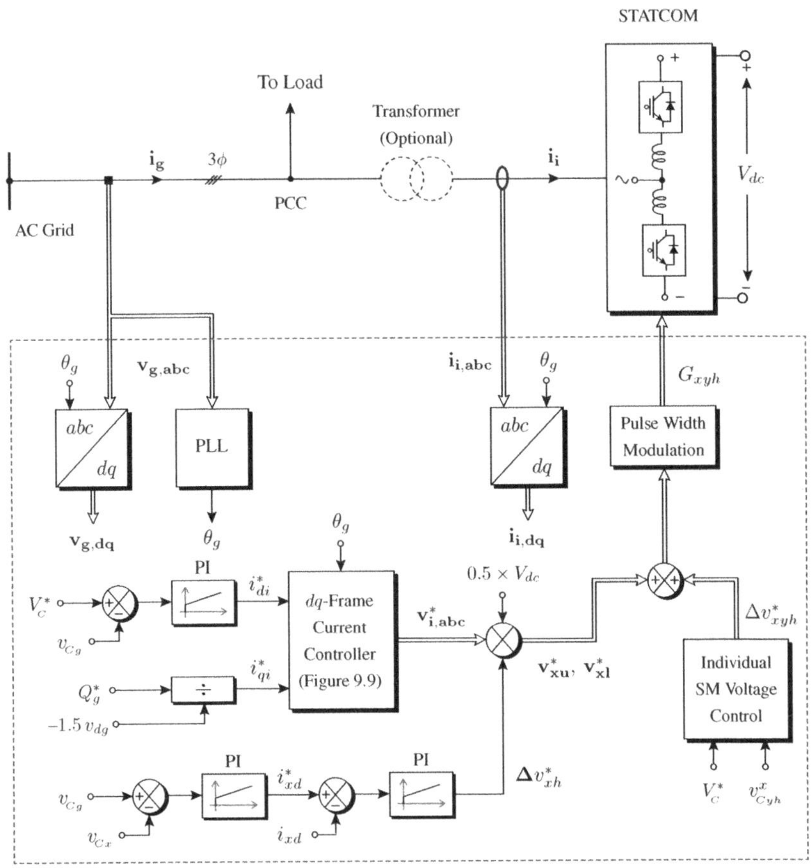

Figure 9.20 Control of STATCOM under balanced grid conditions.

In addition, the MMC-based STATCOM requires arm capacitor voltage and individual submodule voltage control as well. The arm capacitor voltage control is realized with an outer voltage and inner current control loops as shown in Figure

9.20. The outer voltage loop maintains the average leg voltage at the given reference value of average overall DC voltage. This voltage loop gives the reference DC components in the arm (i^*_{xd}). The reference current is compared with the actual DC component in the arm current (i_{xd}), and its magnitude is given in terms of the upper and lower arm currents as

$$i_{xd} = \frac{\mathbf{i_{xu}} + \mathbf{i_{xl}}}{6} \tag{9.12}$$

The inner current control loop minimizes the current error and generates the compensating voltage signal Δv^*_{xh}. This voltage signal is used along with the fundamental voltage signals and a fixed DC offset to form the upper and lower arm modulation signals. The reference arm modulation signals are

$$\begin{aligned} v^*_{xu} &= 0.5\,V_{dc} - v^*_{xi} - \Delta v^*_{xh} \\ v^*_{xl} &= 0.5\,V_{dc} + v^*_{xi} - \Delta v^*_{xh} \end{aligned} \tag{9.13}$$

The reference arm modulation signals are further modified by using the individual submodule voltage control output so that, each submodule capacitor voltage is maintained at their nominal value along with other control objectives of the STATCOM. The details of individual submodule capacitor voltage control are presented in Chapter 5.

CASE STUDY 9.4 Reactive Power Compensation with MMC-STATCOM

Objective: In this case study, the performance of an MMC-based STATCOM under balanced grid condition is analyzed using MATLAB simulation.

Parameters: The system parameters are as follows: rated reactive power $Q_g = \pm$ 10 MVAr, grid line-to-line voltage $V_g = 10$ kV, grid frequency $f_g = 60$ Hz, number of SMs per arm $N = 10$, SM capacitor $C = 4000$ μF, SM capacitor voltage $v_C = 2$ kV, arm inductor $L = 5$ mH, and carrier frequency $f_c = 500$ Hz.

Analysis:
In this case study, the MMC-based STATCOM is designed to supply the reactive power demanded by the AC grid. The grid operator sets the reference reactive power. The performance of the STATCOM is shown in Figure 9.21. From $t = 0.44$ s to $t = 0.5$ s, the STATCOM is operating in the capacitive mode and supplying a reactive power of $Q_g = -10$ MVAr. At $t = 0.5$ s, the STATCOM operation is switched to the inductive mode and absorbing a reactive power of $Q_g = +10$ MVAr as shown in Figure 9.21(a). The reference d- and q-axis current components are shown in Figure 9.21(b). The magnitude of q-axis current component is changed from $+816$ A to -816 A, corresponding to the reference reactive power. The leg voltage control generates the d-axis current component, and its magnitude varies proportionally to the converter power losses. The STATCOM is controlled to generate these reference currents at their output. The three-phase output currents have the peak value of 816 A as shown in Figure 9.21(c). The phase-a current leads the AC grid voltage during the capacitive mode operation, and it lags the voltage during the inductive mode as shown in Figure 9.21(d). The leg voltage control maintains the average voltage of each converter at 10 kV. This voltage is equally distributed the submodules in each arm by using an individual submodule capacitor voltage control. Each submodule capacitor voltage is regulated at 2 kV as shown in Figure 9.21(e).

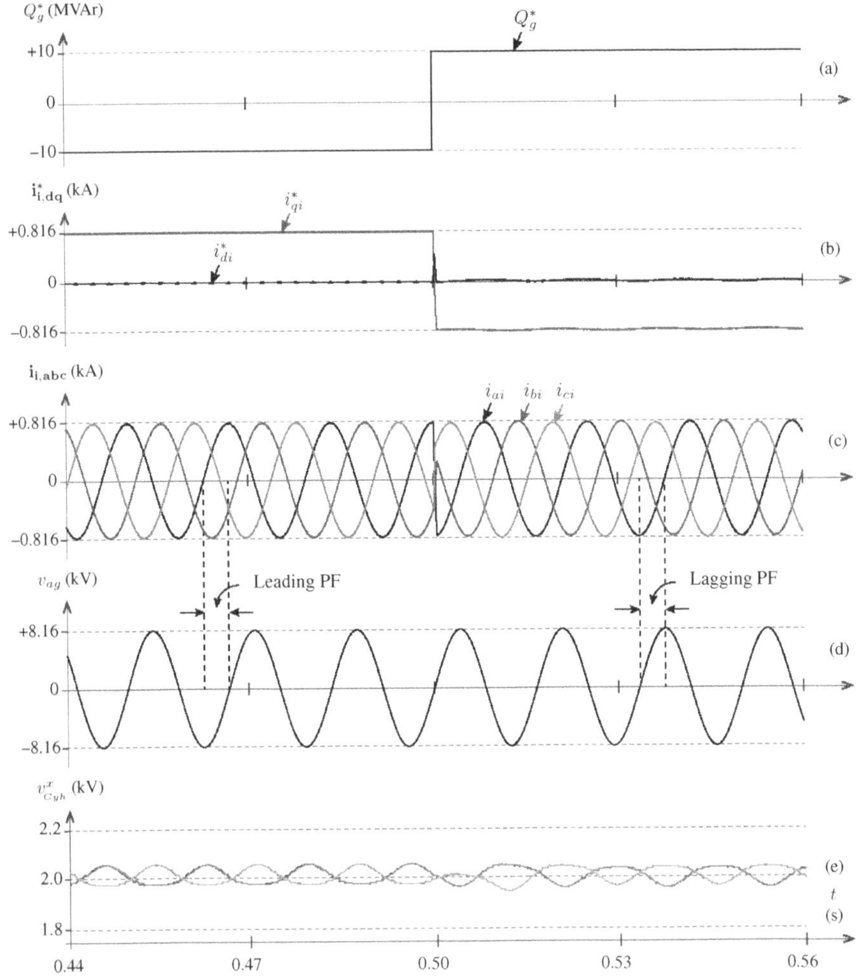

Figure 9.21 Performance of STATCOM under balanced grid conditions.

■

9.5.3 Compensation of Unbalanced AC-Grid Currents

The AC grid currents are unbalanced in nature due to asymmetrical operating conditions like the AC grid fault and unbalanced loads [53]. During the asymmetrical conditions, the AC grid currents consist of positive and negative sequence components in the three-phase three-wire system. These sequence components should be supplied by the STATCOM to balance the AC grid currents. The control block diagram of MMC-based STATCOM under unbalanced conditions is shown in Figure 9.22. It consists of a leg, arm, and individual submodule capacitor voltage controllers along with the dq-frame current control. The reference d-axis current component consists of a feed-forward term i^f_{xd}, converter power loss component Δi_{di}, and oscillating component in the load d-axis current i_{do}. The converter power loss component is obtained from the leg voltage control as shown in Figure 9.22.

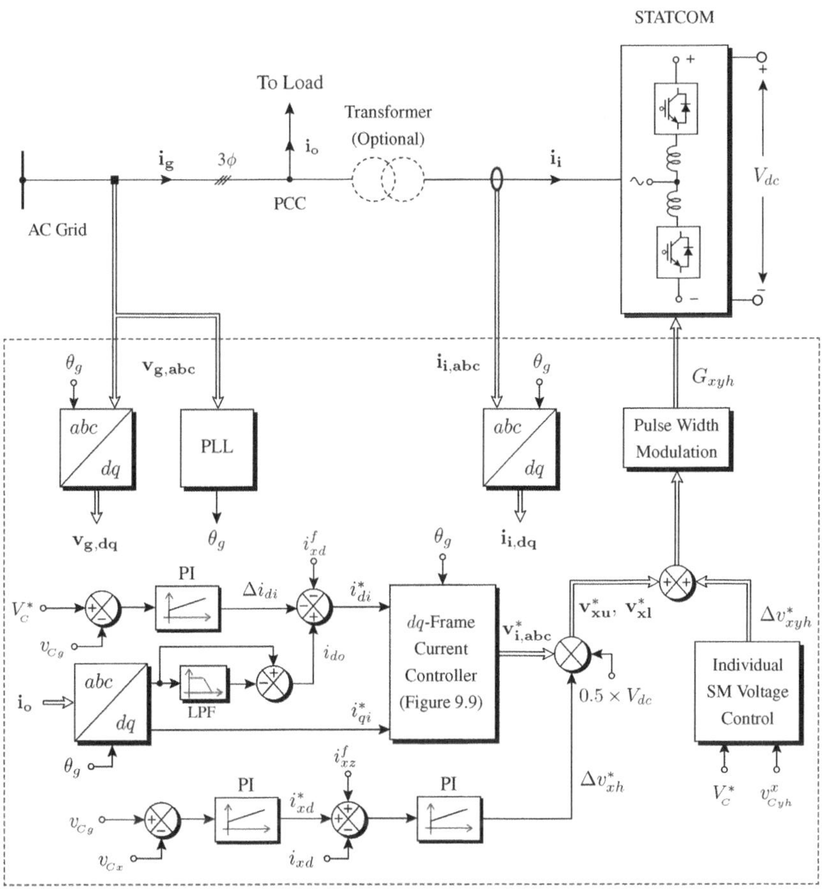

Figure 9.22 Control of STATCOM under unbalanced grid conditions.

The oscillating component in the load d-axis current is obtained by transforming the stationary-abc frame load currents into the synchronous-dq frame. Due to the unbalance (negative sequence components), the transformed d-axis current component consists of a second harmonic component along with a DC signal. The DC-signal represents the positive sequence active power, and the AC grid supplies it. On the other hand, the second harmonic component is supplied by the STATCOM, and it is extracted by using a low-pass filter (LPF). The LPF output is a DC signal, and it is subtracted from the actual signal to obtain the second harmonic component in the d-axis current as shown in Figure 9.22.

Similarly, the q-axis current consists of both DC and oscillating signals. These signals are given to the STATCOM as a reference so that, the STATCOM supplies both positive and negative sequence reactive power to the load. The reference d- and q-axis current components are applied to the dq-frame current controller, and its structure is shown in Figure 9.9. The current controller minimizes the error between the reference and actual current components and generates the fundamental modulation signals.

Finally, the arm modulation signals are obtained by adding the fundamental modulation signals to the arm voltage control output and a fixed DC-offset. The arm voltage control is further improved by adding a feed-forward current component (i_{xz}^f) corresponding to the unbalance operation as shown in Figure 9.22. Finally, the individual SM voltage control output is added to the arm modulation signals. So that, each SM capacitor voltage is regulated at their nominal value.

(1) Extraction of Sequence Components: The positive and negative sequence components of the AC grid voltage and STATCOM output currents are used to estimate the feed-forward current components (i_{xd}^f and i_{xz}^f) of the d-axis and circulating current control loops [53]. The extraction process of positive and negative sequence components is shown in Figure 9.23. The positive sequence components are obtained by using a grid angle θ_g^+ in the abc to dq transformation matrix while the θ_g^- is used to extract the negative sequence components as shown in Figure 9.23. Due to the unbalance, the transformed positive and negative sequence dq components (voltage and current) consist of both DC and oscillating signals. From them, the DC signals are extracted by using a low-pass filter (LPF). The resultant signals are used in the estimation process of the feed-forward current components.

The extraction process is further analyzed with the following mathematical equations. In this analysis, the three-phase unbalanced AC grid voltages are defined as

$$\begin{aligned} v_{ag} &= V_p \cos(\omega_g t) + V_n \cos(\omega_g t + \delta) \\ v_{bg} &= V_p \cos(\omega_g t - \frac{2\pi}{3}) + V_n \cos(\omega_g t + \delta - \frac{2\pi}{3}) \\ v_{cg} &= V_p \cos(\omega_g t - \frac{4\pi}{3}) + V_n \cos(\omega_g t + \delta - \frac{4\pi}{3}) \end{aligned} \tag{9.14}$$

where V_p and V_n are the peak values of the positive and negative sequence voltage components, ω_g is the AC grid angular frequency (rad/sec), and δ is the phase angle between the positive and negative sequence voltage components.

The three-phase voltages in the stationary-abc frame are transformed to the synchronous-$dq0$ frame using the following transformation matrix.

$$\mathbf{T_{abc \Rightarrow dq0}} = \frac{2}{3} \begin{bmatrix} \cos\theta & \cos(\theta - \frac{2\pi}{3}) & \cos(\theta - \frac{4\pi}{3}) \\ -\sin\theta & -\sin(\theta - \frac{2\pi}{3}) & -\sin(\theta - \frac{4\pi}{3}) \\ 0.5 & 0.5 & 0.5 \end{bmatrix} \tag{9.15}$$

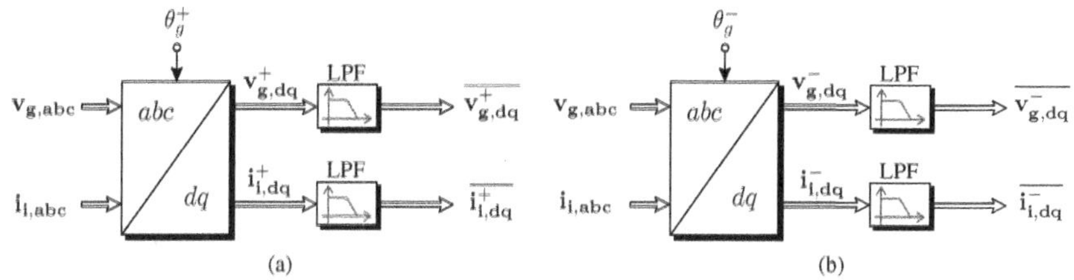

Figure 9.23 Extraction of sequence components: (a) positive sequence, and (b) negative sequence.

The positive sequence $dq0$ voltage components are obtained by replacing the θ in equation (9.15) with the θ_g^+, and are given by

$$\begin{bmatrix} v^+_{dg} \\ v^+_{qg} \\ v^+_{0g} \end{bmatrix} = \mathbf{T_{abc \Rightarrow dq0}} \begin{bmatrix} v_{ag} \\ v_{bg} \\ v_{cg} \end{bmatrix} = \begin{bmatrix} V_p + V_n \cos(2\,\omega_g t + \delta) \\ -V_n \sin(2\,\omega_g t + \delta) \\ 0 \end{bmatrix} \tag{9.16}$$

where $\mathbf{v}^+_{\mathbf{g,dq0}} \in \{v^+_{dg}, v^+_{qg}, v^+_{0g}\}$ are the positive sequence $dq0$ voltage components.

Similarly, the negative sequence $dq0$ voltage components are obtained by using $\theta = \theta_g^-$ in equation (9.15), and are given by

$$\begin{bmatrix} v^-_{dg} \\ v^-_{qg} \\ v^-_{0g} \end{bmatrix} = \mathbf{T_{abc \Rightarrow dq0}} \begin{bmatrix} v_{ag} \\ v_{bg} \\ v_{cg} \end{bmatrix} = \begin{bmatrix} V_n \cos(\delta) + V_p \cos(2\,\omega_g t) \\ -V_n \sin(\delta) + V_p \sin(2\,\omega_g t) \\ 0 \end{bmatrix} \tag{9.17}$$

where $\mathbf{v}^-_{\mathbf{g,dq0}} \in \{v^-_{dg}, v^-_{qg}, v^-_{0g}\}$ are the negative sequence $dq0$ voltage components.

According to equations (9.16) and (9.17), the positive and negative sequence dq components consist of DC and second harmonic oscillating components. The oscillating components are eliminated by using low-pass filters (LPF), and the resultant sequence components are given by

$$\begin{aligned} \overline{v^+_{dg}} &= V_p, \qquad & \overline{v^-_{dg}} &= V_n \cos(\delta) \\ \overline{v^+_{qg}} &= 0.0, \qquad & \overline{v^-_{qg}} &= -V_n \sin(\delta) \end{aligned} \tag{9.18}$$

where $\overline{v^+_{dg}}$ and $\overline{v^+_{qg}}$ are the DC value of positive sequence d- and q-axis voltage components, and $\overline{v^-_{dg}}$ and $\overline{v^-_{qg}}$ are the DC value of negative sequence d- and q-axis voltage components.

Similarly, the positive and negative sequence compensating currents are obtained from the stationary-abc frame currents. The three-phase compensating currents are defined as

$$\begin{aligned} i_{ai} &= I_p \cos(\omega_g t + \alpha) + I_n \cos(\omega_g t + \delta + \beta) \\ i_{bi} &= I_p \cos(\omega_g t + \alpha - \frac{2\pi}{3}) + I_n \cos(\omega_g t + \delta + \beta - \frac{2\pi}{3}) \\ i_{ci} &= I_p \cos(\omega_g t + \alpha - \frac{4\pi}{3}) + I_n \cos(\omega_g t + \delta + \beta - \frac{4\pi}{3}) \end{aligned} \tag{9.19}$$

where I_p and I_n are the peak values of the positive and negative sequence current components, ω_g is the AC grid angular frequency (rad/sec), δ is the phase angle between the positive and negative sequence components, α is the phase angle between the positive sequence voltage and current components, and β is the phase angle between the negative sequence voltage and current components.

The positive sequence voltages in the $dq0$-frame are given by

$$\begin{bmatrix} i_{di}^+ \\ i_{qi}^+ \\ i_{0i}^+ \end{bmatrix} = \mathbf{T_{abc \Rightarrow dq0}} \begin{bmatrix} i_{ai} \\ i_{bi} \\ i_{ci} \end{bmatrix} = \begin{bmatrix} I_p \cos(\alpha) + I_n \cos(2\,\omega_g t + \delta + \beta) \\ I_p \sin(\alpha) - I_n \sin(2\,\omega_g t + \delta + \beta) \\ 0 \end{bmatrix} \tag{9.20}$$

where $\mathbf{i^+_{i,dq0}} \in \{i_{di}^+, i_{qi}^+, i_{0i}^+\}$ are the positive sequence $dq0$ current components.

Similarly, the positive sequence voltages in the $dq0$-frame are given by

$$\begin{bmatrix} i_{di}^- \\ i_{qi}^- \\ i_{0i}^- \end{bmatrix} = \mathbf{T_{abc \Rightarrow dq0}} \begin{bmatrix} i_{ai} \\ i_{bi} \\ i_{ci} \end{bmatrix} = \begin{bmatrix} I_n \cos(\delta + \beta) + I_p \cos(2\,\omega_g t + \alpha) \\ -I_n \sin(\delta + \beta) + I_p \sin(2\,\omega_g t + \alpha) \\ 0 \end{bmatrix} \tag{9.21}$$

where $\mathbf{i^-_{i,dq0}} \in \{i_{di}^-, i_{qi}^-, i_{0i}^-\}$ are the negative sequence $dq0$ current components.

From equations (9.20) and (9.21), the DC value in the positive and negative sequence dq components are given by

$$\begin{aligned} \overline{i_{di}^+} &= I_p \cos(\alpha), & \overline{i_{di}^-} &= I_n \cos(\delta + \beta) \\ \overline{i_{qi}^+} &= I_p \sin(\alpha), & \overline{i_{qi}^-} &= -I_n \sin(\delta + \beta) \end{aligned} \tag{9.22}$$

where $\overline{i_{di}^+}$ and $\overline{i_{qi}^+}$ are the DC value of positive sequence d- and q-axis current components, and $\overline{i_{di}^-}$ and $\overline{i_{qi}^-}$ are the DC value of negative sequence d- and q-axis current components.

(2) Feed-Forward Current Components: From equations (9.14) and (9.19), the phase-x voltage and current signals are defined as

$$\begin{aligned} v_{xg} &= V_p \cos(\omega_g t + \theta_x) + V_n \cos(\omega_g t - \theta_x + \delta) \\ i_{xi} &= I_p \cos(\omega_g t + \theta_x + \alpha) + I_n \cos(\omega_g t - \theta_x + \delta + \beta) \end{aligned} \tag{9.23}$$

where $\theta_x \in \{0, -\frac{2\pi}{3}, -\frac{4\pi}{3}\}$ is phase angle between the three-phase signals.

The definition of positive and negative sequence components given in equations (9.18) and (9.22) are substituted in equation (9.23), and results in

$$\begin{aligned} v_{xg} &= \overline{v_{dg}^{+}} \cos(\omega_g t + \theta_x) + \overline{v_{dg}^{-}} \cos(\omega_g t - \theta_x) + \overline{v_{qg}^{-}} \sin(\omega_g t - \theta_x) \\ i_{xi} &= \overline{i_{di}^{+}} \cos(\omega_g t + \theta_x) - \overline{i_{qi}^{+}} \sin(\omega_g t + \theta_x) + \overline{i_{di}^{-}} \cos(\omega_g t - \theta_x) \\ &\quad + \overline{i_{qi}^{-}} \sin(\omega_g t - \theta_x) \end{aligned} \tag{9.24}$$

According to the fundamental model of MMC (Figure 2.9), the phase-x upper arm voltage and current signals are given by

$$\begin{aligned} v_{xu} &= \frac{V_{dc}}{2} - v_{xg} - L\,\frac{di_{xu}}{dt} - r\,i_{xu} \\ i_{xu} &= i_{xzf} + \frac{1}{2} i_{xi} \end{aligned} \tag{9.25}$$

From equations (9.24) and (9.25), the average power of upper arm is given by

$$\begin{aligned} p_{xu} &= \frac{1}{T_s} \int v_{xu}\, i_{xu}\, dt \\ &= \underbrace{\overline{v_{dg}^{+}}\,\overline{i_{di}^{+}} + \overline{v_{dg}^{-}}\,\overline{i_{di}^{-}} + \overline{v_{qg}^{-}}\,\overline{i_{qi}^{-}}}_{\text{Term-1}} \\ &\quad + \underbrace{\frac{V_{dc}\, i_{xzf}}{2} - \frac{\cos(2\theta_x)}{4} (\overline{v_{dg}^{+}}\,\overline{i_{di}^{-}} + \overline{v_{dg}^{-}}\,\overline{i_{di}^{+}} - \overline{v_{qg}^{-}}\,\overline{i_{qi}^{+}})}_{\text{Term-2}} \\ &\quad + \underbrace{\frac{\sin(2\theta_x)}{4} (\overline{v_{dg}^{+}}\,\overline{i_{qi}^{-}} + \overline{v_{dg}^{-}}\,\overline{i_{qi}^{+}} + \overline{v_{qg}^{-}}\,\overline{i_{di}^{+}})}_{\text{Term-2}} \end{aligned} \tag{9.26}$$

The "term-1" in the above equation represents the overall active power flowing into the MMC-based STATCOM, and it is caused by the positive and negative sequence voltage and current signals. The "term-2" appears due to the interaction of DC-bus voltage and circulating currents, as well as the positive and negative sequence voltage and current signals. At steady state, these two terms should be zero and are given by

$$\begin{aligned} \overline{v_{dg}^{+}}\,\overline{i_{di}^{+}} + \overline{v_{dg}^{-}}\,\overline{i_{di}^{-}} + \overline{v_{qg}^{-}}\,\overline{i_{qi}^{-}} &= 0 \\ \frac{V_{dc}\, i_{xzf}}{2} - \frac{\cos(2\theta_x)}{4} (\overline{v_{dg}^{+}}\,\overline{i_{di}^{-}} + \overline{v_{dg}^{-}}\,\overline{i_{di}^{+}} - \overline{v_{qg}^{-}}\,\overline{i_{qi}^{+}}) & \\ + \frac{\sin(2\theta_x)}{4} (\overline{v_{dg}^{+}}\,\overline{i_{qi}^{-}} + \overline{v_{dg}^{-}}\,\overline{i_{qi}^{+}} + \overline{v_{qg}^{-}}\,\overline{i_{di}^{+}}) &= 0 \end{aligned} \tag{9.27}$$

In the above equation, the positive and negative sequence grid voltages are determined by the AC grid, and the negative sequence q-axis current component is determined by the load. Therefore, the positive-sequence d-axis current ($\overline{i_{di}^{+}}$) and circulating current (i_{xzf}) components are considered as the feed-forward components. These components help to balance the converter input and output powers. The magnitude of feed-forward current components are given by

$$
\begin{aligned}
i^f_{xd} = \overline{i^+_{di}} &= -\frac{\overline{v^-_{dg}}\,\overline{i^-_{di}} + \overline{v^-_{qg}}\,\overline{i^-_{qi}}}{\overline{v^+_{dg}}} \\
i^f_{xz} = i_{xzf} &= \frac{\cos(2\theta_x)}{2V_{dc}}(\overline{v^+_{dg}}\,\overline{i^-_{di}} + \overline{v^-_{dg}}\,\overline{i^+_{di}} - \overline{v^-_{qg}}\,\overline{i^+_{qi}}) \\
&- \frac{\sin(2\theta_x)}{2V_{dc}}(\overline{v^+_{dg}}\,\overline{i^-_{qi}} + \overline{v^-_{dg}}\,\overline{i^+_{qi}} + \overline{v^-_{qg}}\,\overline{i^+_{di}})
\end{aligned}
\tag{9.28}
$$

It is worth mentioning that the intermediate variables in equation (9.28) can be derived from the measured AC grid voltages and the MMC output currents.

CASE STUDY 9.5 Analysis of MMC-STATCOM with AC Grid Fault

Objective: In this case study, the performance of an MMC-based STATCOM under the AC grid fault is analyzed using MATLAB simulation.

Parameters: The system parameters are as follows: grid line-to-line voltage $V_g = 10$ kV, grid frequency $f_g = 60$ Hz, load resistance $R_o = 4.9\ \Omega$, load inductance $L_o = 17.3$ mH, number of SMs per arm $N = 10$, SM capacitor $C = 4000\ \mu$F, SM capacitor voltage $v_C = 2$ kV, arm inductor $L = 5$ mH, and carrier frequency $f_c = 500$ Hz.

Analysis:

In this case study, the three-phase passive load of resistance $R_o = 4.9\ \Omega$ and an inductance $L_o = 17.3$ mH is connected to the AC grid. The MMC-based STATCOM is connected at the PCC to compensate the AC grid unbalance caused by the asymmetrical fault. The performance of MMC-based STATCOM is shown in Figure 9.24. From $t = 0.44$ s to $t = 0.5$ s, the overall system performance is quite similar to the balance grid conditions.

At $t = 0.5$ s, the fault occurs on the AC grid, and the phase-c voltage becomes zero as shown in Figure 9.24(a). The AC grid feeds the power to the passive load with a power factor of 0.6. Due to the fault, the load currents become unbalanced as shown in Figure 9.24(b). From the unbalanced currents, the positive and negative sequence d- and q-axis current components (i_{do} and i_{qo}) are extracted using the transformation theory. These components consist of double line frequency components as shown in Figure 9.24(b). The extracted d-axis current component (i_{do}) is added to the leg voltage controller output (Δi_{di}) and the feed-forward component (i^f_{ad}). The resultant current component becomes the reference d-axis current (i^*_{di}) of the STATCOM as shown in Figure 9.24(c). Before the fault, the feed-forward component (i^f_{ad}) has a zero magnitude. After the fault, it has a finite magnitude and helps to maintain the average leg voltage at their nominal during the fault. Similarly, the q-axis current component (i_{qo}) is extracted from the unbalanced load currents, and it becomes the reference q-axis current (i^*_{qi}) of the STATCOM. The MMC-based STATCOM generates the current waveforms corresponding to the reference current signals (i^*_{di} and i^*_{qi}), and their waveforms are shown in Figure 9.24(d). It is observed that the STATCOM supplies only reactive power before the fault. After the fault, the STATCOM supplies negative sequence current components along with the reactive power.

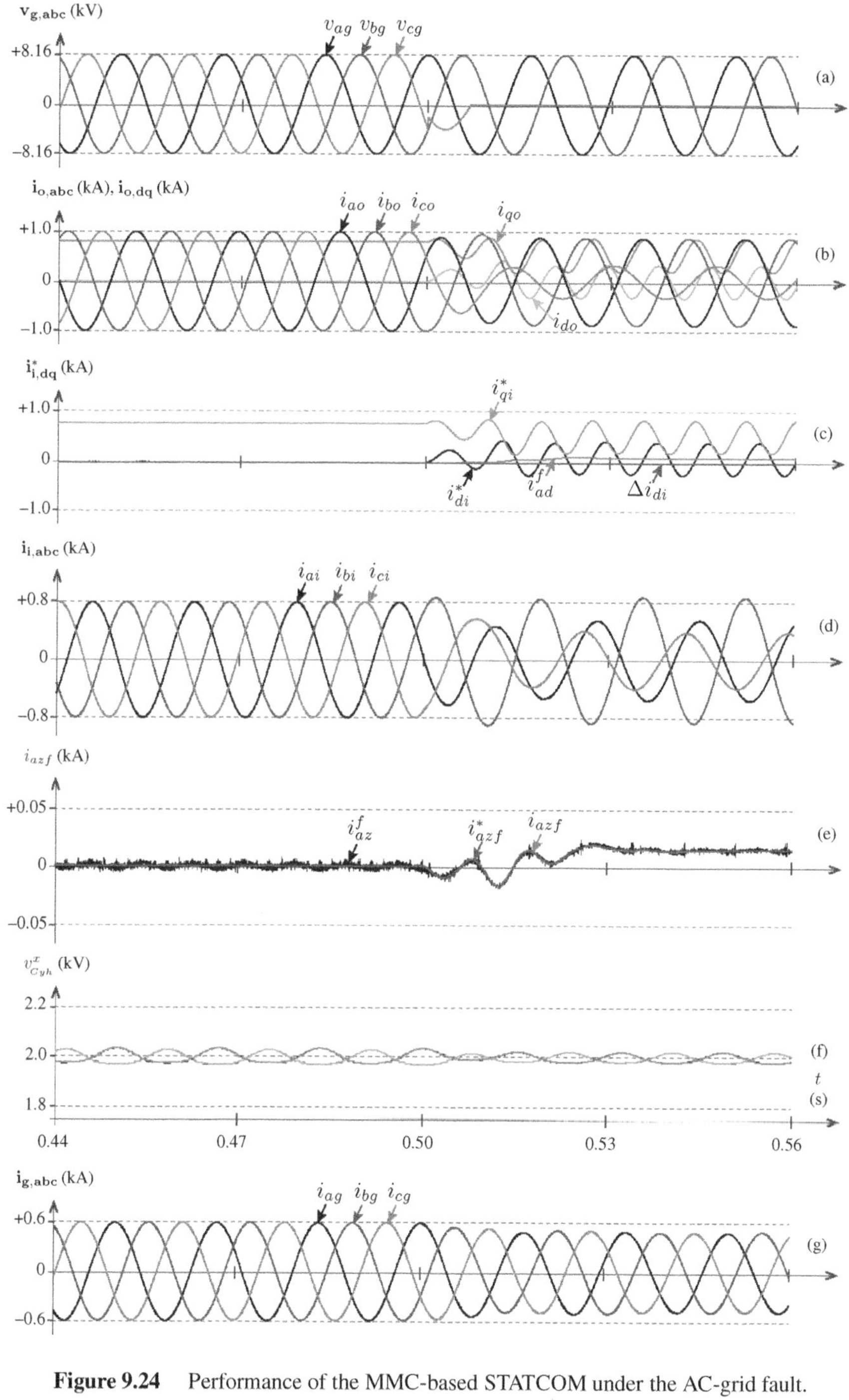

Figure 9.24 Performance of the MMC-based STATCOM under the AC-grid fault.

The negative sequence current components interact with the inner converter voltage and distributes the power among the converter legs. To balance the power distribution among the converter legs, a feed-forward DC current component (i_{az}^{f}) is added to the arm voltage control loop of each leg. The control signals of the arm voltage control loop are shown in Figure 9.24(e). These components are close to zero during the balanced grid conditions and have finite magnitude during the fault conditions. The total DC-bus voltage is equally distributed among the submodules using the individual SM voltage control. Each submodule capacitor voltage is regulated at 2 kV as shown in Figure 9.24(f). The three-phase AC grid currents are shown in Figure 9.24(g). These currents are perfectly balanced and in-phase with their respective voltages even under the AC grid fault. ■

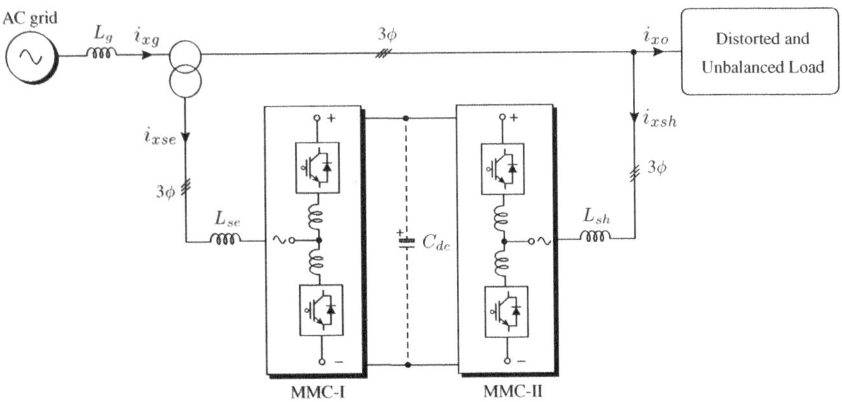

Figure 9.25 System configuration of the MMC-based UPQC.

9.6 MMC-BASED UNIFIED POWER QUALITY CONDITIONER

The unified power quality conditioner (UPQC) is another potential application of the MMC [54–56]. The UPQC is capable of solving both voltage and current-related power quality problems simultaneously. The system configuration of MMC-based UPQC is shown in Figure 9.25. It consists of two modular multilevel converters and is connected in the back-to-back (BTB) as shown in Figure 9.25. In the MMC-based UPQC, the DC-bus capacitor is optional and can be eliminated in practical installations. Among the two MMCs, one MMC output is connected in series with the AC grid through a transformer. This converter is referred to as a "series compensator," and solves the voltage quality problems such as voltage sag/swell in the power system. Other MMC is connected in parallel to the AC grid. This converter is referred to as a "shunt compensator" and solves the current quality problems such as harmonics, reactive power, and grid unbalances in the power system. The shunt compensator can be either directly connected to the grid or through a step-up transformer, depending on the system configuration and its requirements. Along with the power quality problems, the shunt compensator should regulate the total DC-bus voltage (V_{dc}) of the UPQC system.

9.7 SUMMARY

In this chapter, the potential applications of the modular multilevel converter including the high-voltage direct current (HVDC) transmission systems, static synchronous compensators (STATCOM), and unified power quality conditioners (UPQC) are discussed. The MMC-based HVDC systems can be configured in the two-terminal and multi-terminal layout. The operation and control of two-terminal and multi-terminal HVDC transmission systems are presented. The MMC-based HVDC systems are highly vulnerable to the DC-side short-circuit faults. The fault protection schemes including the submodule based approach and HVDC circuit breakers are discussed.

The MMC is also used in the power quality technologies such as STATCOM and UPQC systems. The MMC-based STATCOM is widely used to compensate the reactive power during the balanced grid conditions. Also, it can be used in the compensation of AC grid unbalance during the asymmetrical conditions (grid faults and load unbalance). The operation and control of MMC-based STATCOM during the symmetrical and asymmetrical AC grid conditions are presented. Several case studies are presented to analyze the STATCOM performance during the symmetrical and asymmetrical operating conditions. Finally, the system configuration of MMC-based UPQC is briefly introduced in this chapter.

REFERENCES

1. H. J. Knaak, "Modular multilevel converters and hvdc/facts: A success story," in *Power Electronics and Applications (EPE 2011), Proceedings of the 2011-14th European Conference on*, Aug 2011, pp. 1–6.

2. S. Allebrod, R. Hamerski, and R. Marquardt, "New transformerless, scalable modular multilevel converters for hvdc-transmission," in *Power Electronics Specialists Conference, 2008. PESC 2008. IEEE*, 2008, pp. 174–179.

3. R. Marquardt, "Modular multilevel converter: An universal concept for hvdc-networks and extended dc-bus-applications," in *Power Electronics Conference (IPEC), 2010 International*, Jun 2010, pp. 502–507.

4. S. Debnath and M. Saeedifard, "A new hybrid modular multilevel converter for grid connection of large wind turbines," *IEEE Trans. Sustainable Energy*, vol. 4, no. 4, pp. 1051–1064, Oct 2013.

5. G. P. Adam, O. Anaya-Lara, and G. Burt, "Multi-terminal dc transmission system based on modular multilevel converter," in *2009 44th International Universities Power Engineering Conference (UPEC)*, Sep 2009, pp. 1–5.

6. H. Mohammadi and M. Bina, "A transformerless medium-voltage statcom topology based on extended modular multilevel converters," *IEEE Trans. Power Electron.*, vol. 26, no. 5, pp. 1534–1545, May 2011.

7. M. Hagiwara, R. Maeda, and H. Akagi, "Negative-sequence reactive-power control by a pwm statcom based on a modular multilevel cascade converter (mmcc-sdbc)," *IEEE Trans. Ind. Appl.*, vol. 48, no. 2, pp. 720–729, Mar 2012.

8. J. Beerten, S. Cole, and R. Belmans, "Modeling of multi-terminal vsc hvdc systems with distributed dc voltage control," *IEEE Trans. Power Syst.*, vol. 29, no. 1, pp. 34–42, Jan 2014.

9. M. P. Bahrman, "Hvdc transmission overview," in *2008 IEEE/PES Transmission and Distribution Conference and Exposition*, Apr 2008, pp. 1–7.

10. A. Nami, J. Liang, F. Dijkhuizen, and P. Lundberg, "Analysis of modular multilevel converters with dc short circuit fault blocking capability in bipolar hvdc transmission systems," in *2015 17th European Conference on Power Electronics and Applications (EPE'15 ECCE-Europe)*, Sep 2015, pp. 1–10.

11. S. Teeuwsen, "Modeling the trans bay cable project as voltage-sourced converter with modular multilevel converter design," in *2011 IEEE Power and Energy Society General Meeting*, Jul 2011, pp. 1–8.

12. T. Westerweller, K. Friedrich, U. Armonies, A. Orini, D. Parquet, and S. Wehn, "Trans bay cable-worldâĂŹs first hvdc system using multilevel voltage-sourced converter," *Proc. 2010 CIGRE, Paris*, 2010.

13. Y. Wang, Z. Yuan, and J. Fu, "A novel strategy on smooth connection of an offline mmc station into mtdc systems," *IEEE Trans. Power Del.*, vol. 31, no. 2, pp. 568–574, Apr 2016.

14. G. Tang, Z. He, H. Pang, X. Huang, and X. p. Zhang, "Basic topology and key devices of the five-terminal dc grid," *CSEE Journal of Power and Energy Systems*, vol. 1, no. 2, pp. 22–35, Jun 2015.

15. J. Hu, K. Xu, L. Lin, and R. Zeng, "Analysis and enhanced control of hybrid-mmc-based hvdc systems during asymmetrical dc voltage faults," *IEEE Trans. Power Del.*, vol. 32, no. 3, pp. 1394–1403, Jun 2017.

16. R. Li, J. E. Fletcher, L. Xu, and B. W. Williams, "Enhanced flat-topped modulation for mmc control in hvdc transmission systems," *IEEE Trans. Power Del.*, vol. 32, no. 1, pp. 152–161, Feb 2017.

17. U. N. Gnanarathna, S. K. Chaudhary, A. M. Gole, and R. Teodorescu, "Modular multi-level converter based hvdc system for grid connection of offshore wind power plant," in *9th IET International Conference on AC and DC Power Transmission (ACDC 2010)*, Oct 2010, pp. 1–5.

18. R. Sharma, T. W. Rasmussen, K. H. Jensen, and V. Akamatov, "Modular vsc converter based hvdc power transmission from offshore wind power plant: Compared to the conventional hvac system," in *2010 IEEE Electrical Power Energy Conference*, Aug 2010, pp. 1–6.

19. A. G. Endegnanew, K. Uhlen, G. Bergna-Diaz, and J. Beerten, "Small-signal modelling of an mmc-based hvdc link interfacing large ac systems," in *2016 IEEE 17th Workshop on Control and Modeling for Power Electronics (COMPEL)*, Jun 2016, pp. 1–9.

20. R. T. Pinto, P. Bauer, S. F. Rodrigues, E. J. Wiggelinkhuizen, J. Pierik, and B. Ferreira, "A novel distributed direct-voltage control strategy for grid integration of offshore wind energy systems through mtdc network," *IEEE Trans. Ind. Electron.*, vol. 60, no. 6, pp. 2429–2441, Jun 2013.

21. R. L. Vaughan, J. P. Bowles, and J. Dalzell, "The control and performance of a series connected multiterminal hvdc transmission system," *IEEE Trans. Power App. Syst.*, vol. 94, no. 5, pp. 1868–1877, Sep 1975.

22. J. Arrillaga, Y. H. Liu, and N. R. Watson, *Flexible power transmission: the HVDC options*. John Wiley & Sons, 2007.

23. C. Li, C. Zhao, J. Xu, Y. Ji, F. Zhang, and T. An, "A pole-to-pole short-circuit fault current calculation method for dc grids," *IEEE Trans. Power Syst.*, vol. PP, no. 99, pp. 1–1, 2017.

24. W. Leterme, P. Tielens, S. D. Boeck, and D. V. Hertem, "Overview of grounding and configuration options for meshed hvdc grids," *IEEE Trans. Power Del.*, vol. 29, no. 6, pp. 2467–2475, Dec 2014.

25. G. Liu, F. Xu, Z. Xu, Z. Zhang, and G. Tang, "Assembly hvdc breaker for hvdc grids with modular multilevel converters," *IEEE Trans. Power Electron.*, vol. 32, no. 2, pp. 931–941, Feb 2017.

26. R. Zeng, L. Xu, L. Yao, and D. J. Morrow, "Precharging and dc fault ride-through of hybrid mmc-based hvdc systems," *IEEE Trans. Power Del.*, vol. 30, no. 3, pp. 1298–1306, Jun 2015.

27. J. Zhang and C. Zhao, "The research of sm topology with dc fault tolerance in mmc-hvdc," *IEEE Trans. Power Del.*, vol. 30, no. 3, pp. 1561–1568, Jun 2015.

28. G. P. Adam and I. E. Davidson, "Robust and generic control of full-bridge modular multilevel converter high-voltage dc transmission systems," *IEEE Trans. Power Del.*, vol. 30, no. 6, pp. 2468–2476, Dec 2015.

29. G. Tang, Z. Xu, and Y. Zhou, "Impacts of three mmc-hvdc configurations on ac system stability under dc line faults," *IEEE Trans. Power Syst.*, vol. 29, no. 6, pp. 3030–3040, Nov 2014.

30. X. Li, Q. Song, W. Liu, H. Rao, S. Xu, and L. Li, "Protection of nonpermanent faults on dc overhead lines in mmc-based hvdc systems," *IEEE Trans. Power Del.*, vol. 28, no. 1, pp. 483–490, Jan 2013.

31. T. Modeer, H. P. Nee, and S. Norrga, "Loss comparison of different sub-module implementations for modular multilevel converters in hvdc applications," in *Proceedings of the 2011 14th European Conference on Power Electronics and Applications*, Aug 2011, pp. 1–7.

32. H. Kim, J. Kang, S. Kim, C. K. Kim, and K. Hur, "Dc fault protection for modular multilevel converter hvdc using asymmetrical unipolar full-bridge submodule," in *2015 9th International Conference on Power Electronics and ECCE Asia (ICPE-ECCE Asia)*, Jun 2015, pp. 1083–1089.

33. R. Oliveira and A. Yazdani, "A modular multilevel converter with dc fault handling capability and enhanced efficiency for hvdc system applications," *IEEE Trans. Power Electron.*, vol. 32, no. 1, pp. 11–22, Jan 2017.

34. R. Lizana, A. Dekka, S. Rivera, and B. Wu, "Modular multilevel converter based on 5-level submodule with dc fault blocking capability," in *IECON 2016 – 42nd Annual Conference of the IEEE Industrial Electronics Society*, Oct 2016, pp. 3482–3487.

35. E. Kontos, R. T. Pinto, S. Rodrigues, and P. Bauer, "Impact of hvdc transmission system topology on multiterminal dc network faults," *IEEE Trans. Power Del.*, vol. 30, no. 2, pp. 844–852, Apr 2015.

36. M. Barnes, D. V. Hertem, S. P. Teeuwsen, and M. Callavik, "Hvdc systems in smart grids," *IEEE Proceeding of the.*, vol. PP, no. 99, pp. 1–17, 2017.

37. M. Callavik, A. Blomberg, J. Häfner, and B. Jacobson, "The hybrid hvdc breaker," *ABB Grid Systems Technical Paper*, 2012.

38. B. Parkhideh and S. Bhattacharya, "Vector-controlled voltage-source-converter-based transmission under grid disturbances," *IEEE Trans. Power Electron.*, vol. 28, no. 2, pp. 661–672, Feb 2013.

39. J. Beerten and R. Belmans, "Modeling and control of multi-terminal vsc hvdc systems," *Energy Procedia*, vol. 24, pp. 123–130, 2012.

40. S. Henry, O. Despouys, R. Adapa, L. Barthold, C. Bayfield, K. Bell, J.-L. Binard, A. Edris, P. Egrot, W. Hung *et al.*, "Influence of embedded hvdc transmission on system security and ac network performance," *Electra*, no. 267, pp. 79–86, 2013.

41. I. Etxeberria-Otadui, U. Viscarret, M. Caballero, A. Rufer, and S. Bacha, "New optimized pwm vsc control structures and strategies under unbalanced voltage transients," *IEEE Trans. Ind. Electron.*, vol. 54, no. 5, pp. 2902–2914, Oct 2007.

42. W. Wang, A. Beddard, M. Barnes, and O. Marjanovic, "Analysis of active power control for vsc-hvdc," *IEEE Trans. Power Del.*, vol. 29, no. 4, pp. 1978–1988, Aug 2014.

43. J. Liang, T. Jing, O. Gomis-Bellmunt, J. Ekanayake, and N. Jenkins, "Operation and control of multiterminal hvdc transmission for offshore wind farms," *IEEE Trans. Power Del.*, vol. 26, no. 4, pp. 2596–2604, Oct 2011.

44. T. M. Haileselassie and K. Uhlen, "Impact of dc line voltage drops on power flow of mtdc using droop control," *IEEE Trans. Power Syst.*, vol. 27, no. 3, pp. 1441–1449, Aug 2012.

45. W. Wang and M. Barnes, "Power flow algorithms for multi-terminal vsc-hvdc with droop control," *IEEE Trans. Power Syst.*, vol. 29, no. 4, pp. 1721–1730, Jul 2014.

46. C. Dierckxsens, K. Srivastava, M. Reza, S. Cole, J. Beerten, and R. Belmans, "A distributed dc voltage control method for vsc mtdc systems," *Electric Power Systems Research*, vol. 82, no. 1, pp. 54–58, 2012.

47. T. K. Vrana, J. Beerten, R. Belmans, and O. B. Fosso, "A classification of dc node voltage control methods for hvdc grids," *Electric power systems research*, vol. 103, pp. 137–144, 2013.

48. A. Beddard and M. Barnes, "Modelling of mmc-hvdc systems–an overview," *Energy Procedia*, vol. 80, pp. 201–212, 2015.

49. R. E. Torres-Olguin, A. R. Årdal, H. Støylen, A. G. Endegnanew, K. Ljøkelsøy, and J. O. Tande, "Experimental verification of a voltage droop control for grid integration of offshore wind farms using multi-terminal hvdc," *Energy Procedia*, vol. 53, pp. 104–113, 2014.

50. M. Pereira, D. Retzmann, J. Lottes, M. Wiesinger, and G. Wong, "Svc plus: An mmc statcom for network and grid access applications," in *2011 IEEE Trondheim PowerTech*, Jun 2011, pp. 1–5.

51. H. M. Pirouz and M. T. Bina, "Extended modular multilevel converters suitable for medium-voltage and large-current statcom applications," in *2010 Conference Proceedings IPEC*, Oct 2010, pp. 487–492.

52. X. Yu, Y. Wei, and Q. Jiang, "Statcom operation scheme of the cdsm-mmc during a pole-to-pole dc fault," *IEEE Trans. Power Del.*, vol. 31, no. 3, pp. 1150–1159, Jun 2016.

53. S. Du and J. Liu, "A study on dc voltage control for chopper-cell-based modular multilevel converters in d-statcom application," *IEEE Trans. Power Del.*, vol. 28, no. 4, pp. 2030–2038, Oct 2013.

54. X. Xiao, J. Lu, C. Yuan, and Y. Yang, "A 10kv 4mva unified power quality conditioner based on modular multilevel inverter," in *2013 International Electric Machines Drives Conference*, May 2013, pp. 1352–1357.

55. Z. Liu, J. Lu, Z. Ou, M. Ma, C. Yuan, X. Xiao, and H. Wang, "The start control strategy design of unified power quality conditioner based on modular multilevel converter," in *Electric Machines Drives Conference (IEMDC), 2013 IEEE International*, May 2013, pp. 933–937.

56. Y. Long, X. Xiao, Y. Xu, B. Yu, Y. Xu, and J. Hao, "Mmc-upqc: Application of modular multilevel converter on unified power quality conditioner," in *2013 IEEE Power Energy Society General Meeting*, Jul 2013, pp. 1–5.

APPENDIX A

MATLAB DEMO PROJECTS

To help the readers with the implementation of voltage balancing algorithm for an modular multilevel converter (MMC) in MATLAB/Simulink environment, the following demo project is posted in the companion website of this book.

- M2014Ra_Demo Project-I

The demo project has been developed using SimPowerSystems toolbox in MATLAB /Simulink 2014a release software. The demo project consists of the following three files.

- Initial_Parameters.m
- MMC_Voltage_Balance_Demo1.slx
- ploting_results.m

The file Initial_Parameters.m consists of the MMC design specifications and other system parameters. The file MMC_Voltage_Balance_Demo1.slx consists of the main simulink model of the demo project I. In the simulink model, the three-phase MMC is designed with the HB-SMs for DC–AC power conversion. The phase-disposition pulse width modulation (PD-PWM) is employed to generate the submodule gating signals. To achieve the voltage balancing in an MMC, the gating signals are rotated between the submodules. This approach is also referred to as a gating signal rotation. For more details about the voltage balancing approach, the readers could read the following reference article [1].

The readers should first run the Initial_Parameters.m file to load the system parameters into the MATLAB workspace. Later, the readers needs to run the simulink model (.slx) file. At the end of the simulation, the readers can use the ploting_results.m file to observe the important results such as voltage, current, and submodule capacitor voltage waveforms of an MMC. This simulation provides a basic understanding on the operation of an MMC and its control scheme

Modular Multilevel Converters: Analysis, Control, and Applications, First Edition.
Sixing Du, Apparao Dekka, Bin Wu, and Navid Zargari.

implementation. The readers can use this demo project as a starting point to build their own simulation models or to implement other concepts presented in this book.

In the due time, the authors going to provide few more demo projects and other teaching materials in the companion website.

REFERENCES

1. G. Adam, O. Anaya-Lara, G. M. Burt, D. Telford, B. Williams, and J. McDonald, "Modular multilevel inverter: Pulse width modulation and capacitor balancing technique," *IET Power Electron.*, vol. 3, no. 5, pp. 702–715, Sep 2010.

INDEX

Modular Multilevel Converters: Analysis, Control, and Applications, First Edition.
Sixing Du, Apparao Dekka, Bin Wu, and Navid Zargari.

IEEE Press Series on Power Engineering

Series Editor: M. E. El-Hawary, Dalhousie University, Halifax, Nova Scotia, Canada

The mission of IEEE Press Series on Power Engineering is to publish leading-edge books that cover the broad spectrum of current and forward-looking technologies in this fast-moving area. The series attracts highly acclaimed authors from industry/academia to provide accessible coverage of current and emerging topics in power engineering and allied fields. Our target audience includes the power engineering professional who is interested in enhancing their knowledge and perspective in their areas of interest.

1. *Principles of Electric Machines with Power Electronic Applications, Second Edition*
M. E. El-Hawary

2. *Pulse Width Modulation for Power Converters: Principles and Practice*
D. Grahame Holmes and Thomas Lipo

3. *Analysis of Electric Machinery and Drive Systems, Second Edition*
Paul C. Krause, Oleg Wasynczuk, and Scott D. Sudhoff

4. *Risk Assessment for Power Systems: Models, Methods, and Applications*
Wenyuan Li

5. *Optimization Principles: Practical Applications to the Operations of Markets of the Electric Power Industry*
Narayan S. Rau

6. *Electric Economics: Regulation and Deregulation*
Geoffrey Rothwell and Tomas Gomez

7. *Electric Power Systems: Analysis and Control*
Fabio Saccomanno

8. *Electrical Insulation for Rotating Machines: Design, Evaluation, Aging, Testing, and Repair, Second Edition*
Greg Stone, Edward A. Boulter, Ian Culbert, and Hussein Dhirani

9. *Signal Processing of Power Quality Disturbances*
Math H. J. Bollen and Irene Y. H. Gu

10. *Instantaneous Power Theory and Applications to Power Conditioning*
Hirofumi Akagi, Edson H. Watanabe, and Mauricio Aredes

11. *Maintaining Mission Critical Systems in a 24/7 Environment*
Peter M. Curtis

12. *Elements of Tidal-Electric Engineering*
Robert H. Clark

13. *Handbook of Large Turbo-Generator Operation and Maintenance, Second Edition*
Geoff Klempner and Isidor Kerszenbaum

14. *Introduction to Electrical Power Systems*
Mohamed E. El-Hawary

15. *Modeling and Control of Fuel Cells: Distributed Generation Applications*
M. Hashem Nehrir and Caisheng Wang

16. *Power Distribution System Reliability: Practical Methods and Applications*
Ali A. Chowdhury and Don O. Koval

17. *Introduction to FACTS Controllers: Theory, Modeling, and Applications*
Kalyan K. Sen and Mey Ling Sen

18. *Economic Market Design and Planning for Electric Power Systems*
James Momoh and Lamine Mili

19. *Operation and Control of Electric Energy Processing Systems*
James Momoh and Lamine Mili

20. *Restructured Electric Power Systems: Analysis of Electricity Markets with Equilibrium Models*
Xiao-Ping Zhang

21. *An Introduction to Wavelet Modulated Inverters*
S.A. Saleh and M.A. Rahman

22. *Control of Electric Machine Drive Systems*
Seung-Ki Sul

23. *Probabilistic Transmission System Planning*
Wenyuan Li

24. *Electricity Power Generation: The Changing Dimensions*
Digambar M. Tigare

25. *Electric Distribution Systems*
Abdelhay A. Sallam and Om P. Malik

26. *Practical Lighting Design with LEDs*
Ron Lenk and Carol Lenk

27. *High Voltage and Electrical Insulation Engineering*
Ravindra Arora and Wolfgang Mosch

28. *Maintaining Mission Critical Systems in a 24/7 Environment, Second Edition*
Peter Curtis

29. *Power Conversion and Control of Wind Energy Systems*
Bin Wu, Yongqiang Lang, Navid Zargari, and Samir Kouro

30. *Integration of Distributed Generation in the Power System*
Math H. Bollen and Fainan Hassan

31. *Doubly Fed Induction Machine: Modeling and Control for Wind Energy Generation Applications*
Gonzalo Abad, Jesus Lopez, Miguel Rodrigues, Luis Marroyo, and Grzegorz Iwanski

32. *High Voltage Protection for Telecommunications*
Steven W. Blume

33. *Smart Grid: Fundamentals of Design and Analysis*
James Momoh

34. *Electromechanical Motion Devices, Second Edition*
Paul C. Krause, Oleg Wasynczuk, and Steven D. Pekarek

35. *Electrical Energy Conversion and Transport: An Interactive Computer-Based Approach, Second Edition*
George G. Karady and Keith E. Holbert

36. *ARC Flash Hazard and Analysis and Mitigation*
J. C. Das

37. *Handbook of Electrical Power System Dynamics: Modeling, Stability, and Control*
Mircea Eremia and Mohammad Shahidehpour

38. *Analysis of Electric Machinery and Drive Systems, Third Edition*
Paul Krause, Oleg Wasynczuk, S. D. Sudhoff, and Steven D. Pekarek

39. *Extruded Cables for High-Voltage Direct-Current Transmission: Advances in Research and Development*
Giovanni Mazzanti and Massimo Marzinotto

40. *Power Magnetic Devices: A Multi-Objective Design Approach*
S. D. Sudhoff

41. *Risk Assessment of Power Systems: Models, Methods, and Applications, Second Edition*
Wenyuan Li

42. *Practical Power System Operation*
Ebrahim Vaahedi

43. *The Selection Process of Biomass Materials for the Production of Bio-Fuels and Co-Firing*
Najib Altawell

44. *Electrical Insulation for Rotating Machines: Design, Evaluation, Aging, Testing, and Repair, Second Edition*
Greg C. Stone, Ian Culbert, Edward A. Boulter, and Hussein Dhirani

45. *Principles of Electrical Safety*
Peter E. Sutherland

46. *Advanced Power Electronics Converters: PWM Converters Processing AC Voltages*
Euzeli Cipriano dos Santos Jr. and Edison Roberto Cabral da Silva

47. *Optimization of Power System Operation, Second Edition*
Jizhong Zhu

48. *Power System Harmonics and Passive Filter Designs*
J. C. Das

49. *Digital Control of High-Frequency Switched-Mode Power Converters*
Luca Corradini, Dragan Maksimovic, Paolo Mattavelli, and Regan Zane

50. *Industrial Power Distribution, Second Edition*
Ralph E. Fehr, III

51. *HVDC Grids: For Offshore and Supergrid of the Future*
Dirk Van Hertem, Oriol Gomis-Bellmunt, and Jun Liang

52. *Advanced Solutions in Power Systems: HVDC, FACTS, and Artificial Intelligence*
Mircea Eremia, Chen-Ching Liu, and Abdel-Aty Edris

53. *Power Grid Operation in a Market Environment: Economic Efficiency and Risk Mitigation*
Hong Chen

54. *Current Signature Analysis for Condition Monitoring of Cage Induction Motors: Industrial Application and Case Histories*
William T. Thomson and Ian Culburt

55. *Model Predictive Control of Wind Energy Conversion Systems*
Venkata Yaramasu and Bin Wu

56. *Electric Power System Basics for the Nonelectrical Professional, Second Edition*
Steven W. Blume

57. *Energy Production Systems Engineering*
Thomas H. Blair

58. *Understanding Symmetrical Components for Power System Modeling*
J. C. Das

59. *High-Power Converters and AC Drives, Second Edition*
Bin Wu and Mehdi Narimani

60. *Practical Lighting Design with LEDs, Second Edition*
Ron Lenk and Carol Lenk

61. *Introduction to Electric Power and Drive Systems*
Paul Krause, Oleg Wasynczuk, Timothy O'Connell, and Maher Hasan

62. *Instantaneous Power Theory and Applications to Power Conditioning, Second Edition*
Hirofumi Akagi, Edson Hirokazu Watanabe, and Mauricio Aredes

63. *Introduction to AC Machine Design*
Thomas A. Lipo

64. *Advances in Electric Power and Energy Systems: Load and Price Forecasting*
Mohamed E. El-Hawary

65. *Electricity Markets: Theories and Applications*
Jeremy Lin and Fernando H. Magnago

66. *Multiphysics Simulation by Design for Electrical Machines, Power Electronics, and Drives*
Marius Rosu, Ping Zhou, Dingsheng Lin, Dan Ionel, Mircea Popescu, Frede Blaabjerg, Vandana Rallabandi, and David Staton

67. *Modular Multilevel Converters: Analysis, Control, and Applications*
Sixing Du, Apparao Dekka, Bin Wu, and Navid Zargari

Printed and bound by CPI Group (UK) Ltd, Croydon, CR0 4YY
17/09/2026
14964317-0005